APPROXIMATIONS WITH SPECIAL EMPHASIS ON SPLINE FUNCTIONS

Publication No. 23
of the Mathematics Research Center
United States Army
The University of Wisconsin

Approximations with Special Emphasis on Spline Functions

Edited by I. J. Schoenberg

Proceedings of a Symposium
Conducted by the Mathematics Research Center,
United States Army, at the University
of Wisconsin, Madison
May 5-7, 1969

Academic Press
New York • London 1969

ACADEMIC PRESS, INC.
111 Fifth Avenue, New York, New York 10003

United Kingdom Edition published by
ACADEMIC PRESS, INC. (LONDON) LTD.
Berkeley Square House, London W1X6BA

LIBRARY OF CONGRESS CATALOG CARD NUMBER: 71-86364

PRINTED IN THE UNITED STATES OF AMERICA

Foreword

The present volume contains the proceedings of a symposium on approximations with special emphasis on spline functions held in Madison, Wisconsin, on May 5–7, 1969, and sponsored by the Mathematics Research Center, U.S. Army, University of Wisconsin.

There were five sessions. The following persons acted as chairmen:

Professor J.L. Walsh, University of Maryland
Professor Arthur Sard, Queens College
Dr. E.N. Nilson, Pratt and Whitney Aircraft Company
Professor A.M. Ostrowski, Universität Basel
Professor Salomon Bochner, Rice University

To retain in these proceedings something of the atmosphere of the actual meetings, as well as the interrelation of subject matter, the papers are printed in this volume in the order in which they appeared on the program. The symposium committee consisted of Professor R. Creighton Buck, T.N.E. Greville, and Seymour Parter, all of the University of Wisconsin, with the editor as chairman. The success of the symposium, attended by 162 registrants, and the prompt appearance of these proceedings were made possible by the superb services of Mrs. Gladys Moran as Symposium Secretary, of Mrs. Doris Whitmore who prepared the manuscript, and of the staff of the Academic Press.

I wish to thank the speakers and chairman and all others who made the symposium possible.

Madison, Wisconsin
August 1969

I. J. Schoenberg

Preface

During the last year the Mathematics Research Center has actively sponsored and encouraged interest in the field of spline functions and their applications. In October 1968 it sponsored an Advanced Seminar organized by T.N.E. Greville. Its proceedings appeared under the title "Theory and Applications of Spline Functions" (Academic Press, 1969). In May 1969 the Mathematics Research Center organized a symposium whose proceedings are herewith presented to the public.

Is the recent interest in spline functions due to the current trend, also in the Arts and in Literature, toward the fragmented, the piecewise, the disjointed? In the editor's opinion it is rather due to the fact that spline functions are a good tool for the numerical approximation of functions on the one hand, and that they suggest new, challenging, and rewarding problems on the other. We should also realize that the piecewise and fragmented is not a new idea. With the exception of Queen Dido's, the shapes of plots of land were never circular but rather polygonal in shape. The Greeks used polygons in their method of exhaustion. Johann Bernoulli and Euler used polygons for the approximate solution of differential equations.

Piecewise linear functions, as well as step functions, have long been an important theoretical tool of approximation of functions as long as the step size was made to converge to zero. However, to obtain useful approximations without this limiting process, we need the higher degree analogs of piecewise linear functions, i.e. the functions obtained by integrating step functions not only once but a certain number of times. These have, on the whole, been largely neglected until about 25 years ago. A notable exception was the work done by actuarial mathematicians on so-called "osculatory interpolation" that began soon after Hermite's work on interpolation. In particular, the writings of Greville on this subject suggested to the editor his early work on spline functions during World War II. In 1949 A. Sard started his work on best approximation of linear functionals pointing in an only apparently different direction. During the years from 1957 to 1964 the various optimal properties of spline functions were discovered and their relationships clarified by Holladay, Golomb, and Weinberger; Ahlberg, Nilson, and Walsh; C. de Boor and the editor. The bivariate problems were first attacked by G. Birkhoff and Garabedian in 1960. In 1964 Greville initiated the long list of generalizations of spline functions. Already the year 1967 saw the appearance of the first book on spline functions by Ahlberg, Nilson, and Walsh.

The number of mathematicians, in this country and abroad, actively engaged in practical or theoretical spline analysis has considerably increased since 1964. The mere size of the present proceedings indicates why we were unable to invite more of these active workers to contribute to the symposium. These proceedings show some of the current developments and trends. The editor believes that a mathematical paper should either be beautiful, or else useful, preferably both, but at least one. He hopes that most readers will find that all these papers qualify on at least one of the counts, and perhaps many of them on both.

Madison, Wisconsin
August 1969

I. J. Schoenberg

Contents

Splines in the Complex Plane

J. H. AHLBERG

Introduction

This paper gives a general presentation of a relatively new area of spline research. It brings together in one paper a number of earlier results along with a number of still unpublished results. These results were obtained in part in conjunction with E. N. Nilson and J. L. Walsh, which material, to a large degree, is contained in references [1] and [2]. The remainder is due to the author and E. N. Nilson and, with the exception of reference [3], is as yet unpublished.

The main objective is to give a unified summary of our present knowledge of splines in the complex plane; the ideas underlying methods of proof are indicated, but for the most part detailed proofs are not given. It is hoped that the results obtained so far serve to put complex spline theory on a relatively solid foundation; their ramifications remain to be explored.

Complex Splines

Suppose that N distinct points $t_0, t_1, \ldots, t_N = t_0$ lying on a Jordan curve Γ in counterclockwise order are given. Let Δ denote this set of points. We shall refer to the points t_j in Δ as <u>nodal points</u>. Now, let Γ_j denote the subarc of Γ joining t_{j-1} to t_j. If n is a positive integer and if we are given N complex polynomials, $p_j(z)$, $(j = 1, 2, \ldots, N)$, of degree n with the property

(1) $p_{j-1}^{(\alpha)}(t_{j-1}) = p_j^{(\alpha)}(t_{j-1})$, $(\alpha = 0, 1, \ldots, n-1;\ j = 1, 2, \ldots, N)$,

then these polynomials define a complex polynomial spline $q_\Delta(t)$ of degree n on Γ. On the subarc Γ_j we identify $q_\Delta(t)$ with $p_j(t)$. It is immediate from (1) that $q_\Delta(t)$ is in $C^{n-1}(\Gamma)$. The continuity properties of $q_\Delta(t)$ could be relaxed, if desired, but we shall not do so. The splines $q_\Delta(t)$, as defined, we shall call simple splines.

Since $q_\Delta(t)$ is integrable on Γ, we can now define, in a natural way, an associated analytic spline, $S_\Delta(z)$, in the interior of Γ by the Cauchy integral

(2) $$S_\Delta(z) = \frac{1}{2\pi i} \int_\Gamma \frac{q_\Delta(t)}{t-z}\, dt .$$

It can be established that $S_\Delta(z)$ depends only on the point set Δ and not on the particular Jordan curve Γ used in its definition. It is actually analytic at all points in the complex plane with the exception of the point set Δ. The points in Δ are logarithmic branch points.

Periodicity Conditions

Let the polynomials defining $q_\Delta(t)$ on Γ have the representation

(3) $$p_j(t) = \sum_{k=0}^{n} a_{jk}(t-z)^k, \qquad (j = 1, 2, \ldots, N) ,$$

where z is a fixed, but otherwise arbitrary, point in the complement of Δ. Since t_N and t_0 denote the same point, we have the identities

(4) $$\sum_{j=1}^{N} (a_{jk} - a_{j-1,k}) = 0, \qquad (k = 0, 1, \ldots, n) .$$

Let

(5) $$\tau_j = a_{j+1,n} - a_{j,n} , \quad (j = 1, 2, \ldots, N) .$$

We now can rewrite the last of equations (4) as

(6) $$\sum_{j=1}^{N} \tau_j = 0 ;$$

here τ_j is the saltus (if any) in the n-th derivative of $q_\Delta(t)$ at the point t_j .

From the continuity conditions

(7) $$p_{j+1}^{(n-1)}(t_j) = p_j^{(n-1)}(t_j) , \quad (j = 1, 2, \ldots, N) .$$

we have

(8) $$-n!\,(a_{j+1,n} - a_{jn})\,(t_j - z) = (n-1)!\,(a_{j-1,n-1} - a_{j,n-1}) .$$

It now follows from (4), with $k = n-1$, that

(9) $$\sum_{j=1}^{N} \tau_j (t_j - z) = 0 .$$

This argument can be repeated using the continuity conditions

$$p_{j+1}^{(n-2)}(t_j) = p_j^{(n-2)}(t_j) , \quad (j = 1, 2, \ldots, N) ,$$

equations (4) with $k = n-2$, and equations (8) to establish

(10) $$a_{j+1,n-2} - a_{j,n-2} = C_{n-2}\,\tau_j (t_j - z)^2 , \quad (j = 1, 2, \ldots, N) ,$$

where C_{n-2} is a constant independent of j, and

(11) $$\sum_{j=1}^{N} \tau_j (t_j - z)^2 = 0 .$$

The general result is

$$(12)\quad a_{j+1,n-\alpha} - a_{j,n-\alpha} = C_{n-\alpha}\tau_j(t_j-z)^\alpha, \quad (j=1,2,\ldots,N; \ \alpha = 0,1,\ldots,n) .$$

and

$$(13)\quad \sum_{j=1}^{N} \tau_j(t_j-z)^\alpha = 0, \qquad (\alpha = 0,1,\ldots,n) .$$

Equations (13) we shall call the periodicity conditions for reasons that will become evident later.

If a set of polynomials $p_j(t)$ of degree n define a spline $q_\Delta(t)$ on Γ, equations (12) and (13) must be satisfied; conversely, it is not hard to see that given a set of complex numbers $\tau_1, \tau_2, \ldots, \tau_N$ satisfying equations (13) and (n+1) arbitrary complex coefficients $\alpha_{10}, \alpha_{11}, \ldots, \alpha_{1n}$ defining a complex polynomial $p_1(t)$, then these quantities define a unique polynomial spline $q_\Delta(t)$ of degree n on Γ. Since there are N degrees of freedom in the quantities $\tau_1, \tau_2, \ldots, \tau_N, \alpha_{10}, \ldots, \alpha_{1n}$, a polynomial spline has N degrees of freedom. Later we shall examine two fruitful methods of using this freedom to uniquely determine $q_\Delta(t)$ and, hence, $S_\Delta(z)$.

Representation of Analytic Splines

Suppose that we substitute equations (3) into (2) and integrate. Then

$$S_\Delta(z) = \frac{1}{2\pi i}\sum_{j=1}^{N}\int_{\Gamma_j}\frac{p_j(t)}{t-z}\,dt$$

$$= \frac{1}{2\pi i}\sum_{j=1}^{N}\left\{\sum_{k=1}^{n} a_{jk}\frac{(t-z)^k}{k} + p_j(z)\ln(t-z)\right\}\Bigg|_{t_{j-1}}^{t_j}$$

$$S_\Delta(z) = \frac{1}{2\pi i} \sum_{k=1}^{n} \{-\frac{1}{k} \sum_{j=1}^{N} (a_{j+1,k} - a_{jk})(t_j - z)^k\}$$

$$+ \frac{1}{2\pi i} \sum_{j=1}^{N} p_j(z) \ln\left(\frac{t_j - z}{t_{j-1} - z}\right) .$$

But, in view of equations (12) and (13), the double summation vanishes and we have

$$S_\Delta(z) = \frac{1}{2\pi i} \sum_{j=1}^{N} p_j(z) \ln\left(\frac{t_j - z}{t_{j-1} - z}\right) . \tag{14}$$

From this representation our earlier assertion that $S_\Delta(z)$ is analytic at all points in the complex plane with the exception of logarithmic branch points at the points t_j is clear. A discussion of selecting proper branches of the logarithmic terms is contained in [2].

An alternate representation for $S_\Delta(z)$ is

$$S_\Delta(z) = \frac{1}{2\pi i} \sum_{j=1}^{N} \{p_{j+1}(z) - p_j(z)\} \ln(t_j - z) + p_1(z) . \tag{15}$$

Since, however,

$$p_{j+1}(z) - p_j(z) = \frac{(-1)^n (t_j - z)^n \tau_j}{n!} ,$$

equation (15) can be expressed as

$$S_\Delta(z) = \frac{1}{2\pi i} \frac{(-1)^n}{n!} \sum_{j=1}^{N} \tau_j (t_j - z)^n \ln(t_j - z) + p_1(z) . \tag{16}$$

With the help of the periodicity conditions, we obtain by repeated differentiation of (16) the useful representations

$$(17)\quad S_{\Delta}^{(\alpha)}(z) = \frac{1}{2\pi i}\,\frac{(-1)^{n-\alpha}}{(n-\alpha)!}\sum_{j=1}^{N}\tau_j (t_j - z)^{n-\alpha}\ln(t_j - z) + p_1^{(\alpha)}(z)$$

$$(\alpha = 0, 1, \ldots, n),$$

and

$$(18)\quad S_{\Delta}^{(\alpha)}(z) = -\frac{(\alpha-n-1)!}{2\pi i}\sum_{j=1}^{N}\frac{\tau_j}{(t_j - z)^{\alpha-n}}, \quad (\alpha = n+1, n+2, \ldots).$$

Splines of Multiple Interpolation

Let z_0 be a fixed point in the complex plane not in Δ. Suppose we wish to use the N degrees of freedom inherent in an analytic spline $S_{\Delta}(z)$ so that $S_{\Delta}(z)$ and its first $N-1$ derivatives interpolate to prescribed quantities A_{α} $(\alpha = 0, 1, \ldots, N-1)$ at z_0, i.e.,

$$S_{\Delta}^{(\alpha)}(z_0) = A_{\alpha} \qquad (\alpha = 0, 1, \ldots, N-1).$$

Then, using the periodicity conditions and equations (18) we are led to the equations

$$(19)\quad \begin{bmatrix} 1 & \cdots & 1 \\ (t_1-z_0)^{-1} & \cdots & (t_N-z_0)^{-1} \\ \vdots & & \vdots \\ (t_1-z_0)^{-N+1} & \cdots & (t_N-z)^{-N+1} \end{bmatrix} \begin{bmatrix} \tau_1 (t_1-z_0)^n \\ \\ \tau_{n+1}(t_{n+1}-z_0)^n \\ \tau_{n+2}(t_{n+2}-z_0)^n \\ \vdots \\ \tau_N (t_N-z_0)^n \end{bmatrix} = \begin{bmatrix} 0 \\ \\ 0 \\ \dfrac{2\pi i A_{n+1}}{0!} \\ \vdots \\ \dfrac{2\pi i A_{N-1}}{(N-n-2)!} \end{bmatrix}$$

The matrix appearing in the left member of (19) is a Vandermonde matrix which is non-singular since the t_j are distinct. Thus $\tau_1, \tau_2, \ldots, \tau_N$ are uniquely determined. Equations (17) now permit us to obtain $p_1(z)$. Thus an analytic spline $S_\Delta(z)$ is uniquely determined by prescribing $S_\Delta^{(\alpha)}(z_0)$ $(\alpha = 0, 1, \ldots, N-1)$ at a non-nodal point z_0 . We call splines determined in this manner <u>splines of multiple interpolation</u>.

The Unit Circle

We now consider the interesting situation where the nodal points are the N-th roots of unity, i.e., $t_j = e^{j\frac{2\pi i}{N}}$ $(j = 0, 1, \ldots, N-1)$. This choice of nodal points leads to considerable simplifications. In particular, if $A_0, A_1, \ldots, A_{N-1}$ are prescribed values of $S_\Delta(0), S'_\Delta(0), \ldots, S_\Delta^{(N-1)}(0)$, it can be shown that

$$(20) \qquad S_\Delta(z) = \sum_{p=0}^{n} \frac{A_p}{p!} z^p + \frac{1}{n!} \sum_{p=n+1}^{N-1} \frac{A_p}{(p-n+1)!} \int_0^z \frac{(z-t)^n}{1-t^N} dt .$$

If $f(z)$ is a function holomorphic in $|z| < r \leq 1$ and r is its radius of convergence, the analytic spline of multiple interpolation to $f(z)$ at $z = 0$ behaves much like the truncated power series

$$\sum_{p=0}^{N-1} \frac{A_p}{p!} z^p .$$

We have, in fact, the following convergence theorems:

<u>Theorem</u>. <u>Let</u> $f(z)$ <u>be holomorphic in</u> $|z| < r \leq 1$. <u>Let</u> $\{\Delta_k\}$ <u>be a sequence of meshes on</u> $|z| = 1$ <u>with</u>

$$\Delta_k = \{t_{k,1}, \ldots, t_{k,N_k}\}$$

where

$$t_{kj} = \exp[j2\pi i/N_k]$$

and $N_{k'} > N_k$ if $k' > k$. Then for $r' < r$,

(21) $$\limsup_{N\to\infty}\left[\sup_{|z|=r'} |f^{(n+1)}(z) - S_{\Delta_k}^{(n+1)}(z)|\right]^{1/N_k} = \frac{r'}{r} .$$

For $r \le r' < 1$

(22) $$\limsup_{N\to\infty}\left[\sup_{|z|\le r'} |S_{\Delta_k}^{(n+1)}(z)|\right]^{1/N_k} = r'/r .$$

Theorem. Under the conditions of the preceding theorem, for $r' < r$ and arbitrary positive $\varepsilon < r - r'$, we have

(23) $$\sup_{|z|\le r'} |f^{(p)}(z) - S_{\Delta_k}^{(p)}(z)| \le \text{const.} \left(\frac{r'}{r-\varepsilon}\right)^{N_k} (r')^{n+1-p}$$

$$(p = 0,1, \ldots, n) ,$$

and for $r \le r' < 1$ and arbitrary $\varepsilon > 0$

(24) $$\sup_{|z|\le r'} |S_{\Delta_k}^{(p)}(z)| \le \text{const.} \left(\frac{r'+\varepsilon}{r}\right)^{N_k} (r')^{n+1-p}$$

$$(p = 0, 1, \ldots, n) .$$

Nodal Point Interpolation

An alternate method of using the N degrees of freedom inherent in an analytic spline $S_\Delta(z)$ to advantage is to cause the associated polynomial spline $q_\Delta(t)$ to interpolate to prescribed values $f_0, f_1, \ldots, f_N = f_0$ at the nodal points. The theory becomes especially interesting when the spline is of odd degree and the nodes are uniformly spaced on the circumference of a circle in the complex plane. There is no loss in generality in assuming that $t_j = \exp[j2\pi i/N]$, $(j = 0, 1, \ldots, N)$, i.e., the nodes are the N-th roots of unity. Most of the succeeding formulae do not, however, depend on this choice of nodes. The basic formulae hold for non-uniform spacing and apply to periodic splines on the unit interval as well. If the nodes are equally-spaced with respect to arc length (but with t still the independent variable) along a smooth Jordan curve Γ, the existence and uniform boundedness property which we shall establish are still valid. If, however, Γ is a circle, the crucial matrix that arises is a circulant, a fact which we shall take advantage of later.

It happens that not only do the equations which we shall give apply to the periodic spline on the unit interval, but, in addition, the matrix applicable to $q_\Delta(t)$ with equally-spaced nodes approaches this matrix in the row-max norm as $N \to \infty$. It is this property that can be exploited to show that the n-th derivatives of $q_\Delta(t)$ are uniformly bounded as $N \to \infty$ if $q_\Delta(t)$ interpolates at the points of Δ to a function $f(t)$ in $C^n(\Gamma)$. This limiting property also establishes the existence of $q_\Delta(t)$ for large N. If $N \geq n+1$, methods analogous to those in [4] (Chapter IV) can be used to establish the existence of $q_\Delta(t)$.

Basic Equations

Let $\bar{n} = (n-1)/2$. Then the central divided difference $f[t_{k-\bar{n}}, \ldots, t_{k+\bar{n}}, \ldots, t_{k+\bar{n}}]$ is given by

$$(25)\quad f[t_{k-\bar{n}}, \ldots, t_{k+\bar{n}}] = C_{-\bar{n}}(k) f_{k-\bar{n}} + C^{(k)}_{-\bar{n}+1} f_{k-\bar{n}+1} + \ldots + C^{(k)}_{n} f_{k+\bar{n}}$$

where $f_j = f(t_j)$ and

$$(26)\quad C_j = [(t_{k+j} - t_{k-\bar{n}})(t_{k+j} - t_{k-\bar{n}+1}) \cdots (t_{k+j} - t_{k+j-1})$$

$$\cdots (t_{k+j} - t_{k+j+1}) \cdots (t_{k+j} - t_{k+\bar{n}})]^{-1}$$

$$(j = -\bar{n}, -\bar{n}+1, \ldots, \bar{n}) .$$

Now, let $q_\Delta(t)$ have the representation

$$(27)\quad q_\Delta(t) = q_\Delta(t_k) + q'_\Delta(t_k)(t-t_k) + \ldots + \frac{1}{(n-2)!} q_\Delta^{(n-2)}(t_k)(t-t_k)^{n-2} + \frac{1}{(n-2)!} \int_{t_k}^{t} (t-s)^{n-2} q_\Delta^{(n-1)}(s)\,ds$$

where the path of integration is a subarc of Γ. We let $\Delta_{t_k}^{n-1}$ denote the $(n-1)$th divided difference operator centered at t_k and based on the points $t_{k-\bar{n}}, t_{k-\bar{n}+1}, \ldots, t_{k+\bar{n}}$. it follows that

$$(28)\quad (n-1)!\,\Delta_{t_k}^{n-1} q_\Delta(t) = (n-1)\,\Delta_{t_k}^{n-1} \int_{t_k}^{t} (t-s)^{n-2} q_\Delta^{(n-1)}(s)\,ds$$

$$= (n-1) \sum_{j=-n}^{\bar{n}} C_j(k) \int_{t_k}^{t_{k+j}} (t_{k+j}-s)^{n-2} q_\Delta^{(n-1)}(s)\,ds$$

If we let

$$(29)\quad A_{\alpha\beta} = \frac{1}{t_\beta - t_{\beta-1}} \left[\frac{1}{n-1}(t_\alpha - t_{\beta-1})^{n-1}(t_\beta - t_{\beta-1}) + \frac{1}{(n-1)n}(t_\alpha - t_\beta)^n - \frac{1}{(n-1)n}(t_\alpha - t_{\beta-1})^n\right],$$

$$(30)\quad B_{\alpha\beta} = \frac{1}{t_\beta - t_{\beta-1}} \left[\frac{-1}{n-1}(t_\alpha - t_\beta)^{n-1}(t_\beta - t_{\beta-1}) - \frac{1}{(n-1)n}(t_\alpha - t_\beta)^n + \frac{1}{(n-1)n}(t_\alpha - t_{\beta-1})^n\right],$$

$$
(31)\quad \left.\begin{aligned} R_{k,\ell} &= -(n-1)\sum_{j=\ell}^{-\bar{n}} C_j(k)\,A_{k+j,\,k+\ell+1} \\ S_{k,\ell} &= -(n-1)\sum_{j=\ell}^{-\bar{n}} C_j(k)\,B_{k+j,\,k+\ell+1} \end{aligned}\right\}\ \ell = -1, -2, \ldots, -\bar{n}\ ,
$$

$$
(32)\quad \left.\begin{aligned} R_{k,\ell} &= (n-1)\sum_{j=\ell+1}^{\bar{n}} C_j(k)\,A_{k+j,\,k+j,\,k+\ell+1} \\ S_{k,\ell} &= (n-1)\sum_{j=\ell+1}^{\bar{n}} C_j(k)\,B_{k+j,\,k+\ell+1} \end{aligned}\right\}\ \ell = 0, 1, \ldots, \bar{n}-1;
$$

it can be shown, [2] that

$$
(33)\quad (n-1)!\,\Delta_{t_k}^{n-1} q_\Delta(t) = R_{k,-\bar{n}} M_{k-\bar{n}} + \sum_{\ell=-\bar{n}}^{\bar{n}-2} (R_{k,\ell+1} + S_{k,\ell}) M_{k+\ell+1} + S_{k,\bar{n}-1} M_{k+\bar{n}} \qquad (k = 1, 2, \ldots, N)
$$

where $M_j = q_\Delta^{(n-1)}(t_j)$ $(j = 1, 2, \ldots, N)$.

Equations (33) are applicable to a periodic spline on a Jordan arc and, in particular, cover both the case of a periodic spline on the unit interval and of a complex polynomial spline on the unit circle in the complex plane. Let the nodes, in the latter case, be at the N-th roots of unity, and let

$$
(34)\qquad h_j = t_j - t_{j-1} = e^{j2\pi i/N} - e^{(j-1)2\pi i/N} = \omega_N h_{j-1}
$$

where $\omega_N = e^{2\pi i/N}$. The coefficients in (33) can be shown to be functions of ω_N alone and not of h_j $(j = 1, 2, \ldots, N)$. Near $\omega = 1$, they are continuous functions of ω and since the matrix associated with equations (33) is banded, they approach the equations for $\omega = 1$ in the row max-max norm. For $\omega = 1$, however, equations (33) represent the case of the periodic spline on the unit interval. Thus, for large N the matrix associated with equations (33) is invertible and its inverse is uniformly bounded with respect to N. These properties follow from the analogous properties known to hold for the periodic spline on the unit interval.

On the subarc Γ_1 of Γ let $q_\Delta(t)$ have the representation

$$q_\Delta(t) = Q_1(t) + p(t) \qquad (t \in \Gamma_1) \tag{35}$$

where

$$Q_1(t) = \frac{1}{n!}(t - t_0)^n \frac{M_1 - M_0}{h_1} \tag{36}$$

and $p(t)$ is a polynomial of degree $n-1$. Since (36) gives the correct value for the n-th derivative of $q_\Delta(t)$ on Γ_1, the polynomial $p(t)$ can be chosen so that (35) is valid. Now let

$$Q_2(t) = \frac{1}{n!}(t - t_1)^n \left[\frac{M_2 - M_1}{h_2} - \frac{M_1 - M_0}{h_1}\right] = \frac{1}{n!}(t - t_1)^n \tau_1 . \tag{37}$$

Then

$$q_\Delta(t) = Q_1(t) + Q_2(t) + p(t) \qquad (t \in \Gamma_2) . \tag{38}$$

Continuing in this fashion, we define

$$Q_j(t) = \frac{1}{n!}(t - t_{j-1})^n \left[\frac{M_j - M_{j-1}}{h_j} - \frac{M_{j-1} - M_{j-2}}{h_{j-1}}\right]$$

$$= \frac{1}{n!}(t - t_{j-1})^n \tau_{j-1} ;$$

then

$$q_\Delta(t) = Q_1(t) + Q_2(t) + \ldots + Q_j(t) + p(t) \quad (t \in \Gamma_j) \tag{40}$$
$$(j = 1, 2, \ldots, N) \ .$$

Observe that formula (40) holds with $j = N+1$ since

$$\sum_{j=2}^{N+1} Q_j(t) = \frac{1}{n!}(-1)^n \sum_{j=1}^{N} (t-t_j)^n \tau_j = 0 \tag{41}$$

by the last ($\alpha = n$, $z = t$) of the periodicity conditions; thus, $q_\Delta(t)$ is periodic and in $C^n(\Gamma)$. We also point out that the remaining periodicity conditions are obtainable from this one by differentiation and they guarantee the periodicity of $q_\Delta^{(\alpha)}(t)$, $\alpha = 1, 2, \ldots, n-1$.

Since $M_0, M_1, \ldots, M_N = M_0$ can be obtained by means of equations (33), the polynomial spline $q_\Delta(t)$ is completely determined once $p(t)$ is known. But, if we require

$$\begin{aligned} p(t_1) &= f_1 - Q_1(t_1) \\ p(t_2) &= f_2 - Q_1(t_2) - Q_2(t_2) \\ &\cdots\cdots\cdots\cdots\cdots\cdots \\ p(t_n) &= f_n - Q_1(t_n) - Q_2(t_n) - \ldots - Q_n(t_n) \ , \end{aligned} \tag{42}$$

these equations uniquely determine $p(t)$ such that $q_\Delta(t)$ will interpolate to $f_0, f_1, \ldots, f_n$ at $t_0, t_1, \ldots, t_n$, respectively. It can be shown by separate argument that $q_\Delta(t)$ also interpolates to f_j at t_j $(j = n+1, n+2, \ldots, N)$. Now that $q_\Delta(t)$ is determined, $S_\Delta(t)$ is also determined.

Convergence Properties of $q_\Delta(t)$

Let $f^{(\alpha+1)}$ be continuous on Γ and let p be a point on Γ. Then

$$f(t) = f(p) + f'(p)(t-p) + \dots + \frac{1}{\alpha!} f^{(\alpha)}(p)(t-p)^{\alpha} + \int_p^t (t-s)^{\alpha} f^{(\alpha+1)}(s)\,ds \quad . \tag{43}$$

Now, let

$$\Delta^{\alpha}_{t_k} f(t) = f[t_{k-\frac{\alpha}{2}}, \dots, t_{k+\frac{\alpha}{2}}] \qquad (\alpha \text{ even})$$

(44)

$$\Delta^{\alpha}_{t_k} f(t) = f[t_{k-\frac{\alpha-1}{2}}, \dots, t_{k+\frac{\alpha+1}{2}}] \qquad (\alpha \text{ odd})$$

$$(k = 1, 2, \dots, N)$$

It follows that

$$f^{(\alpha)}(p) - \alpha!\,\Delta^{\alpha}_{t_k} f(t) = \Delta^{\alpha}_{t_k} \int_p^t (t-s)^{\alpha} f^{(\alpha+1)}(s)\,ds \quad . \tag{45}$$

Now impose the requirement $|t_j - t_{j-1}| = h$ for all j and observe that we then can always choose t_k such that $|t_k - p| < h$. With these restrictions

$$|f^{(\alpha)}(p) - \alpha!\,\Delta^{\alpha}_{t_k} f(t)| = O(\|f^{(\alpha+1)}\|_{\infty} h) \tag{46}$$

since the coefficients $C_j(k)$ in (25) have the property

$$|C_j(k)| = O(h^{-\alpha}) \quad , \tag{47}$$

and

(48) $$\left| \int_{p}^{t_{k+j}} (t_{k+j}-s)^{\alpha} f^{(\alpha+1)}(s)\,ds \right| = O(h^{\alpha+1} \| f^{(\alpha+1)} \|_{\infty})$$

$$(j = 0, \pm 1, \ldots, \pm n;\ |t_k - p| < h) \ .$$

Here we have adopted the notation

(49) $$\| g \|_{\infty} = \sup_{\Gamma} |g(t)| \ .$$

If we choose p between t_{k-1} and t_k on Γ, it follows that

(50) $$|\alpha!\, \Delta^{\alpha}_{t_k} f(t) - \alpha!\, \Delta^{\alpha}_{t_{k-1}} f(t)| = O(\| f^{(\alpha+1)} \|_{\infty}\, h) \ .$$

We now infer from (50) that

(51) $$\max_{k} |(\alpha+1)\, \Delta^{\alpha+1}_{t_k} f(t)| = O(\| f^{(\alpha+1)} \|_{\infty}) \ .$$

Assume now that Γ is either the unit interval with $t_j = j/N$, $(j = 0, 1, \ldots, N)$ or the unit circle in the complex plane with $t_j = e^{j2\pi i/N}$, $(j = 0, 1, \ldots, N)$. In either of these two cases the matrix associated with equations (33) is a circulant. Consequently, its inverse, which exists for large N, is also a circulant. Moreover, we know that these inverse matrices are uniformly bounded with respect to N. With the aid of these properties we can show that

(52) $$\left| \frac{M_j - M_{j-1}}{h_j} \right| = O(\| f \|_{\infty}\, h^{-n}), \quad (j = 1, 2, \ldots, N)$$

if $f(t)$ is continuous on Γ. If $f^{(\alpha+1)}(t)$ is continuous on Γ, we can use (51) to show

(53) $$\left| \frac{M_j - M_{j-1}}{h_j} \right| = O(\| f^{(\alpha+1)} \|_{\infty}\, h^{-n+\alpha+1}), \quad (j = 1, 2, \ldots, N) \ .$$

But from (36) and (39) we have

$$(54)\quad |Q_j^{(\alpha+1)}(t)| = O(h^{n-\alpha-1}O(\left|\frac{M_j-M_{j-1}}{h_j}\right|)),\ (j=1, 2, \ldots, N)\ ;$$

thus

$$(55)\quad |Q_j^{(\alpha+1)}(t)| = O(\|f^{(\alpha+1)}\|_\infty),\qquad (\alpha = -1, 0, \ldots, n-1\ ;\ j = 1, 2, \ldots, N)\ .$$

We also have

$$(56)\quad |Q_j(t)| = O(\|f^{(\alpha+1)}\|_\infty\ h^{\alpha+1})$$

$$(\alpha = -1, 0, \ldots, n-1; j = 1, 2, \ldots, N)$$

Furthermore, we can employ the interpolation conditions (42) imposed on $p(t)$ to show

$$(57)\quad |p^{(\alpha+1)}(t)| = O(\|f^{(\alpha+1)}\|_\infty),\quad (t \in \Gamma_1 \cup \Gamma_2 \cup \ldots \cup \Gamma_{\alpha+1})$$

if $f^{(\alpha+1)}(t)$ is continuous on Γ. Equations (56) and (57) combine to give

$$(58)\quad |q_\Delta^{(\alpha+1)}(t)| = O(\|f^{(\alpha+1)}\|_\infty),\quad (t \in \Gamma_1 \cup \Gamma_2 \cup \ldots \cup \Gamma_{\alpha+1})$$

if $q_\Delta(t)$ interpolates to $f(t)$ at the points t_j $(j=0, 1, \ldots, N)$ and $f^{(\alpha+1)}(t)$ is continuous on Γ. Here as elsewhere the constant implied by the right side of (58) is independent of $f(t)$ and its derivatives. Furthermore, if a different initial point is chosen, it remains unchanged. Thus (58) holds for all t on Γ.

Now let $0 \leq \alpha \leq n-1$ and $f^{(\alpha+1)}(t)$ be continuous on Γ. Apply (46) to the function $f(t) - q_\Delta(t)$ and note that $\Delta_{t_k}^{\alpha} f(t) = \Delta_{t_k}^{\alpha} q_\Delta(t)$. It follows that

$$(59)\quad |f^{(\alpha)}(t) - q_\Delta^{(\alpha)}(t)| = O(\|f^{(\alpha+1)}\|_\infty h + \|q_\Delta^{(\alpha+1)}\|_\infty h)$$

$$= O(\|f^{(\alpha+1)}\|_\infty h) \ .$$

Again, by using $f(t) - q_\Delta(t)$ in (43) in place of $f(t)$ we have

$$(60)\quad O = (\alpha-1)! \{\Delta_{t_k}^{\alpha-1} f(t) - \Delta_{t_k}^{\alpha-1} q_\Delta(t)\}$$

$$= \{f^{(\alpha-1)}(p) - q_\Delta^{(\alpha-1)}(p) + \frac{1}{\alpha}\{f^{(\alpha)}(p) - q_\Delta^{(\alpha)}(p)\}\Delta_{t_k}^{\alpha-1}(t-p)^\alpha$$

$$+ \frac{1}{\alpha}\Delta_{t_k}^{\alpha-1}\int_p^t \frac{f^{(\alpha-1)}(s) - q_\Delta^{(\alpha+1)}(s)\}ds}{(t-s)^\alpha} \ .$$

Consequently,

$$(61)\qquad |f^{(\alpha-1)}(p) - q^{(\alpha-1)}(p)| = O(\|f^{(\alpha+1)}\|_\infty h^2) \ .$$

By further repetition of this argument we obtain the general result

$$(62)\quad |f^{(\alpha-\beta)}(p) - q_\Delta^{(\alpha-\beta)}(p)| = O(\|f^{(\alpha+1)}\|_\infty h^{\beta+1}), \quad (\beta = 0, 1, \dots, \alpha) \ .$$

If, in particular, $\alpha+1 = n$, then (62) and (58) can be combined and rewritten in the form

$$(63)\quad |f^{(\beta)}(t) - q_\Delta^{(\beta)}(t)| = O(\|f^{(n)}\|_\infty h^{n-\beta}), \ (\beta = 0, 1, \dots, n; t \in \Gamma) \ .$$

Now suppose that instead of assuming $f^{(n)}(t)$ is continuous on Γ, we assume $f^{(n)}(t)$ is absolutely continuous on Γ. Then for $|t-t_k| < h$,

$$(64)\quad |f^{(n)}(t_k) - f^{(n)}(t)| = O(\sup_{t_k} \int_{t_{k-n}}^{t_{k+n}} |f^{(n+1)}(s)|\,|ds|)\ ,$$

and

$$(65)\quad |\Delta^n_{t_k} f(t) - \Delta^n_{t_k} f(t)| = O(\sup_{t_k} \int_{t_{k-n}}^{t_{k+n}} |f^{(n+1)}(s)|\,|ds|)\ .$$

It follows that

$$(66)\quad \left|\frac{M_{k+1}-M_k}{h_{k+1}} - \frac{M_k-M_{k-1}}{h_k}\right| = O(\sup_{t_k} \int_{t_{k-n}}^{t_{k+n}} |f^{(n+1)}(s)|\,|ds|)\ .$$

If we write equations (33) as

$$(67)\quad (n-1)!\,\Delta^{n-1}_{t_k} f(t) = \sum_{j=-\bar{n}}^{\bar{n}} a_j M_{k+j} \quad (k = 1, 2, \ldots, N)\ ,$$

then

$$(68)\quad (n-1)!\,\frac{\Delta^{n-1}_{t_k} f(t) - \Delta^{n-1}_{t_{k-1}} f(t)}{h_k} = \frac{1}{h_k} \sum_{j=-n}^{\bar{n}} h_{k+j} a_j \frac{M_{k+j}-M_{k+j-1}}{h_{k+j}}$$

$$(k = 1, 2, \ldots, N)\ .$$

But, it is known that $\sum_{j=-\bar{n}}^{\bar{n}} a_j = 1$, i.e. $a_0 = -a_{-\bar{n}} - \ldots -a_{\bar{n}}$.
Using this fact and equations (66) and (68), we can establish that

$$(69)\quad |f^{(n)}(t) - q^{(n)}_\Delta(t)| = O(\sup_{t_k} \int_{t_{k-n}}^{t_{k+n}} |f^{(n+1)}(s)|\,|ds|)\ .$$

The same methods used to obtain (63) can now be employed to show

$$(70)\quad |f^{(\beta)}(t) - q_{\Delta}^{(\beta)}(t)| = O(\sup_{t_k} \int_{t_{k-n}}^{t_{k+n}} |f^{(n+1)}(s)|\,|ds| \cdot h^{n-\beta})$$

$$(\beta = 0, 1, \ldots, n) \ .$$

If $f^{(n+1)}(t)$ is in $C(\Gamma)$, we can replace (70) by

$$(71)\qquad |f^{(\beta)}(t) - q^{(\beta)}(t)| = O(\|f^{(n+1)}\|_{\infty}\, h^{n+1-\beta})$$

$$(\beta = 0, 1, \ldots, n) \ .$$

Convergence of $S_{\Delta}(z)$

From the Cauchy integral

$$(72)\qquad S_{\Delta}(z) - f(z) = \frac{1}{2\pi i} \int_{\Gamma} \frac{q_{\Delta}(t) - f(t)}{t - z}\, dt$$

we have

$$(73)\quad |S_{\Delta}(z) - f(z)| = O(\|f^{(\alpha+1)}\|_{\infty}\, h^{\alpha+1}) \qquad (|z| \le r < 1)$$

if $f^{(\alpha+1)}(t)$ is in $C(\Gamma)$ and Γ is the unit circle. The constant implied by the right side of (73) deteriorates as $r \to 1$, however. We can go further and assert that

$$(74)\qquad |S_{\Delta}^{(\beta)}(z) - f^{(\beta)}(z)| = O(\|f^{(\alpha+1)}\|_{\infty}\, h^{(\alpha+1)})$$

$$(|z| \le r < 1;\ \beta = 0, 1, \ldots,) \ .$$

In this case the constant deteriorate with increasing β as well as with increasing r .

If a function g(t) satisfies a Hölder condition

(75) $$|g(t') - g(t'')| \leq A|t' - t''|^{\alpha} \quad (0 < \alpha \leq 1)$$

for any pair of points t', t'' on Γ and if

(76) $$G(z) = \frac{1}{2\pi i} \int_{\Gamma} \frac{g(t)}{t-z} dt ,$$

then the Plemelj formulas assert

(77) $$G^{+}(t_0) = \frac{1}{2} g(t_0) + \frac{1}{2\pi i} \int_{\Gamma} \frac{g(t)}{t-t_0} dt$$

$$= g(t_0) + \frac{1}{2\pi i} \int_{\Gamma} \frac{g(t)-g(t_0)}{t-t_0} dt$$

where t_0 is on Γ, $G^{+}(z)$ is the limit of $G(z)$ as $t \to t_0$ from within Γ, and the integrals are interpreted as Cauchy Principal Values. If $g^{(\beta)}(t)$ satisfies a similar Hölder condition, a similar formula is applicable. These formulas allow us to relate the convergence of $S_{\Delta}(z)$ to that of $q_{\Delta}(t)$ in a fashion such that the convergence does not deteriorate as z approach Γ . The integrals in (77) cause trouble and if simple argument is used, lead to a loss of one in the exponent of h in the rate of convergence. A more careful argument is possible which will reduce this loss.

Power Series

The ensuing discussion is given explicitly for analytic splines of degree three with their nodes at the N-th roots of unity. As long, however, as the nodes are equally-spaced on the circumference of a circle in the complex plane, the discussion carries over with minor modification. More important, the analysis also applies to analytic splines of degree n; the restriction to cubic splines is for notational convenience and greater simplicity in presentation. We do relate the convergence properties of analytic splines with multiple-interpolation at $z = 0$, to corresponding properties of analytics splines with nodal interpolation. Here the results are most significant when n is odd since it is in this case that we have adequate knowledge of the existence and convergence properties of analytic splines with nodal interpolation.

Let a cubic analytic spline have the power series representation

$$(78) \qquad S_\Delta(z) = S_\Delta(0) + S'_\Delta(0)z + \frac{1}{2}S''_\Delta(0)z^2 + \dots \ .$$

But, we know

$$(79) \quad S_\Delta^{(\alpha)}(z) = \frac{(\alpha-4)!}{2\pi i}\sum_{j=1}^{N}\frac{\tau_j}{(t_j-z)^{\alpha-3}}, \qquad (\alpha = 4, 5, \dots) \ .$$

Now let $\alpha = \omega_\alpha N + k$ where $0 \le k < N$. Then, since $t_j^N = 1$ for all j,

$$(80) \qquad S_\Delta^{(\alpha)}(0) = \frac{(\omega_\alpha N + k - 4)!}{2\pi i}\sum_{j=1}^{N}\frac{\tau_j}{t_j^{k-3}} \ .$$

For $k = 0, 1, 2, 3$ the periodicity conditions imply that $(\omega_\alpha > 0)$

$$(81) \qquad S_\Delta^{(\omega_\alpha N+k)}(0) = 0 \ ,$$

and for $k \ge 4$

$$(82) \qquad S_\Delta^{(\omega_\alpha N+k)}(0) = \frac{(\omega_\alpha N+k-r)!}{(k-4)!} S_\Delta^{(k)}(0) \ .$$

Thus, (78) can be written as

$$(83)\quad S_\Delta(z) = S_\Delta(0) + S'_\Delta(0)z + \frac{1}{2}S''_\Delta(0)z^2 + \frac{1}{6}S_\Delta'''(0)z^3$$

$$+ \sum_{k=4}^{N} \frac{S_\Delta^{(k)}(0)}{(k-4)!} z^k \left\{ \sum_{\omega_\alpha=1}^{\infty} \frac{(\omega_\alpha N+k-4)!}{(\omega_\alpha N+k)!} z^{\omega_\alpha N} \right\} .$$

Now, let $C_{k,\Delta}(z)$ be the analytic spline determined by

$$(84)\quad C_{k,\Delta}^{(\alpha)}(z) = \delta_{\alpha k}, \quad (\alpha = 0, 1, \ldots, N-1; \; k = 0, \ldots, N-1)$$

where $\delta_{\alpha j}$ is the Kronecker delta. In particular,

$$(85)\quad C_{k,\Delta}(z) = \frac{z^k}{k!} \quad (k = 0, 1, 2, 3) \; .$$

For $4 \leq k < N$ the splines $C_{k,\Delta}(z)$ are non trivial and can be shown to be given by (in fact, for $k = 0, 1, 2, 3$ as well)

$$(86)\, C_{k,\Delta}(z) = \frac{z^k}{(k-4)!} \sum_{r=0}^{\infty} \frac{1}{(rN+k)(rN+k-1)(rN+k-2)(rN+k-3)} z^{rN}$$

or

$$(87)\quad C_{k,\Delta}(z) = \frac{z^k}{k!}\left\{1 + R_k(z)\right\}$$

where

$$(88)\quad R_k(z) = \sum_{r=1}^{\infty} \frac{k(k-1)(k-2)(k-3)}{(rN+k)(rN+k-1)(rN+k-2)(rN+k-3)} z^{rN} ,$$

If $S_\Delta^{(k)}(z)$ interpolates to A_k $(k = 0, 1, \ldots, N-1)$ at $z = 0$, we have the representation

(89) $$S_\Delta(z) = \sum_{k=0}^{N-1} A_k C_{k,\Delta}(z)$$

for an analytic spline of multiple interpolation at $z = 0$.

Generalized Hypergeometric Functions

The function ${}_AF_B\,(a_1, a_2, \ldots, a_A; b_1, b_2, \ldots, b_B; w)$ defined by

(90) $${}_AF_B\,(a_1, a_2, \ldots, a_A; b_1 b_2, \ldots, b_B; w) = \sum_{n=0}^{\infty} \frac{(a_1)_n \ldots (a_A)_n}{(b_1)_n \ldots (b_B)_n} \frac{w^n}{n!}$$

where

(91) $$(C)_0 = 1, \quad (C)_n = C(C+1) \ldots (C+n-1) \qquad n > 1$$

is known as a generalized hypergeometric function. It satisfies the linear differential equation $(D_w = \frac{d}{dw})$

(92) $$\{wD_w(wD_w + b_1 - 1) \ldots (wD_w + b_B - 1) - wD_w(wD_w + a_1 - 1) \ldots$$
$$(wD_w - a_A - 1)\}y = 0 .$$

Considered as a power series in w, the function ${}_AF_B(w)$ has $|w| = 1$ as its circle of convergence. If $A + B = 1$, the differential equation (92) is Fuchsian and has regular singularities at $w = 0, 1, \infty$. In this case (92) is replaced by

(93) $$\sum_{\ell=1}^{B} w^{\ell-1}(a_\ell w - b_\ell)\frac{d^\ell y}{dw^\ell} + a_0 y + w^B(1-w)\frac{d^{B+1} y}{dw^{B+1}} = 0 ,$$

and the power series (90), which always converges for $|w| < 1$, converges for $w = 1$, and hence absolutely, if

$$\text{Re}\left(\sum_{\ell=1}^{B} b_\ell - \sum_{\ell=1}^{A} a\right) > 0 \quad . \tag{94}$$

Now let $w = z^N$ and consider the general term $T_r(w)$ in the power series (88). This term can be shown to be given by

$$T_r(w) = \frac{(\frac{k}{N})_r (\frac{k-1}{N})_r (\frac{k-2}{N})_r (\frac{k-3}{N})_r (1)_r}{(\frac{k}{N}+1)_r (\frac{k-1}{N}+1)_r (\frac{k-2}{N}+1)_r (\frac{k-3}{N}+1)_r} \frac{w^r}{r!} \quad . \tag{95}$$

Thus

(96) $C_{k,\Delta}(z) =$

$$= \frac{z^k}{k!} \, {}_5F_4\left(1, \frac{k-3}{N}, \frac{k-2}{N}, \frac{k-1}{N}, \frac{k}{N}, \frac{k-3}{N}+1, \frac{k-2}{N}+1, \frac{k-1}{N}+1 ; \frac{k}{N}+1; z^N\right) \quad .$$

It is easily verified that the inequality (94) is satisfied. Consequently $C_{k,\Delta}(z)$ converges absolutely on $|z| = 1$. This fact is easily seen directly by examining (86). For simplicity, we write (96) as

$$C_{k,\Delta}(z) = \frac{z^k}{k!} F_k(z) \quad . \tag{97}$$

If $|z| \leq 1$ we can use (86) to show

$$|F_k^{(\alpha)}(z)| < k^\alpha G_\alpha < \infty \qquad (\alpha = 0, 1, 2) \tag{98}$$

where G_α is independent of k. For $\alpha = 3$ the quantity G_α depends on $|z|$ and approaches ∞ as $|z| \to 1$.

Convergence of Splines of Multiple Interpolation

If $f(z)$ is holomorphic in $|z| < 1$, the spline of multiple interpolation to $f(z)$ at $z = 0$ converges to $f(z)$ uniformly on any circle $|z| \leq r < 1$. We can go further and show that $S_\Delta(z)$ converges to $f(z)$ in essentially the same manner as the analytic spline of nodal interpolation to $f(z)$. Suppose, for instance, that $f^{(4)}(z)$ is continuous on $|z| = 1$. Let $S_\Delta(z)$ be the spline of multiple interpolation to $f(z)$ at $z = 0$ and $\hat{S}_\Delta(z)$ the spline of nodal interpolation to $f(z)$ on $|z| = 1$. Then

$$(99) \quad |f^{(\alpha)}(z) - S_\Delta^{(\alpha)}(z)| \leq |f^{(\alpha)}(z) - \hat{S}_\Delta^{(\alpha)}(z)| + |\hat{S}_\Delta^{(\alpha)}(z) - S_\Delta^{(\alpha)}(z)| .$$

But

$$(100) \quad |f^{(\alpha)}(z) - \hat{S}_\Delta^{(\alpha)}(z)| = O(\|f^{(n+1)}\|_\infty h^{4-\alpha}) \quad (\alpha = 0, 1, 2, 3, \ldots, n).$$

Let us consider

$$(101) \qquad \varepsilon_\Delta(z) = \hat{S}_\Delta(z) - S_\Delta(z) .$$

From (89) we have

$$(102) \qquad \varepsilon^{(\alpha)}(z) = \sum_{n=0}^{N-1} \{\hat{S}_\Delta^{(k)}(0) - S_\Delta^{(k)}(0)]C_{k,\Delta}^{(\alpha)}(z)$$

$$= \sum_{k=0}^{N-1} [\hat{S}_\Delta^{(k)}(0) - f^{(k)}(0)]C_k^{(\alpha)}(z)$$

$$(\alpha = 0, 1, 2, 3) .$$

But

$$(103) \quad |\hat{S}_\Delta^{(k)}(0) - f^{(k)}(0) = \frac{k!}{2\pi} \left| \int_{|t|=1} \frac{q_\Delta(t) - f(t)}{t^{k+1}} dt \right| = k!\, O(\|f^{(4)}\|_\infty h^4) .$$

We can show, however, that

$$(104) \qquad |C_{k,\Delta}^{(\alpha)}(z)| \leq \frac{|z|^{k-\alpha}}{(k-\alpha)!} H_{\alpha}(|z|) \quad (\alpha = 0, 1, 2, 3; k \geq \alpha)$$

and

$$C_{k,\Delta}^{(\alpha)}(z) = 0, \quad 0 < k < \alpha .$$

where $H_\alpha(|z|)$ is independent of k and N. The function $H_\alpha(|z|)$ is bounded by constants k_α on $|z| \leq 1$ for $\alpha = 0, 1, 2$; but is unbounded as $|z| \to 1$ for $\alpha = 3$. Now, since $N = 2\pi/h$ and the largest value of k in (102) is $N-1$; we can show that

$$(105) \qquad |\varepsilon_{\Delta}^{(\alpha)}(z)| = O(H_{\alpha}(z)| \, |f^{(4)}||_{\infty} h^{4-\alpha} \sum_{k=\alpha}^{N-1} |z^{k-\alpha}|).$$

Thus, for $0 \leq |z| \leq r < 1$ we have

$$(106) \qquad |f^{(\alpha)}(z) - S_{\Delta}^{(\alpha)}(z)| = O(\|f^{(4)}\|_{\infty} h^{4-\alpha})$$

$$(\alpha = 0, 1, 2, 3)$$

and for $|z| = 1$

$$(107) \qquad |f^{(\alpha)}(z) - S_{\Delta}^{(\alpha)}(z)| = O(\|f^{(4)}\|_{\infty} h^{3-\alpha})$$

$$(\alpha = 0, 1, 2) \quad .$$

Clearly, other convergence results are obtainable under assumptions on $f(t)$ other than the assumption that $f^{(4)}(t)$ is continuous on $|z| = 1$ which we have made here.

REFERENCE

[1] J. H. Ahlberg, E. N. Nilson, and J. L. Walsh, "Complex Cubic Splines", Trans. Amer. Math. Soc. 129 (1967), 391-413.

[2] J. H. Ahlberg, E. N. Nilson, and J. L. Walsh, "Properties of Analytic Splines (1). Complex Polynomial Splines", J. Math. Anal. and Applic., to appear.

[3] J. H. Ahlberg and E. N. Nilson, "Existence of Complex Polynomial Splines on Fine Uniform Meshes", Brown University Report, Feb. 1969 (Contract Nonr 562(36).

[4] J. H. Ahlberg, E. N. Nilson, and J. L. Walsh, The Theory of Splines and Their Applications, Academic Press (New York) 1967.

The author's research was supported by the Office of Naval Research Contract Nonr 562(36) with Brown University.

Prolongement d'une Fonction en une Fonction Différentiable. Diverses Majorations sur le Prolongement

CHRISTIAN COATMELEC

I - INTRODUCTION

Soit E un compact de $\mathbb{R}^n$ et E_1 un ouvert (de $\mathbb{R}^n$) contenant E. Pour $f \in C^m(E_1)$ nous appellerons $T_A[f]$ le polynôme de Taylor, de degré m, de f au point $A \in E_1$. Une fonction $f \in C^m(E_1)$ induit sur E un champ de polynômes de degrés $\leq m$:

$E \ni A \to T_A[f] \in \Pi_m$ où Π_m est l'espace vectoriel, de dimension $\binom{m+n}{n}$, des polynômes à n indéterminées et de degrés $\leq m$.

Ce champ est appelé champ taylorien associé à f sur E. Réciproquement donnons nous un champ $A \xrightarrow{T} T_A$ défini sur E et à valeurs dans Π_m. Introduisons comme dans (Coatmélec I): AB distance euclidienne dans $\mathbb{R}^n$

$$\frac{\partial^K g}{\partial X^K} = \frac{\partial^{k_1+\dots+k_n} g}{\partial X_1^{k_1} \dots \partial X_n^{k_n}} \quad \text{avec} \begin{cases} K = (k_1, k_2, \dots, k_n) \\ k = k_1 + \dots + k_n \end{cases}$$

$$\mathfrak{J}_1(A,B,T,m) = \sup_{M \in \mathbb{R}^n} \frac{|T_B(M) - T_A(M)|}{(AM+BM)^m}$$

$$\mathfrak{J}_2(A,B,T,m) = \sup_{0 \leq k \leq m} \frac{\left|\frac{\partial^K T_B}{\partial X^K}(A) - \frac{\partial^K T_A}{\partial X^K}(A)\right|}{AB^{m-k}}$$

$$\mathfrak{J}_3(A,B,T,m) = \sup_{M \in C(AB)} \frac{|T_A(M) - T_B(M)|}{AB^m}$$ où $C(AB)$ est la boule de diamètre AB.

$\mathfrak{J}_2$ a été introduit par Whitney (Whitney II) et $\mathfrak{J}_1$ par Glaeser (Glaeser I). En introduisant $\mathfrak{J}_3$ dans (Coatmélec I) nous avons démontré que : $\mathfrak{J}_1(A,B,T,m) \leq e\, \mathfrak{J}_2(A,B,T,m) \leq n^m e 2^m (m!)^2\, \mathfrak{J}_3(A,B,T,m) \leq n^m 2^{2m} e (m!)^2\, \mathfrak{J}_1(A,B,T,m)$ qui montre l'équivalence des $\mathfrak{J}_i$.

(I,1) Prolongement d'un champ de polynômes

Le théorème de Whitney (Whitney II) peut s'énoncer ainsi (Coatmelec I): Pour qu'il existe une fonction $\hat{f} \in C^m(E_1)$ dont le champ taylorien coincïde en restriction sur E, avec le champ donné : $A \to T_A$, il faut et il suffit qu'il existe un module de continuité ω_E tel que : $\forall (A,B) \in E^2$ on ait $\mathfrak{J}_i(A,B,T,m) \leq \omega_E(AB)$ avec $i = 1$ ou 2 ou 3 .

Soit $W^m(E)$ l'espace des champs T, définis sur E, qui sont restrictions à E de champs tayloriens induits par des fonctions $\hat{f}$ m fois continûment dérivables dans $E_1 \supset E$. Le théoremè de Whitney caractérise donc les éléments T de $W^m(E)$.

Soit T un élément de $W^m(E)$: $A \xrightarrow{T} T_A$. On peut poser

$$\|T\|_E^m = \max_{0 \leq k \leq m} \left(\max_{A \in E} \left| \frac{\partial^K T_A}{\partial X^K}(A) \right| \right)$$

qui correspond à la norme

$$\|g\|_E^m = \max_{0 \leq k \leq m} \left\| \frac{\partial^K g}{\partial X^K} \right\|_E \text{ avec } \|g_1\|_E = \max_{M \in E} |g_1(M)| .$$

$W^m(E)$ est alors normé, malheureusement $W^m(E)$ n'est pas en général complet pour la norme $\| \ \|_E^m$. Au contraire si nous posons

$$|||T|||_E^{m,i} = \max\left(\|T\|_E^m, \max_{A \in E, B \in E} \mathfrak{J}_i(A,B,T,m)\right)$$

$W^m(E)$ est complet pour les trois normes correspondant à $i = 1, 2, 3$.

$W^m(E)$ est isomorphe au quotient de l'ensemble $C^m(C)$ des fonctions m fois continûment dérivables dans un hypercube C contenant E par l'idéal fermé des fonctions de $C^m(C)$ qui sont m-plates en tout point de E. Les 3 normes $||| \;\; |||_E^{m,i}$ sont uniformément équivalentes à la norme quotient, borne inférieure des normes qui rendent $W^m(E)$ complet (Glaeser I).

(I, 2) Cas où $n=1$. Prolongement d'une fonction $f \in C(E)$ en une fonction $\hat{f} \in C^m(E_1)$

Dans un article (Whitney II) Whitney donne une condition nécessaire et suffisante pour qu'une fonction f définie sur un fermé E de $\mathbb{R}$ soit prolongeable en une fonction $\hat{f} \in C^m(\mathbb{R})$ avec $\hat{f}|_E = f$. Merrien dans (Merrien I) a clarifié les démonstrations. Nous reprenons cependant dans (II, 1), (II, 2), (II, 3) des démonstrations contenues à quelques détails près dans (Merrien I) pour préciser les constantes qui interviennent.

En (II,4) nous utilisons (Coatmélec I) pour montrer qu'un champ de polynômes est bien le champ induit sur E par une fonction $\hat{f} \in C^m(E_1)$.

En (III, 1) nous utilisons le prolongement extrémal de Glaeser (Glaeser II) puis nous cherchons en (III, 1), (III, 2) et (III, 3) des majorations concernant le prolongement obtenu. Les méthodes de prolongement du type de celles utilisées par Whitney ne permettent pas d'obtenir commodément des majorations. Nous montrons, en particulier, le résultat annoncé par Glaeser dans (Glaeser I):

$$|||\hat{T}|||_C^{m,i} \leq \Gamma_1(d, m)\, |||T|||_E^{m,i}$$

où Γ_1 dépend de la longueur de l'intervalle C (contenant E) sur lequel on prolonge. Nous explicitons cette constante et nous montrons, en précisant les valeurs de constantes Γ_{ij} ne dépendant que de m, que lorsqu'on prolonge à $\mathbb{R}$ d'une certaine façon on a :

$$\sup_{\substack{u \in \mathbb{R} \\ v \in \mathbb{R}}} \mathfrak{J}_i(u,v,\hat{T},m) \leq \Gamma_{ij} \sup_{\substack{a \in E \\ b \in E}} \mathfrak{J}_j(a,b,T,m)$$

On peut alors donner des majorations qui ne font intervenir que les données du départ.

En particulier, on peut obtenir des majorations en fonction de m, de la borne supérieure des valeurs absolues des différences divisées d'ordre $\leq m$ sur E et de ω module de continuité donné dans les hypothèses de départ.

II - LE THEOREME DE WHITNEY SUR $\mathbb{R}$

Soit E un compact contenant plus de $(m+1)$ points de $\mathbb{R}$, C un intervalle fermé contenant E. Dans toute la suite E, C et l'entier $m \geq 0$ sont fixés. Soit $\mathfrak{A} = (a_0, a_1, \ldots, a_m)$ avec $a_{i-1} \leq a_i$ un ensemble de $(m+1)$ points de $\mathbb{R}$. On a $\mathfrak{A} \in \mathbb{R}_{m+1}$ où $\mathbb{R}_{m+1}$ est le quotient de $\mathbb{R}^{m+1}$ par le groupe des permutations des $(m+1)$ coordonnées de $\mathbb{R}^{m+1}$. Soit $\mathbb{R}^*_{m+1}$ l'ensemble des $\mathfrak{A} = (a_0, a_1, \ldots, a_m)$ tels que $\forall i = 1$ à m on ait $a_{i-1} < a_i$.

Nous définissons sur $\mathbb{R}_{m+1}$ une distance en posant:

$$d(\mathfrak{A}, \mathfrak{B}) = \frac{1}{m+1} \sum_{i=0}^{m} |a_i - b_i| .$$

On définit de même E_{m+1}, E^*_{m+1}, C_{m+1}, C^*_{m+1}. Soit f définie sur E. Pour $\mathfrak{A} \in E^*_{m+1}$ nous désignerons par $\mathfrak{F}(\mathfrak{A})$ la différence divisée, d'ordre m, relative à f sur $\mathfrak{A}$.

(II, 0) Un lemme concernant $d(\mathfrak{A}, \mathfrak{B})$

Soit $\mathfrak{A} \in \mathbb{R}_{m+1}$ et $\mathfrak{B} \in \mathbb{R}^*_{m+1}$.
On a: $a_0 < a_1 \ldots < a_m$ et $b_0 < b_1 \ldots < b_m$.

Lemme: Dans le calcul de la distance $d(\mathfrak{A},\mathfrak{B})$ on peut écrire $d(\mathfrak{A},\mathfrak{B}) = \frac{1}{m+1}\sum_{j=0}^{s}|\alpha_j - \beta_j|$ avec $s \le m$ où les α_j sont obtenus en prenant les a_i non contenus dans $\mathfrak{B}$ et en les indiçant de telle façon que $\alpha_j < \alpha_{j+1}$ et de même pour les β_j.

En effet il suffit de considérer que $\mathfrak{A}$ et $\mathfrak{B}$ ont un point seulement en commun: $a_k = b_p$ et que par exemple $p<k$.

$$\sum_{i=0}^{m}|a_i - b_i| = \sum_{i=0}^{p-1}|a_i - b_i| + \sum_{i=p}^{k}|a_i - b_i| + \sum_{i=k+1}^{m}|a_i - b_i| .$$

Mais $\sum_{i=p}^{k}|a_i - b_i| = \sum_{i=p}^{k}(b_i - a_i) = \sum_{i=p}^{k-1}(b_{i+1} - a_i)$ et les points a_k et b_p ont été supprimés.

(II.1) Théorème 1

Si f définie sur E est la restriction à E d'une fonction $\hat{f} \in C^m(E_1)$ alors il existe un module de continuité ω tel que

$$\forall \mathfrak{A} \in E^*_{m+1},\ \forall \mathfrak{B} \in E^*_{m+1}:\ |\mathfrak{F}(\mathfrak{A}) - \mathfrak{F}(\mathfrak{B})| \le \omega(d(\mathfrak{A},\mathfrak{B})) .$$

Démonstration: Soit ω_m le module de continuité de $\hat{f}^{(m)}$ sur $C \supset E$:

$$\omega_m(\delta) = \max_{\substack{|x_1 - x_2| \le \delta \\ x_1 \in C,\ x_2 \in C}} |\hat{f}^{(m)}(x_1) - \hat{f}^{(m)}(x_2)| .$$

Par (Guelfond I) on a:

$$\mathfrak{F}(\mathfrak{A}) - \mathfrak{F}(\mathfrak{B}) = \int_0^1 dt_1 \int_0^{t_1} \cdots \int^{t_{m-1}} \Big(\hat{f}^{(m)}\big(a_0 + \sum_{i=1}^{m}(a_i - a_{i-1})t_i\big) -$$

$$- \hat{f}^{(m)}\big(b_0 + \sum_{i=1}^{m}(b_i - b_{i-1})t_i\big)\Big)\, dt_m$$

$$\text{et}\quad |\mathfrak{F}(\mathfrak{A})-\mathfrak{F}(\mathfrak{B})| \leq \int_0^1 dt_1 \int_0^{t_1} \cdots \int_0^{t_{m-1}} \omega_m\Big(2\sum_{i=0}^{m} |a_i - b_i|\Big)\, dt_m$$

$$|\mathfrak{F}(\mathfrak{A})-\mathfrak{F}(\mathfrak{B})| \leq \frac{1}{m!}\,\omega_m(2(m+1)\,d(\mathfrak{A},\mathfrak{B})) \leq \frac{2(m+1)}{m!}\,\omega_m\, d(\mathfrak{A},\mathfrak{B})$$

et on a le résultat avec $\omega = \dfrac{2(m+1)}{m!}\,\omega_m$.

(II, 2) Théorème 2

Réciproquement, une fonction f étant définie sur E, pour qu'il existe $\hat{f} \in \mathcal{C}^m(E_1)$ telle que $\hat{f}|_E = f$ il suffit qu'il existe un module de continuité ω tel que:

$$(\text{II}, 2, 1):\ \forall \mathfrak{A} \in E^*_{m+1},\ \forall \mathfrak{B} \in E^*_{m+1}:\ |\mathfrak{F}(\mathfrak{A})-\mathfrak{F}(\mathfrak{B})| \leq \omega(d(\mathfrak{A},\mathfrak{B}))\ .$$

Avant de démontrer ce théorème nous démontrerons en (II, 3) quelques lemmes fondamentaux. Remarquons tout d'abord que, si (II, 2, 1) est satisfaîte, $\mathfrak{F}$ est une fonction définie et uniformément continue sur E^*_{m+1}. On peut donc prolonger $\mathfrak{F}$ en une fonction (que nous appellerons toujours $\mathfrak{F}$), uniformément continue sur l'adhérence $\bar{E}^*_{m+1}$ de E^*_{m+1} dans E_{m+1}, et de même module de continuité. Le prolongement se fait de la manière suivante: l'adhérence $\bar{E}^*_{m+1}$ de E^*_{m+1} dans E_{m+1} se compose d'ensembles $\mathfrak{A}$ de $(m+1)$ points distincts ou confondus correspondant à des points isolés de E et à des points d'accumulation de E, ces derniers intervenant dans $\mathfrak{A}$ avec une multiplicité $\leq m$ tandis que les points isolés de E interviennent chacun avec une multiplicité égale à l'unité.

On pose alors:

$$\text{pour } \mathfrak{A} \notin E^*_{m+1} \text{ et } \mathfrak{A} \in \bar{E}^*_{m+1}:\ \mathfrak{F}(\mathfrak{A}) = \lim_{\substack{d(\mathfrak{A},\mathfrak{B}) \to 0 \\ \mathfrak{B} \in E^*_{m+1}}} \mathfrak{F}(\mathfrak{B})$$

ce qui permet de donner un sens a $\mathfrak{F}(\mathfrak{A})$ pour tout $\mathfrak{A} \in \bar{E}^*_{m+1}$.

Soit E_a l'ensemble des points d'accumulation de E et E_i l'ensemble des points isolés de E: $E = E_a \cup E_i$.

$$\text{Si } x \in E_a \quad \text{et si } X = (x, x, \dots, x) \in \bar{E}^*_{m+1}$$

$$\text{on a } \mathfrak{F}(X) = \lim_{\substack{d(X,\mathfrak{B}) \to 0 \\ \mathfrak{B} \in E^*_{m+1}}} \mathfrak{F}(\mathfrak{B})$$

et pour toute fonction $\hat{f} \in C^m(E_1)$ telle que $\hat{f}|_E = f$ on doit avoir l'égalité $\hat{f}^{(m)}(x) = m!\ \mathfrak{F}(X)$.

De plus pour toutes les différences divisées (d'ordre $p<m$) $\mathfrak{F}_p$ relatives à f et définies sur les ensembles $\mathfrak{A}_p$ de $(p+1)$ points distincts dans E on a une inégalité encore plus forte que (II, 2, 1). En utilisant (II, 0), en posant $\alpha_q = \beta_q$ pour $s+1 \leq q \leq p$ pour les points communs à $\mathfrak{A}_p$ et $\mathfrak{B}_p$ et en utilisant le fait que la valeur d'une différence divisée ne dépend pas de l'ordre des points on a :

$$\mathfrak{F}_p(\mathfrak{A}_p) - \mathfrak{F}_p(\mathfrak{B}_p) =$$

$$\mathfrak{F}_p(\alpha_0, \alpha_1, \dots, \alpha_s, \alpha_{s+1}, \dots, \alpha_p) - \mathfrak{F}_p(\beta_0, \beta_1, \dots, \beta_s, \alpha_{s+1}, \dots, \alpha_p)$$

$$= \mathfrak{F}_p(\alpha_0, \alpha_1, \dots, \alpha_s, \alpha_{s+1}, \dots, \alpha_p) - \mathfrak{F}_p(\beta_0, \alpha_1, \dots, \alpha_s, \alpha_{s+1}, \dots, \alpha_p)$$

$$+ \mathfrak{F}_p(\beta_0, \alpha_1, \dots, \alpha_s, \alpha_{s+1}, \dots, \alpha_p) - \mathfrak{F}_p(\beta_0, \beta_1, \alpha_2, \dots, \alpha_s, \alpha_{s+1}, \dots, \alpha_p)$$

+ -

$$+ \mathfrak{F}_p(\beta_0, \beta_1, \dots, \beta_{s-1}, \alpha_s, \alpha_{s+1}, \dots, \alpha_p) - \mathfrak{F}_p(\beta_0, \dots, \beta_s, \alpha_{s+1}, \dots, \alpha_p)$$

$$\text{et donc } \quad |\mathfrak{F}_p(\mathfrak{A}_p) - \mathfrak{F}_p(\mathfrak{B}_p)| \leq \left(\sum_{i=0}^{s} |\alpha_i - \beta_i|\right) \sup_{V_{p+1} \in E^*_{p+2}} |\mathfrak{F}_{p+1}(V_{p+1})| .$$

Pour $p = m-1$ et comme $\mathfrak{F}_m = \mathfrak{F}$ est uniformément continue sur E^*_{m+1} $\mathfrak{F}$ est majoré par un nombre μ_{m+1} et donc on a (avec $d_{p+1}(\mathfrak{A}_p, \mathfrak{B}_p) = \frac{1}{p+1} \sum_{i=0}^{p} |a_i - b_i|$)

$$|\mathfrak{F}_{m-1}(\mathfrak{A}_{m-1}) - \mathfrak{F}_{m-1}(\mathfrak{B}_{m-1})| \leq m\, d_m(\mathfrak{A}_{m-1}, \mathfrak{B}_{m-1}) \cdot \mu_{m+1} .$$

Par récurrence on peut donc prolonger chaque $\mathfrak{F}_p$ pour $p = 0$ à m aux points $X_p = (x, \dots, x) \in \bar{E}^*_{p+1}$ par $\mathfrak{F}_p(X_p) = \lim_{\substack{d_{p+1}(X_p, \mathfrak{B}_p) \to 0 \\ \mathfrak{B}_p \in E^*_{p+1}}} \mathfrak{F}_p(\mathfrak{B}_p)$.

Pour toute fonction $\hat{f} \in C^m(E_1)$ telle que $\hat{f}|_E = f$ on doit avoir l'égalité $f^{(p)}(x) = p!\ \mathfrak{F}_p(X_p)$ pour $x \in E_a$ et pour $p = 0$ à m .

Définissons sur E_a le champ T de polynômes de degrés $\leq m$: $x \xrightarrow{T} T_x \in \Pi_m$ par $T_x^{(p)}(x) = p!\ \mathfrak{F}_p(X_p)$ pour $p = 0$ à m puis cherchons à définir sur E_i un champ de polynômes de degrés $\leq m$ de telle façon que sur $E = E_a \cup E_i$ on obtienne un champ T satisfaisant à la condition du théorème (I, 1).

(II, 3) THEOREMES PRELIMINAIRE

(II, 3, 1) : Q-projections de Whitney

Soit $D(A, B)$ la "distance" habituelle de 2 ensembles A et B compacts de $\mathbb{R}$. $D(A, B) = \min_{\substack{x \in A \\ y \in B}} |x-y|$.

Soit $a = a_0$ un point isolé de E: nous allons associer à a_0 un ensemble $\mathfrak{A} = (a_0, a_1, \dots, a_k, a_k, \dots, a_k)$ contenu, à une permutation près sur les a_i de $\mathfrak{A}$, dans $\bar{E}^*_{m+1}$.

Nous posons $a = a_0 = S_0(a) = S_0$ et nous définissons $S_k = (a_0, \dots, \dots, a_k)$ pour $k \leq m$ par récurrence:

Si S_k a été obtenu nous appellerons a_{k+1} le point de E (il peut y en avoir deux, nous prenons alors celui de gauche) tel que:

$a_{k+1} = a_k$ si a_k est point d'accumulation de E .
$D(a_{k+1}, S_k) = D(S_k, E-S_k)$ si a_k n'est pas point d'accumulation de E .

On pose $S_m = Q(a)$.

<u>(II, 3, 2) Lemme</u>

<u>Si</u> $a \in E$ <u>et</u> $b \in E$ <u>et si</u> δ <u>désigne le diamètre sur</u> $\mathbb{R}$ <u>pour la distance euclidienne</u>:

$$\delta(Q(a)) > m|a-b| > 0 \Longrightarrow Q(a) = Q(b)$$

<u>et donc</u> $\quad Q(a) \neq Q(b) \Longrightarrow \delta(Q(a)) \leq m|a-b|$.

Soit $a \neq b$. Par l'hypothèse $\delta(Q(a)) > m|a-b|$ il exsite un premier indice $k \leq m-1$ tel que $D(a_{k+1}, S_k(a)) > a-b$.

D'où $S_k(a)$ est formé de points isolés de E et $\forall x \in E - S_k(a)$ et $\forall i \leq k$ on a : $|x-a_i| \geq |a_{k+1} - a_i| > |a-b|$. D'où $b \in S_k(a)$.

Supposons que dans $S_k(b)$ il existe $b_{p+1} \notin S_k(a)$ avec $p < k$ et $S_p(b) \subset S_k(a)$; $D(b_{p+1}, S_p(b)) \geq D(b_{p+1}, S_k(a)) > |a-b|$. Mais dans $S_k(a) - S$ il existerait aussi a_i, avec $i \leq k$, tel que $D(a_i, S_p(b)) \leq |a-b|$ et cet a_i devrait être b_{p+1} .

On a donc $S_k(a) = S_k(b)$.

<u>(II, 3, 3) Lemme</u>

<u>Pour</u> $(a, b) \in E^2$ <u>on a</u>: $d(Q(a), Q(b)) \leq (2m+1) \cdot |a-b|$

$$: d(Q(a), Q(b)) = \frac{1}{m+1} \sum_{i=0}^{m} |a_i - b_i|$$

Mais si $Q(a) \neq Q(b)$ on a $\delta(Q(a)) \leq m|a-b|$ et $\delta(Q(b)) \leq m|a-b|$ et donc:

$$|a_i - b_i| \leq |a_i - a| + |a-b| + |b-b_i|$$

$$|a_i - b_i| \leq \delta(Q(a)) + |a-b| + \delta(Q(b))$$

$$|a_i - b_i| \leq (2m+1)|a-b| \quad .$$

Remarque: On a aussi $\delta(Q(a)\ \ Q(b)) \leq \delta(Q(a)) + |a-b| + \delta(Q(b)) \leq (2m+1)|a-b|$.

(II, 4) DEMONSTRATION DU THEOREME (II, 2)

Définissons notre champ de polynômes sur E_i par: pour $x \in E_i$: $T_x = \mathcal{P}_{Q(x)}$ où $\mathcal{P}_{\mathcal{A}}$ désigne le polynôme d'interpolation généralisée de Lagrange de f sur $\mathcal{A}$. Le champ T a été défini sur E_a dans (II, 2).

Théorème:

Pour $x \in [a,b]$ avec $a \in E$ et $b \in E$ on a $\dfrac{|T_a(x)-T_b(x)|}{|a-b|^m} \leq \varphi(m)\omega(|a-b|)$ où ω est le module de continuité (donné) qui figure dans (II, 2, 1) et $\varphi(m)$ ne dépend que de m .

Soit $\mathcal{A} = Q(a)$ et $\mathcal{B} = Q(b)$. $\mathcal{A}$ et $\mathcal{B}$ sont dans $\bar{E}^*_{m+1}$

Soit $\mathcal{A} = (a_0, a_1, \ldots, a_m)$ et $\mathcal{B} = (b_0, \ldots, b_m)$.

Mais par (II, 0) on a (en posant $\alpha_{s+1} = \beta_{s+1}, \ldots, \alpha_m = \beta_m$ pour les points communs à $\mathcal{A}$ et $\mathcal{B}$):

$$\mathcal{P}_{\mathcal{A}} - \mathcal{P}_{\mathcal{B}} = \mathcal{P}_{(\alpha_0, \alpha_1, \ldots, \alpha_m)} - \mathcal{P}_{(\beta_0, \alpha_1, \ldots, \alpha_m)} + \mathcal{P}_{(\beta_0, \alpha_1, \ldots, \alpha_m)}$$

$$-\mathcal{P}_{(\beta_0, \beta_1, \alpha_2, \ldots, \alpha_m)} + \ldots + \mathcal{P}_{(\beta_0, \beta_1, \ldots, \beta_{s-1}, \ldots, \alpha}$$

$$- \mathcal{P}_{(\beta_0, \beta_1, \ldots, \beta_s, \alpha_{s+1}, \ldots, \alpha_s)} \ .$$

Par la forme de Newton du polynôme d'interpolation généralisé $\mathcal{P}_{\mathcal{D}} - \mathcal{P}_{\mathcal{E}}$, où $\mathcal{D}$ et $\mathcal{E}$ ne diffèrent que par un point, est égal à $\mathcal{P}_{\mathcal{D}} - \mathcal{P}_{\mathcal{E}} = [\mathcal{F}(\mathcal{D}) - \mathcal{F}(\mathcal{E})]\pi(x-d_i)$ où les d_i (i=1 à m) sont les points communs à $\mathcal{D}$ et $\mathcal{E}$.

D'où pour $x \in [a,b]$

$$|P_{\mathcal{A}}(x) - P_{\mathcal{B}}(x)| \leq \left(\sum_{i=0}^{s} \omega\left(\frac{|\alpha_i - \beta_i|}{m+1}\right)\right) \quad (\delta(Q(a) \cup Q(b)))^m$$

et $|P_{\mathcal{A}}(x) - P_{\mathcal{B}}(x)| \leq$

$$\sum_{i=0}^{s} \left(1 + \frac{|\alpha_i - \beta_i|}{\sum_{j=0}^{s} |\alpha_j - \beta_j|} \right) \omega \left(\frac{\sum_{j=0}^{s} |\alpha_j - \beta_j|}{m+1} \right) \quad (2m+1)^m |a-b|^m$$

car pour un module de continuité $\omega(\lambda\delta) \leq (\lambda + 1)\,\omega(\delta)$ pour $\delta \geq 0$ et $\lambda \geq 0$ (avec d'ailleurs $\omega(\lambda\delta) \leq \lambda\omega(\delta)$ si λ est entier).

Par (II, 0) on obtient l'inégalité (pour $x \in [a,b]$):

$$|P_{\mathcal{A}}(x) - P_{\mathcal{B}}(x)| \leq (m+2)\,(2m+1)^m \omega\,(d(\mathcal{A}, \mathcal{B})) \cdot |a-b|^m .$$

Par le lemme (II, 3, 3) le résultat final est:

$$\frac{|T_a(x) - T_b(x)|}{|a-b|^m} \leq (m+2)\,(2m+1)^{m+1} \omega(|a-b|) .$$

Par (Coatmélec [I]) le champ taylorien défini sur E peut être prolongé de telle sorte qu'il existe $\hat{f} \in C^m(E_1)$ telle que $\hat{f}|_E = f$.

III - PROLONGEMENT ET MAJORATIONS SUR LE PROLONGEMENT $\hat{T}$ DE T

(III, 1) Prolongement de f à $\mathbb{R}$

Soit $\omega_E = (m+2)\,(2m+1)^{m+1}\omega$ où ω est le module de continuité qui intervient dans l'hypothèse (II, 2, 1). Nous venons de voir que si (II, 2, 1) est satisfaite, on peut construire un champ $a \to T_a$ défini sur E tel que:

$$\forall a \in E, \forall b \in E: \mathfrak{I}_3(a,b,T,m) = \sup_{x \in [a,b]} \frac{|T_b(x) - T_a(x)|}{|b-a|^m} \leq$$

$$\leq \omega_E(|b-a|) \quad .$$

(III, 1, 0) Définition du prolongement:

Soit $]a,b[$ un des intervalles ouverts dont la réunion constitue le complémentaire de E dans $\mathbb{R}$.

1er Cas $a = -\infty$, $b = \min_{x \in E} x$. On peut prendre le prolongement $\hat{f}$ défini sur $]-\infty, b[$ par $\hat{f}(x) = T_b(x)$.

2e Cas $a = \max_{x \in E} x$, $b = +\infty$. On prend $\hat{f}(x) = T_a(x)$ sur $[a, +\infty[$.

3e Cas a et b sont finis, a et b sont dans E et on prend a <

Soit $\{a,b\}$ le couple formé par a et b et soit g une fonction, absolument continue sur $[a,b]$, qui induit sur $\{a,b\}$ le champ obtenu précedémment T_a, T_b .

Un théorème de Glaeser (Glaeser II) permet de prendre pour g une "spline parfaite" c'est-à-dire une fonction m fois continûment dérivable sur $[a,b]$ dont la dérivée $(m+1)^e$ ne prend que 2 valeurs λ et $-\lambda$ sauf en m points au plus de $[a,b]$ où la dérivée $(m+1)^e$ est discontinue et $|\lambda|$ est la borne inférieure des normes dans $L^\infty([a,b])$ des dérivées d'ordre $(m+1)$ des fonctions g qui induisent sur $\{a,b\}$ le champ T .

Soit $g = T_a + g_1$. g_1 doit satisfaire aux conditions aux limites $g_1^{(m-k)}(a) = 0$ pour $k = 0$ à m et $g_1^{(m-k)}(b) = T_b^{(m-k)}(b) - T_a^{(m-k)}(a)$.

Par (Louboutin I) on a:

$$|\lambda| = \max_{Q \in \Pi_m} \frac{\left|\sum_{k=0}^{m} (-1)^k Q^{(k)}(b)\, g_1^{(m-k)}(b)\right|}{\int_a^b |Q(t)|\,dt} .$$

On en déduit le théorème:

<u>(III, 1, 1) Théorème</u>: En utilisant le prolongement extrémal de Glaeser on a:

$$|\lambda| = |g^{(m+1)}(x)| \leq \frac{(m+1)^2\, 2^{m+1}\,(2m!)}{b-a} \sup_{x \in [a,b]} \frac{|T_b(x) - T_a(x)|}{(b-a)^m} .$$

En effet par l'inégalité de Markoff on a:

$$|Q^{(k)}(b)| \leq \frac{2^k\, m^2 (m-1)^2 \dots (m-k+1)^2}{(b-a)^k} \max_{x \in [a,b]} |Q(x)|$$

et

$$|g_1^{(m-k)}(b)| \leq \frac{2^{m-k}\, m^2 (m-1)^2 \dots (k+1)^2}{(b-a)^{m-k}} \sup_{x \in [a,b]} |T_b(x) - T_a(x)| .$$

Il en résulte que:

$$|\lambda| < 2^m \left(\sum_{k=0}^{m} \frac{(m!)^4}{(k!)^2 ((m-k)!)^2} \right) \frac{\max_{x \in [a,b]} |Q(x)|}{\int_a^b |Q(t)|\,dt} \sup_{x \in [a,b]} \frac{|T_b(x) - T_a(x)|}{(b-a)^m}$$

mais $\sum_{k=0}^{m} \binom{m}{k}^2 = \binom{2m}{m} = \frac{(2m!)}{(m!)^2}$ et d'autre part par (Louboutin II) :

$$\max_{x\in[a,b]} |Q(x)| \leq \frac{2(m+1)^2}{b-a} \int_a^b |Q(t)|\,dt$$

et donc

$$|\lambda| \leq \frac{2^{m+1}(2m!)\,(m+1)^2}{b-a} \sup_{x\in[a,b]} \frac{|T_b(x) - T_a(x)|}{(b-a)^m} .$$

On a donc $|\lambda| \leq \varphi_1(m)\dfrac{\omega_E(b-a)}{b-a}$ avec $\varphi_1(m) = 2^{m+1}(m+1)^2(2m!)$.

(III, 1, 2) Majorations concernant le prolongement de T .

Soit $\hat{f}$ le prolongement obtenu par la méthode précédente et cherchons des majorations pour:

$$A^3_{\mathbb{R}}(\hat{T}) = \sup_{\substack{u\in\mathbb{R}\\ v\in\mathbb{R}}} \max_{x\in[u,v]} \frac{|\hat{T}_v(x) - \hat{T}_u(x)|}{|v-u|^m} = \sup_{\substack{u\in\mathbb{R}\\ v\in\mathbb{R}}} \mathfrak{T}_3(u,v,\hat{T},m)$$

où $\hat{T}$ désigne le champ induit sur $\mathbb{R}$ par $\hat{f}$.

Posons de même $A^3_E(T) = \sup\limits_{\substack{a\in E\\ b\in E}} \mathfrak{T}_3(a,b,T,m)$.

Théorème: On a l'inégalité suivante:

$$A^3_{\mathbb{R}}(\hat{T}) \leq \varphi_2(m)\; A^3_E(T)$$

et si $\mathfrak{T}_3(a,b,T,m) \leq \omega_E(b-a)$ pour tout $a\in E$ et tout $b\in E$ alors

$$\forall u \in \mathbb{R} \text{ et } \forall v \in \mathbb{R} : \mathfrak{T}_3(u,v,\hat{T},m) \leq \varphi_3(m)\,\omega_E(v-u) = \omega_{\mathbb{R}}(v-u) .$$

Démonstration:

Soit tout d'abord u et v dans le même intervalle ouvert borné $]a,b[$ du complémentaire de E dans $\mathbb{R}$ (avec

$u < v$ et $a \in E$, $b \in E$, $a < b$).

Par la formule de Taylor

$$\hat{T}_v(x) - \hat{T}_u(x) = \int_u^v \frac{(x-t)^m}{m!} \hat{f}^{(m+1)}(t)\, dt .$$

En prenant pour $\hat{f}$ la fonction correspondant au prolongement extrémal de Glaeser on a:

$$\max_{x \in [u,v]} |\hat{T}_v(x) - \hat{T}_u(x)| \leq |\lambda| \int_u^v \frac{(v-t)^m}{m!} dt \leq \frac{|\lambda|(v-u)^{m+1}}{(m+1)!}$$

et donc

$$\frac{|\hat{T}_v(x) - \hat{T}_u(x)|}{(v-u)^m} \leq \frac{(m+1)^2\, 2^{m+1}(2m!)}{(m+1)!} \frac{v-u}{b-a} \sup_{x \in [a,b]} \frac{|T_a(x) - T_b(x)|}{(b-a)^m} .$$

Il en résulte tout d'abord que pour u et v dans $[a,b]$

$$\sup_{x \in [u,v]} \frac{|\hat{T}_v(x) - \hat{T}_u(x)|}{(v-u)^m} \leq \frac{(m+1)^2\, 2^{m+1}(2m!)}{(m+1)!} \times \mathfrak{J}_3(a,b,T,m) .$$

D'autre part

$$\sup_{x \in [u,v]} \frac{|\hat{T}_v(x) - \hat{T}_u(x)|}{(v-u)^m} \leq \frac{\varphi_1(m)}{(m+1)!} \frac{v-u}{b-a} \omega_E(b-a)$$

mais $\frac{v-u}{b-a} \omega_E(b-a) \leq \frac{v-u}{b-a} (\frac{b-a}{v-u} + 1)\, \omega_E(v-u)$ par la propriété $\omega(\lambda\delta) \leq (\lambda+1)\, \omega(\delta)$ du module de continuité.

et donc $$\sup_{x \in [u,v]} \frac{|\hat{T}_v(x) - \hat{T}_u(x)|}{(v-u)^m} \leq \frac{2\varphi_1(m)}{(m+1)!} \omega_E(v-u) .$$

Supposons maintenant que entre u et v il existe un point de E au moins. Soit $u<v$ et soit a le point, situé à droite de u, le plus proche de u dans E et de même soit b le point, situé à gauche de v, le plus proche de v dans E .

$$\hat{T}_v(x)-\hat{T}_u(x) = \hat{T}_v(x)-T_b(x) + T_b(x)-T_a(x) + T_a(x)-\hat{T}_u(x)$$

$$\frac{|\hat{T}_v(x)-\hat{T}_u(x)|}{(v-u)^m} \leq \left(\frac{v-b}{v-u}\right)^m \frac{|\hat{T}_v(x)-T_b(x)|}{(v-b)^m} + \left(\frac{b-a}{v-u}\right)^m \frac{|T_b(x)-T_a(x)|}{(b-a)^m} + \left(\frac{a-u}{v-u}\right)^m \frac{|T_a(x)-\hat{T}_u(x)|}{(a-u)^m} .$$

(Le cas où $a=b$ se traite encore plus simplement)

mais $\sup_{x\in[u,v]} |T_b(x)-T_a(x)| \leq (v-u) \sup_{x\in[a,b]} |T'_b(x)-T'_a(x)|$ par la formule des accroissements finis.

Donc $\sup_{x\in[u,v]} |T_b(x)-T_a(x)| \leq 2m^2 \frac{v-u}{b-a} \sup_{x\in[a,b]} |T_b(x)-T_a(x)|$.

On peut raisonner de même sur les 2 autres termes et en remarquant que s'il n'y a pas de point de E à gauche de u on a $\hat{T}_u(x) = T_a(x)$, s'il n'y a pas de point de E à droite de v: $\hat{T}_v(x) = T_b(x)$ (par III, 1, 0) . Dans ces 2 cas un terme au moins disparaît. Dans les autres cas soit a_1 le point de E le plus proche de u à gauche et b_1 le point de E le plus proche de v à droite. En prolongeant sur $[a_1, a]$ et $[b, b_1]$ par la méthode du prolongement extrémal de Glaeser on obtient:

$$\sup_{x \in [u,v]} \frac{|\hat{T}_v(x) - \hat{T}_u(x)|}{(v-u)^m} \leq 2m^2 \sup_{x \in [u,a]} \frac{|\hat{T}_u(x) - T_a(x)|}{(a-u)^m} (\frac{a-u}{v-u})^{m-1} +$$

$$+ 2m^2 \sup_{x \in [a,b]} \frac{|T_a(x) - T_b(x)|}{(b-a)^m} \times (\frac{b-a}{v-u})^{m-1}$$

$$+ 2m^2 \sup_{x \in [b,v]} \frac{|\hat{T}_v(x) - T_b(x)|}{(v-b)^m} \times (\frac{v-b}{v-u})^{m-1} .$$

En utilisant le résultat concernant 2 points dans le même intervalle $[a,b]$ on obtient pour $m \geq 1$ et pour $u \in \mathbb{R}$, $v \in \mathbb{R}$:

$$\sup_{x \in [u,v]} \frac{|\hat{T}_v(x) - \hat{T}_u(x)|}{(v-u)^m} \leq \frac{6\,\varphi_1(m)}{(m+1)!} m^2 A_E(T) .$$

Finalement nous obtenons: $A^3_{\mathbb{R}}(\hat{T}) \leq \varphi_2(m) A^3_E(T)$ avec

$$\varphi_2(m) = \frac{6m^2 \varphi_1(m)}{(m+1)!} = \frac{6m(m+1)(2m!)2^{m+1}}{(m-1)!}$$

D'autre part:

$$\mathfrak{J}_3(u,v,\hat{T},m) \leq \frac{4m^2 \varphi_1(m)}{(m+1)!} [(\frac{a-u}{v-u})^{m-1} \omega_E(a-u) + (\frac{b-a}{v-u})^{m-1} \omega_E(b-a) + $$

$$+ (\frac{v-b}{v-u})^{m-1} \omega_E(v-b)]$$

et

$$\mathfrak{J}_3(u,v,\hat{T},m) \leq \frac{12m^2 \varphi_1(m)}{(m+1)!} \omega_E(v-u) = \varphi_3(m)\omega_E(v-u) = \omega_{\mathbb{R}}(v-u)$$

avec $\varphi_3(m) = \dfrac{12m(m+1)(2m!)2^{m+1}}{(m-1)!}$.

Finalement en revenant aux données initiales on a :

$$\mathfrak{J}_3(u,v,\hat{T},m) \leq \varphi_4(m)\omega(|v-u|) .$$

Remarque: Par suite des équivalences entre les $\mathfrak{J}_i$ données dans (Coatmélec I) on en déduit en remplaçant le nombre 3 par i ou j $(0 \leq i \leq 3;\ 0 \leq j \leq 3)$

$$A^i_{\mathbb{R}}(\hat{T}) \leq \Psi_{i,j}(m)\, A^j_E(T)$$

$$\mathfrak{J}_i(u,v,\hat{T},m) \leq \chi_i(m)\, \omega_E(v-u)$$

où les $\Psi_{i,j}$ et χ_i ne dépendent que de m .

(III, 2) Majorations concernant $\hat{f}$ et ses dérivées

Soit d le diamètre de E et soit C le plus petit intervalle fermé contenant E . Le diamètre de C est d . Avec les notations précédentes nous avons d'abord:

$$A^3_C(\hat{T}) \leq \varphi_2(m)\, A^3_E(T)$$

et $\mathfrak{J}_3(u,v,\hat{T},m) \leq \varphi_3(m)\omega_E(v-u)$ pour $u \in C$ et $v \in C$.

$$\text{Soit } \|\hat{f}\|^m_C = \max_{0 \leq k \leq m} \left(\max_{x \in C} |\hat{f}^{(k)}(x)|\right)$$

$$\text{et } |||\hat{T}|||^{m,3}_C = \max(\|\hat{f}\|^m_C,\ A^3_C(\hat{T})) .$$

Nous définissons de même $|||T|||^{m,3}_E$ en posant

$$\|T\|_E^{m,3} = \max_{0\leq k\leq m} \left(\max_{a\in E} \left|\frac{d^K T_a}{dx^k}(a)\right|\right) \text{ et } |||T|||_E^{m,3} = \max(\|T\|_E^m, A_E^3(T)) .$$

Cette norme $||| \quad |||_E^{m,3}$ rend $W^m(E)$ complet. Nous allons majorer $|||\hat{T}|||_C^{m,3}$ en fonction de $|||T|||_E^{m,3}$.

Tout d'abord on a $\hat{f}^{(m)}(v) - \hat{f}^{(m)}(u) = \frac{d^m}{dx^m}(T_v(x) - T_u(x))$

et donc

$$|\hat{f}^{(m)}(v) - \hat{f}^{(m)}(u)| \leq \frac{2^m (m!)^2}{(v-u)^m} \sup_{x\in[u,v]} |T_v(x) - T_u(x)| .$$

Soit $\omega_C(g, .)$ le module de continuité de g sur C .

On a donc $\omega_C(\hat{f}^{(m)}, \delta) \leq \varphi_5(m)\omega_E(\delta)$ avec $\delta \geq 0$ et

$$\varphi_5(m) = 2^m (m!)^2 \varphi_3(m) = 2^{2m+1}(2m!)12m^2(m+1)(m!) .$$

Soit $u \in C$ et $a \in E$ on a de même:

$|\hat{f}^{(m)}(u) - \hat{f}^{(m)}(a)| \leq \varphi_5(m)\omega_E(d)$ et donc $\|\hat{f}^{(m)}\|_C \leq \|\hat{f}^{(m)}\|_E +$

$+ \varphi_5(m)\omega_E(d)$.

(III, 2, 1) Théorème: On a les inégalités suivantes:

$$\|\hat{f}\|_C^m \leq (1+\frac{d}{1!} + \ldots + \frac{d^m}{m!}) \|\hat{f}\|_E^m + \varphi_5(m)\frac{d^m}{m!}\omega_E(d) \text{ et } \omega_C(f^{(m)}, \delta) \leq$$

$$|||\hat{T}|||_C^{m,3} \leq (1+\frac{d}{1!} + \ldots + \frac{d^m}{m!} + \varphi_6(m)d^m) |||T|||_E^m$$

avec $\varphi_6(m) = 2^m(m!)\varphi_2(m) = 6m^2(m+1)(2m!)2^{2m+1}$.

En effet par la formule des accroissements finis $\|\hat{f}^{(m-1)}\|_C \leq \|\hat{f}^{(m-1)}\|_E + d\|\hat{f}^{(m)}\|_C$ et plus généralement par la formule de Taylor:

$$\|\hat{f}^{(m-k)}\|_C \leq \|\hat{f}^{(m-k)}\|_E + \frac{d}{1!}\|\hat{f}^{(m-k+1)}\|_E + \dots + \frac{d^k}{k!}\|\hat{f}^{(m)}\|_C .$$

On a donc:

$$\|\hat{f}\|_C^m \leq \|\hat{f}\|_E^m(1+\frac{d}{1!}+\dots+\frac{d^m}{m!}) + 2^m(m!)^2\varphi_2(m)A_E^3(T)\times\frac{d^m}{m!} .$$

Il en résulte que:

$$|||\hat{T}|||_C^{m,3} \leq (1+\frac{d}{1!}+\dots+\frac{d^m}{m!}+\varphi_6(m)d^m)|||T|||_E^{m,3} .$$

Pour $d \leq 1$: $|||\hat{T}|||_C^{m,3} \leq \varphi_7(m)\,|||T|||_E^{m,3}$

et pour $d > 1$: $|||\hat{T}|||_C^{m,3} \leq \varphi_8(m)d^m\,|||T|||_E^{m,3}$.

BIBLIOGRAPHIE

COATMELEC (I) Approximation et Interpolation des fonctions différentiables de plusieurs variables. Annales de l'Ecole Normale Supérieure 3e serie t. 83 (1966).

GLAESER (I) Etude de quelques algèbres tayloriennes Journal Analyse Math. Jérusalem t. 6 1958.

(II) Prolongement extrémal de Fonctions différentiables (Publications de la section de Mathématiques de la Faculté des Sciences de Rennes 1967).

PROLONGEMENT D'UNE FONCTION

GUELFOND (I) Calcul des différences finies. Dunod Paris 1963.

LOUBOUTIN (I) Sur une "bonne" partition de l'unité. (Prolongateur de Whitney tome II, Publications de la Section de Mathématiques de la Faculté des Sciences de Rennes) 1967.

(II) Inégalité entre les normes L^1 et L^∞ sur l'espace des polynômes. (Prolongateur de Whitney tome II, Publications de la Section de Mathématiques de la Faculté des Sciences de Rennes) 1967.

MERRIEN (I) Prolongateurs de fonctions différentiables. Journal de Maths pures et appliquées 45 (1966).

WHITNEY (I) Differentiable functions defined (Transactions of the American Math. Soc. Vol. 36 n° 2 April 1934).

(II) Analytic extensions of differentiable functions defined in closed sets. (Trans. Amer. Math. Soc. Vol. 36 n° 1 January 1934).

Spline Interpolation near Discontinuities

MICHAEL GOLOMB

1. Introduction

It is well known that the rate of convergence of spline interpolants to the function interpolated depends on the degree of smoothness of the function. For example, if $x \in \overset{\circ}{\mathfrak{W}}_r$ (this is the class of complex-valued functions of period 1, with an (r-1)th absolutely continuous derivative and an r-th derivative in $\mathcal{L}_2(0, 1)$) and $S_r^n x$ is the 2r-spline (piecewise polynomial of degree $2r-1$) of period 1, with knots at the points ν/n $(\nu = 0, \pm 1, \pm 2, \ldots)$ that interpolates x at the knots: $S_r^n x(\nu/n) = x(\nu/n)$), then

$$\sup_{0 \le t \le 1} |x(t) - S_r^n x(t)| = O(n^{-r+\frac{1}{2}}) \quad \text{as} \quad n \to \infty . \tag{1.1}$$

This is proved in a paper [1; Theorem 9.1], whose results and methods will be extensively used in the following. We refer to it as PSI .

The asymptotic relation (1.1) is the best possible, in the sense that in the presence of a jump discontinuity of $x^{(r-1)}$ at 0,

$$\lim_{n \to \infty} n^{r-1}[x(1/2n)] - S_r^n x(1/2n)] \quad \text{exists,} \neq 0 . \tag{1.2}$$

This will be shown below (Theorem 1). In general, we deal in this paper with asymptotic relations like (1.2), that hold in the presence of jump discontinuities of one or more derivatives of the function x. Except for the last part of the paper we deal only with periodic 2r-splines that interpolate x at the knots ν/n, and denote these by $S_r^n x$, as indicated above. Theorem 1, 2, 3 and 4 deal with the local errors, $x(t) - S_r^n x(t)$ and $x^{(s)}(t) - (S_r^n x)^{(s)}(t)$, $(s = 1, 2, \ldots, r-1)$, either near a point of discontinuity or away from it. In Theorems 5 and 6 we deal with the error

$$(1.3) \quad \|x^{(s)} - (S_r^n x)^{(s)}\|_2 = \{\int_0^1 |x^{(s)}(t) - (S_r^n x)^{(s)}(t)|^2 \, dt\}^{\frac{1}{2}} .$$

The rate of convergence of this error, too, is affected by the presence of discontinuities. In Theorem 7 we deal with the error $x(t) - T_r^n x(t)$, where $T_r^n x$ is the spline that matches the boundary values $x^{(s)}(0+)$, $x^{(s)}(1-)$ $(s = 0, 1, \ldots, r-1)$ of the (nonperiodic) function x.

The limit in (1.2) equals $\gamma_r \cdot [x^{(r-1)}(0+) - x^{(r-1)}(0-)]$, where γ_r is a constant independent of the function x (its value is given below). One may make use of this result to compute the jump $x^{(r-1)}(0+) - x^{(r-1)}(0-)$ at the same time while the spline interpolant $S_r^n x$ is computed and the error $x(1/2n) - S_r^n x(1/2n)$ is checked. By observing the asymptotic behavior of the error $x(\tau + 1/2n) - S_r^n x(\tau + 1/2n)$ at various points τ, the unknown abscissa of a discontinuity can also be determined (see Theorem 4). These are important problems of numerical analysis, but the results can also be used to accelerate the rate of convergence of the approximation errors. Indee if 0 is the only point of discontinuity of $x^{(r-1)}$ then

$$y = x - [x^{(r-1)}(0+) - x^{(r-1)}(0-)] \overset{o}{B}_r$$

is in $\overset{o}{\mathfrak{W}}_r$, thus $y - S_r^n y = O(n^{-r+\frac{1}{2}})$, uniformly. Here $\overset{o}{B}_r$ is the Bernoulli function (the periodic extension of the restriction of the Bernoulli polynomial of degree r to the interval $(0,1)$, whose $(r-1)$th derivative jumps by 1 at the integers).

Most of the results of the paper are derived from the examination of the error $\overset{o}{B}_s - S_r^n \overset{o}{B}_s$ near and away from the discontinuities of $\overset{o}{B}{}_s^{(s-1)}$. This is carried out by use of the Fourier series expansion of $\overset{o}{B}_s - S_r^n \overset{o}{B}$, which is obtained in explicit form from the Fourier series of the basic splines:

(1.4) $$b_\nu(t) = S_r^n(e^{2\pi i\nu t}) ,$$

introduced in PSI. The expansion is (PSI (5.8))

(1.5)
$$b_\nu(t) = \sum_k (k-\nu/n)^{-2r} e^{2\pi i(\nu-kn)t} / \sum (k-\nu/n)^{-2r} \qquad \nu \not\equiv 0 \,(\mathrm{mod}\, n)$$
$$= 1 \qquad , \nu \equiv 0 \,(\mathrm{mod}\ n) .$$

The coefficients γ_r, γ_{rs}, κ_r of the asymptotic expansions that occur in the paper are given as definite integrals which involve the analytic function

$$C_s(z) = z^{-s} + \sum_k{}' (z-k)^{-s}$$

(1.6) $$= [(-1)^{s-1}\pi^s/(s-1)!] \cot^{(s-1)}(\pi z) \qquad s = 1, 2, \ldots$$

$$C_s(\tfrac{1}{2}z) = 2^s \sum_{k \text{ even}} (z-k)^{-s}, \quad C_s(\tfrac{1}{2}z + \tfrac{1}{2}) = 2^s \sum_{k \text{ odd}} (z-k)^{-s} .$$

(Here and in the following, the symbol $\sum_k'$ stands for $\lim_{n\to\infty}$ $(\sum_{k=1,2,\ldots,N} + \sum_{k=-1,-2,\ldots,-N}))$.

In Section 2 all the results are presented. For ease of reading the more extensive proofs are collected in Section 3.

2. Results

Theorem 1. Suppose the function x of period 1 has an absolutely continuous $(r-1)$th $(r \geq 1)$ derivative $x^{(r-1)}$ in the open interval $(0, 1)$, and $x^{(r)} \in \mathcal{L}_2(0, 1)$. Then

$$\text{(i)}\quad x(1/2n) - S_r^n x(1/2n) = \gamma_r[x^{(r-1)}(0+) - x^{(r-1)}(0-)]n^{-r+1} + O(n^{-r+\frac{1}{2}})$$

as $n \to \infty$, and

$$\text{(ii)} \quad x(\tau + 1/2n) - S_r^n x(\tau + 1/2n) = O(n^{-r+\frac{1}{2}})$$

if $\tau = \ell/m$ for some $\ell = 1, \ldots, m-1$; $m = 2, 3, \ldots$, and $n \to \infty$ through the sequence $m, 2m, 3m, \ldots$. The constant γ_r depends on r only (not on n, nor x), and its value is

$$(2.1) \quad \gamma_r = 2^{1-2r}(4\pi i)^{-r} \int_0^1 dt\, e^{\pi i t}[C_r(\tfrac{1}{2}t)C_{2r}(\tfrac{1}{2}t + \tfrac{1}{2}) - C_{2r}(\tfrac{1}{2}t)C_r(\tfrac{1}{2}t + \tfrac{1}{2})]/C_{2r}(t) \; .$$

Remark: For $r = 1$, where $x(0+)$ may be different from $x(0-)$, interpolation by $S_1^n x$ at 0 is understood to mean

$$S_1^n x(0) = \tfrac{1}{2}[x(0+) + x(0-)] \; .$$

Then $S_1^n x(1/2n) = \frac{1}{2}x(1/n) + \frac{1}{4}x(0+) + \frac{1}{4}x(0-)$ and since $\gamma_1 = \frac{1}{4}$, (i) asserts

$$x(1/2n) - \tfrac{1}{2}x(1/n) = \tfrac{1}{2}x(0+) + O(n^{-\frac{1}{2}}) \; .$$

For $r = 2$ (cubic splines), one obtains

$$x(1/2n) - S_2^n x(1/2n) = \frac{3-\sqrt{3}}{16}[x'(0+) - x'(0-)]n^{-1} + O(n^{-3/2}) \; .$$

An immediate consequence of Theorem 1 is the following

Corollary 1.1. Under the hypotheses of Theorem 1 the equation

$$\text{(iii)} \quad x^{(r-1)}(0+) - x^{(r-1)}(0-) = \frac{1}{\gamma_r} \lim_{n\to\infty} n^{r-1}[x(1/2n) - S_r^n x(1/2n)]$$

holds.

By the use of this corollary the jump of $x^{(r-1)}$ at 0 can be computed with little additional work in the same process in which the spline interpolant $S_r^n x$ is computed.

In the following theorem $2 \le m$ and $0 \le \ell_0 < \ell_1 < \ldots < \ell_{N-1} < m$ are given integers. We set $t_i = \ell_i/m$ $(i = 0, 1, \ldots, N-1)$ and $t_N = 1 + \ell_0/m$. We allow the derivative

$x^{(r-1)}$ to have jumps at $t_0, \ldots, t_{N-1}$.

Theorem 2. Suppose the function x of period 1 has an absolutely continuous $(r-1)$th derivative $x^{(r-1)}$ $(r \geq 1)$ in each of the open intervals (t_i, t_{i+1}) $(i = 0, 1, \ldots, N-1)$, and $x^{(r)} \in \mathcal{L}_2(0, 1)$. Then

$$\text{(iv)} \quad x(\tau + 1/2n) - S_r^n x(\tau + 1/2n) = \gamma_r[x^{(r-1)}(\tau+) - x^{(r-1)}(\tau-)]n^{-r+1} + O(n^{-r+\frac{1}{2}})$$

if $m\tau$ is an integer and $n \to \infty$ through the sequence $m, 2m, 3m, \ldots$.

For $x \in \overset{\circ}{\mathfrak{W}}_r$, the s-th derivative $(s = 1, \ldots, r-1)$ of the spline interpolant $S_r^n x$ approximates the s-th derivative of x with an error uniformly of order $O(n^{-r+s+\frac{1}{2}})$ (see PSI, Theorem 9.1). The following theorem shows that if $x^{(r-1)}$ has jump discontinuities the order $O(n^{-r+s+\frac{1}{2}})$ still holds at points away from the discontinuities.

Theorem 3. Suppose the function x of period 1 has an absolutely continuous $(r-1)$th derivative $x^{(r-1)}(r \geq 1)$ in each of the open intervals (t_i, t_{i+1}) $(i = 0, 1, \ldots, N)$, and $x^{(r)} \in \mathcal{L}_2(0, 1)$. Then, for $s = 1, 2, \ldots, r-1$

$$\text{(v)} \qquad x^{(s)}(\tau) - (S_r^n x)^{(s)}(\tau) = O(n^{-r+s+\frac{1}{2}})$$

if $\tau \neq t_i$ $(i = 0, 1, \ldots, N)$, $m\tau$ is an integer, and $n \to \infty$ through the sequence $m, 2m, 3m, \ldots$.

In the following theorem we allow jump discontinuities not only of $x^{(r-1)}$, but also of $x'', x''', \ldots, x^{(r-2)}$. We formulate this theorem only for the case where all the discontinuities occur at one point, $t = 0$. The generalization to the case of a finite number of such discontinuities is easily supplied, following the pattern of Theorem 2 above.

Theorem 4. Suppose the function x of period 1 has absolutely continuous derivatives $x', \ldots, x^{(r-1)}$ in the open interval $(0, 1)$ and $x^{(r)} \in \mathcal{L}_2(0,1)$ $(r \geq 2)$. Then

$$\text{(vi)}\quad x(1/2n) - S_r^n x(1/2n) = \sum_{s=2}^{r} [\gamma_{rs}\xi^{(s-1)} + \delta_{s-1}\xi^{(s-2)}]n^{-s+1} + O(n^{-r+\frac{1}{2}})$$

as $n \to \infty$, and,

$$\text{(vii)}\quad x(\tau + 1/2n) - S_r^n x(\tau + 1/2n) = \sum_{s=3}^{r} \delta_{s-1}\xi^{(s-2)} n^{-s+1} + O(n^{-r+\frac{1}{2}})$$

if $\tau = \ell/m$ for some $\ell = 1, \ldots, m-1$; $m = 2, 3, \ldots$, and $n \to \infty$ through the sequence $m, 2m, 3m, \ldots$. Here

$$\xi^{(j)} = x^{(j)}(0+) - x^{(j)}(0-)$$

$$(2.2)\quad \gamma_{rs} = 2^{1-2r}(4\pi i)^{-s}\int_0^1 dt\, e^{\pi i t}[C_s(\tfrac{1}{2}t)C_{2r}(\tfrac{1}{2}t+\tfrac{1}{2}) - C_{2r}(\tfrac{1}{2}t)C_s(\tfrac{1}{2}t+\tfrac{1}{2})]/C_{2r}(t)$$

$$\delta_s = 0 \text{ if } s \text{ odd}, \quad \delta_s = 2(1-2^{-s})B_s/s! \text{ if } s \text{ even}.$$

It should be observed that if a derivative of order lower than $r-1$ is discontinuous at 0 then the rate of convergence of $x(t) - S_r^n x(t)$ is affected even near points $\tau \neq 0$. Thus, by (vii), $x(\tau + 1/2n) - S_3^n x(\tau + 1/2n) = (1/8)\xi^{(1)} n^{-2} + O(n^{-5/2})$

The asymptotic relations (vi) and (vii) can be used, like the previous ones, to compute the jumps $\xi^{(i)}$ in the same process in which the splines $S_r^n x$ are computed.

The mean-square error $\|x - S_r^n x\|_2$ is known to be of order $O(n^{-r})$ for $x \in \overset{o}{\mathfrak{W}}_r$ (see PSI, Theorem 8.1). The presence of a discontinuity of $x^{(r-1)}$ increases this error by a factor $n^{\frac{1}{2}}$, as the following theorem shows.

<u>Theorem 5.</u> Suppose the function x of period 1 has an absolutely continuous derivative of $(r-1)$th order in the open

interval $(0, 1)$, $x^{(r)} \in \mathcal{L}_2(0, 1)$, and $\xi^{(i)} = x^{(i)}(0+) - x^{(i)}(0-) = 0$ for $i = 0, 1, \ldots, r-1$, except for $i = s-1$, where s is an integer between 1 and r. Then

$$\|x - S_r^n x\|_2 = \kappa_s |\xi^{(s-1)}| n^{-s+\frac{1}{2}} + O(n^{-s}) \tag{viii}$$

as $n \to \infty$. Here

$$\kappa_s = (4\pi)^{-s}\{\int_0^1 dt[C_{2r}^2(t)C_{2s}(t) - 2C_{2r+s}(t)C_{2r}(t)C_s(t) + C_{4r}(t)C_s^2(t)]/C_{2r}^2(t)\}^{\frac{1}{2}} . \tag{2.3}$$

A simple consequence of this theorem is the following.

Corollary 5.1. If the function x of period 1 has a piecewise continuous $(r-1)$th derivative which is absolutely continuous in each of the open intervals between the discontinuities, and if $x^{(r)} \in \mathcal{L}_2(0, 1)$, then

$$\|x - S_r^n x\|_2 = O(n^{-r+\frac{1}{2}}), \qquad n \to \infty . \tag{ix}$$

In the next theorem the rate of convergence of $\|x^{(p)} - (S_r^n x)^{(p)}\|_2$ is established for the case where $x^{(s)}$ $(p \le s \le r-1)$ has a discontinuity.

Theorem 6. Suppose the function x of period 1 has an absolutely continuous $(s-2)$th derivative (s some integer between 2 and r) in $\mathbb{R}$ and absolutely continuous derivatives $x^{(s-1)}, \ldots, x^{(r-1)}$ in the open interval $(0, 1)$, and $x^{(r)} \in \mathcal{L}_2(0, 1)$. Then, for $p = 0, 1, \ldots, s-1$

$$\|x^{(p)} - (S_r^n x)^{(p)}\|_2 = \kappa_{s-p} |\xi^{(s-1)}| n^{-s+p+\frac{1}{2}} + O(n^{-s+p}) \tag{x}$$

as $n \to \infty$.

We formulate the special result where $p = s - 1$ in a slightly more general form.

Corollary 6.1. Suppose the function x of period 1 has an absolutely continuous $(s-2)$th derivative (for some s between 2 and r) in $\mathbb{R}$ and piecewise continuous derivatives $x^{(s-1)}, \ldots, x^{(r-1)}$, which are absolutely continuous in the open intervals between discontinuities, and $x^{(r)} \in \mathcal{L}_2(0, 1)$. Then

$$\text{(xi)} \qquad \|x^{(s-1)} - (S_r^n x)^{(s-1)}\|_2 = O(n^{-\frac{1}{2}}), \quad \text{as } n \to \infty .$$

The last theorem deals with a different kind of spline interpolant. Let $T_r^n x$ be the 2r-spline defined on the interval $[0, 1]$, with knots at the points ν/n $(\nu = 1, \ldots, n)$ that interpolates x at the knots, and satisfies the boundary conditions

$$\text{(2.4)} \qquad \begin{aligned} (T_r^n x)^{(s)}(0+) &= x^{(s)}(0+) \\ (T_r^n x)^{(x)}(1-) &= x^{(s)}(1-) , \qquad s = 0, 1, \ldots, r-1 . \end{aligned}$$

We assume that there are points $t_i = \ell_i/m$ for integers $0 < \ell_1 < \ell_2 < \ldots < \ell_N < m$, where $x^{(r-1)}$ has jump discontinuities, and prove that $x - T_r^n x$ behaves near and away from these points in the same way $x - S_r^n x$ (for periodic x) does.

Theorem 7. Suppose the function x defined on $[0, 1]$ has an absolutely continuous $(r-1)$th derivative $x^{(r-1)}$ $(r \geq 1)$ in each of the open intervals $(0, t_1), (t_1, t_2), \ldots, (t_N, 1)$, and $x^{(r)} \in \mathcal{L}_2(0, 1)$. Then

$$\text{(xii)} \quad x(\tau + 1/2n) - T_r^n x(\tau + 1/2n) = \gamma_r [x^{(r-1)}(\tau+) - x^{(r-1)}(\tau-)] n^{-r+1} + O(n^{-r+\frac{1}{2}})$$

if $\tau \neq 0, 1$; $m\tau$ is an integer, and $n \to \infty$ through the sequence $m, 2m, 3m, \ldots$. The constant γ_r is the same as in Theorem 1.

The proof of this result is based on a construction of $T_r^n x$, which is itself of interest. Let P be the polynomial of degree $\leq 2r - 1$ for which

$$(2.5)\quad P^{(s)}(0) = x^{(s)}(0+),\quad P^{(s)}(1) = x^{(s)}(1-),\quad s = 0, 1, \ldots, r-1 .$$

Then the function

$$(2.6)\qquad y = x - P$$

is of the same kind as the function x in Theorem 7, the jumps are not changed,

$$(2.7)\qquad y^{(r-1)}(\tau+) - y^{(r-1)}(\tau-) = x^{(r-1)}(\tau+) - x^{(r-1)}(\tau-)$$

at any τ, and besides we have

$$(2.8)\quad y^{(s)}(0+) = 0,\quad y^{(s)}(1-) = 0,\qquad s = 0, 1, \ldots, r-1 .$$

Clearly,

$$(2.9)\qquad T_r^n x = T_r^n P + T_r^n y = P + T_r^n y .$$

Thus, it suffices to construct $T_r^n y$. By PSI, Section 2, the periodic spline $S_r^n y$ is explicitly known. Hence, also the numbers

$$(2.10)\qquad \eta_{r,s}^n = \eta_s = (S_r^n y)^{(s)}(0),\quad s = 0, 1, \ldots, r-1$$

are known. We clearly have

$$(2.11)\qquad T_r^n y = S_r^n y - \sum_{s=1}^{r-1} \eta_s h_s ,$$

where $h_s = h^n_{r,s}$ is a (2r-1)-spline on $[0, 1]$, with knots at ν/n $(\nu = 1, \ldots, n-1)$, defined by the data

$$(2.12)\qquad \begin{aligned} &h_s(\nu/n) = 0, \\ &h_s^{(\sigma)}(0+) = h_s^{(\sigma)}(1-) = \delta_{s\sigma}, \quad \sigma = 0, 1, \ldots, r-1 . \end{aligned}$$

With the "boundary splines" $h_1, \ldots, h_{r-1}$ calculated (they depend only on r and n), $T^n_r x$ is explicitly given by (2.9) and (2.11):

$$(2.13)\qquad T^n_r x = P + S^n_r y - \sum_{s=1}^{r-1} \eta_s h_s .$$

In particular,

$$(2.14)\qquad x - T^n_r x = y - S^n_r y + \sum_{s=1}^{r-1} \eta_s h_s .$$

By Theorem 2 and (2.7), we have

$$(2.15)\qquad \begin{aligned} y(\tau + 1/2n) - S^n_r y(\tau + 1/2n) &= \gamma_r [x^{(r-1)}(\tau+) - x^{(r-1)}(\tau-)] n^{-r+1} \\ &+ O(n^{-r+\frac{1}{2}}) . \end{aligned}$$

By Theorem 3, we have

$$(2.16)\qquad \eta_s = (S^n_r y)^{(s)}(0) - y^{(s)}(0+) = O(n^{-r+s+\frac{1}{2}}), \quad s = 1, \ldots, r-1$$

Thus, (2.14), (2.15) and (2.16) constitute a proof of Theorem 7 if we can show that

$$(2.17)\qquad h_s(t) = O(n^{-s}), \quad s = 1, \ldots, r-1$$

uniformly in compact subintervals of $(0, 1)$. This is the content of the following

Lemma. For the boundary splines $h^n_{r,s}$ $(r \geq 2,\ s = 1, \ldots, r-1)$ the asymptotic relation

$$\text{(xiii)} \qquad h^n_{r,s}(t) = O(n^{-s}), \qquad \text{as } n \to \infty$$

holds uniformly in any compact interval in $(0, 1)$.

The proof of the Lemma is given in Section 3.

3. Proofs

Proof of Theorem 1. Set

$$\text{(3.1)} \qquad x^{(r-1)}(0+) - x^{(r-1)}(0-) = \xi \ .$$

The Bernoulli function

$$\text{(3.2)} \qquad \overset{\circ}{B}_r(t) = \frac{1}{(2\pi i)^r} \sum_\nu{}' \nu^{-r} e^{2\pi i \nu t}$$

is in $\overset{\circ}{C}_{r-2}$ and

$$\text{(3.3)} \qquad \begin{aligned} &\overset{\circ}{B}{}^{(r-1)}_r(0+) - \overset{\circ}{B}{}^{(r-1)}_r(0-) = 1 \\ &\overset{\circ}{B}{}^{(r)}_r(0+) - \overset{\circ}{B}{}^{(r)}_r(0-) = 0 \ . \end{aligned}$$

Therefore, the function

$$\text{(3.4)} \qquad y = x - \xi \overset{\circ}{B}_r$$

is in $\overset{\circ}{\mathbb{W}}_r$, and by PSI (Theorem 9.1)

$$\text{(3.5)} \qquad y(t) - S^n_r y(t) = O(n^{-r+\frac{1}{2}}), \qquad n \to \infty$$

for all t. Substitution of (3.4) in (3.5) gives

$$x(t) - S_r^n x(t) = \xi[\overset{\circ}{B}_r(t) - S_r^n \overset{\circ}{B}_r(t)] + O(n^{-r+\frac{1}{2}}) \ . \tag{3.6}$$

Using the Fourier expansion (3.2), we have by (1.4) and (1.5)

$$
\begin{aligned}
(2\pi i)^r S_r^n \overset{\circ}{B}_r(t) &= \sum_\nu{}' \nu^{-r} b_\nu(t) \\
&= \sum_{\nu \not\equiv 0} [\nu^{-r} \sum_k (k-\nu/n)^{-2r} e^{2\pi i(\nu - kn)t} / \sum_k (k-\nu/n)^{-2r}] \qquad (3.7) \\
&\quad + n^{-r} \sum_\nu{}' \nu^{-r}
\end{aligned}
$$

where the first summation is over all integers $\nu \not\equiv 0 \pmod{n}$.

We first consider the case $t = 1/2n$. Then (3.2) and (3.7) give

$$
\begin{aligned}
&\overset{\circ}{B}_r(1/2n) - S_r^n \overset{\circ}{B}_r(1/2n) = (2\pi i)^{-r} \sum_{\nu \not\equiv 0} \nu^{-r} e^{\pi i \nu/n} [1 - \sum_k (-1)^k (k-\nu/n)^{-2r} / \\
&\qquad \sum_k (k-\nu/n)^{-2r}] + (2\pi i n)^{-r} \sum_\nu{}' [(-1)^\nu - 1] \nu^{-r} \qquad (3.8) \\
&= (2\pi i n)^{-r} \sum_{\nu \not\equiv 0} (\nu/n)^{-r} e^{\pi i \nu/n} 2 \sum_{k \text{ odd}} (k-\nu/n)^{-2r} / \sum_k (k-\nu/n)^{-2r} \\
&\quad + O(n^{-r}) \ .
\end{aligned}
$$

With the use of the function (1.6) this last result may be written as

$$
\begin{aligned}
\overset{\circ}{B}_r(1/2n) - S_r^n \overset{\circ}{B}_r(1/2n) &= 2^{1-2r} (2\pi i n)^{-r} \sum_{\nu \not\equiv 0} (\nu/n)^{-r} e^{\pi i \nu/n} \\
&\quad C_{2r}(\nu/2n + \tfrac{1}{2}) / C_{2r}(\nu/n) + O(n^{-r}) \qquad (3.9)
\end{aligned}
$$

and since C_s had period 1,

$$\overset{o}{B}_r(1/2n) - S_r^n\overset{o}{B}_r(1/2n) = 2^{1-2r}(2\pi in)^{-r}\sum_{\nu=1}^{n-1} e^{\pi i\nu/n}[C_{2r}(\nu/2n+\tfrac{1}{2})$$

$$\sum_{k \text{ even}} (\nu/n+k)^{-r} - C_{2r}(\nu/2n)\sum_{k \text{ odd}} (\nu/n+k)^{-r}]/C_{2r}(\nu/n) + O(n^{-r}) \tag{3.10}$$

$$= 2^{1-2r}(4\pi in)^{-r}\sum_{\nu=1}^{n-1} e^{\pi i\nu/n}[C_{2r}(\nu/2n+\tfrac{1}{2})C_r(\nu/2n)$$

$$- C_{2r}(\nu/2n)C_r(\nu/2n+\tfrac{1}{2})]/C_{2r}(\nu/n) + O(n^{-r}) .$$

The function $[C_r(\frac{1}{2}t)C_{2r}(\frac{1}{2}t+\frac{1}{2}) - C_{2r}(\frac{1}{2}t)C_r(\frac{1}{2}t + \frac{1}{2})]/C_{2r}(t)$ occurring in (3.10) is analytic in $[0,1]$. Indeed, it approaches $-2^{2r}C_r(\frac{1}{2})$, $2^{2r}C_r(\frac{1}{2})$ as t approaches 0, 1, respectively. Therefore, in (3.10) we have trapezoidal sums for the integral of an analytic function, and one concludes [2, Theorem 1] that

$$\overset{o}{B}_r(1/2n) - S_r^n\overset{o}{B}_r(1/2n) = 2^{1-2r}(4\pi i)^{-r}n^{-r+1}\int_0^1 dt\, e^{\pi it}[C_r(\tfrac{1}{2}t)C_{2r}(\tfrac{1}{2}t$$

$$+\tfrac{1}{2}) - C_{2r}(\tfrac{1}{2}t)C_r(\tfrac{1}{2}t+\tfrac{1}{2})]/C_{2r}(t) + O(n^{-r}) . \tag{3.11}$$

(3.11) substituted in (3.6) results in (i).

Next, we take $t = \tau + 1/2n$ in (3.2) and (3.7). It is not difficult to see that, instead of (3.10), we obtain

(3.12)

$$\overset{o}{B}_r(\tau+1/2n) - S_r^n\overset{o}{B}_r(\tau+1/2n) = 2^{1-2r}(4\pi in)^{-r}\sum_{\nu=1}^{n-1} e^{\pi i\nu/n+2\pi i\tau\nu}$$

$$[C_r(\nu/2n)C_{2r}(\nu/2n+\tfrac{1}{2}) - C_{2r}(\nu/2n)C_r(\nu/2n+\tfrac{1}{2})]/C_{2r}(\nu/n) + O(n^{-r}) .$$

Now $n = mN$ for some positive integer N. As ν ranges from 0 to $n-1$, we have $\nu = m\lambda + \mu$, where λ ranges from 0 to $N-1$ and μ from 0 to $m-1$. Equation (3.12) becomes

$$\overset{o}{B}_r(\tau + 1/2n) - S_r^n \overset{o}{B}_r(\tau + 1/2n) = 2^{1-2r}(4\pi in)^{-r} \sum_{\lambda=0}^{N-1} e^{\pi i(\lambda/N + \mu/N)}$$

$$(3.13) \quad [C_r(\lambda/2N + \mu/2n)\, C_{2r}(\lambda/2N + \tfrac{1}{2} + \mu/2n) - C_{2r}(\lambda/2N + \mu/2n)$$

$$C_r(\lambda/2N + \tfrac{1}{2} + \mu/2n)]/C_{2r}(\lambda/N + \mu/n) + O(n^{-r}) \; .$$

As $n \to \infty$, $N \to \infty$. Again we have in (3.13) a trapezoidal sum of an analytic function, and one concludes

$$(3.14) \quad \overset{o}{B}_r(\tau + 1/2n) - S_r^n \overset{o}{B}_r(\tau + 1/2n) = \Big(\frac{1}{m} \sum_{\mu=0}^{m-1} e^{2\pi i\mu\tau}\Big) n^{-r+1} \gamma_r + O(n^{-r})$$

Since the factor in parentheses on the right of (3.14) is 0, we obtain

$$(3.15) \qquad \overset{o}{B}_r(\tau + 1/2n) - S_r^n \overset{o}{B}_r(\tau + 1/2n) = O(n^{-r})$$

and (3.15) substituted in (3.6) gives assertion (ii). This completes the proof of the theorem.

Proof of Theorem 2. We set

$$(3.16) \qquad \xi_i = x^{(r-1)}(t_i+) - x^{(r-1)}(t_i-), \quad i = 1, 2, \ldots, N .$$

The function

$$(3.17) \qquad y(t) = x(t) - \sum_{i=1}^{N} \xi_i \overset{o}{B}_r(t - t_i)$$

is in $\overset{o}{\mathfrak{B}}_r$, hence

(3.18) $$y(t) - S_r^n y(t) = O(n^{-r+\frac{1}{2}}), \quad \text{as } n \to \infty$$

for all t.

For the function $x_i(t) = \xi_i \overset{o}{B}_r(t - t_i)$ we have, by Theorem 1,

(3.19) $$x_i(\tau + 1/2n) - S_r^n x_i(\tau + 1/2n) = \gamma_r [x_i^{(r-1)}(\tau_+) - x_i^{(r-1)}(\tau_-)] + O(n^{-r+\frac{1}{2}})$$

valid for any τ for which τ is an integer, if $n = m, 2m, 3m, \ldots$. Observe that we arrive at this conclusion by the use of both assertions (i) and (ii) of Theorem 1. But clearly,

(3.20) $$\sum_{i=1}^{N} [x_i^{(r-1)}(\tau+) - x_i^{(r-1)}(\tau-)] = x^{(r-1)}(\tau_+) - x^{(r-1)}(\tau_-) .$$

By (3.17, 18, 19, 20) we conclude (iv); hence Theorem 2 is proved.

Proof of Theorem 3. The function y defined in (3.17) is in $\overset{o}{\mathfrak{W}}_r$, hence by PSI, Theorem 9.1,

(3.21) $$y^{(s)}(t) - (S_r^n y)^{(s)}(t) = O(n^{-r+s+\frac{1}{2}}), \quad \text{as } n \to \infty$$

for all t. Thus, the theorem is proved if we can show that at any nonintegral t for which mt is an integer

(3.22) $$\overset{o}{B}{}^{(s)}(t) - (S_r^n \overset{o}{B})^{(s)}(t) = O(n^{-r+s})$$

as $n \to \infty$ through the sequence $m, 2m, 3m, \ldots$.

Using (3.2), (3.7) and the functions (1.6), one finds

$$\overset{\circ}{B}{}^{(s)}(t) - (S_r^n \overset{\circ}{B})^{(s)}(t) = (2\pi i n)^{-r+s}\{\sum_{\nu=1}^{n-1} e^{2\pi i\nu t}[C_{r-s}(\nu/n)\,C_{2r}(\nu/n) \tag{3.23}$$

$$- C_r(\nu/n)\,C_{2r-s}(\nu/n)]/C_{2r}(\nu/n) + \sum{}' k^{-r+s}\} .$$

Putting, as before, $n = mN$, $\nu = m\lambda + \mu$, this equation becomes

$$\overset{\circ}{B}{}^{(s)}(t) - (S_r^n \overset{\circ}{B})^{(s)}(t) = (2\pi i)^{-r+s} n^{-r+s+1}\{\frac{1}{m}\sum_{\mu=0}^{m-1} e^{2\pi i\mu t}\} \tag{3.24}$$

$$\{\frac{1}{N}\sum_{\lambda=0}^{N-1}[(C_{r-s}C_{2r} - C_r C_{2r-s})/C_{2r}](\lambda/N + \mu/n)\} + O(n^{-r+s}) .$$

As $n \to \infty$, $N \to \infty$, and in (3.24) we have a trapezoidal sum of an analytic function, which differs from the corresponding integral $\int_0^1 (C_{r-s}C_{2r} - C_r C_{2r-s})/C_{2r}$ certainly by less than $O(n^{-1})$. Since, moreover, the first factor in braces on the right of (3.24) is 0, we obtain (3.22).

Proof of Theorem 4. The function

$$y = x - \sum_{s=2}^{r} \xi^{(s-1)} \overset{\circ}{B}_s \tag{3.25}$$

is in $\overset{\circ}{\mathfrak{W}}_r$, hence

$$y(t) - S_r^n y(t) = O(n^{-r+\frac{1}{2}}), \quad \text{as } n \to \infty . \tag{3.26}$$

Using the Fourier expansion

$$\overset{\circ}{B}_s(t) = \frac{1}{(2\pi i)^s} \sum_{\nu}{}' \nu^{-s} e^{2\pi i\nu t} \qquad s = 2, \ldots, r \tag{3.27}$$

we have

$$(2\pi i)^s \, S_r^n \overset{\circ}{B}_s(t) = \sum_{\nu \neq 0} [\nu^{-s} \sum_k (k-\nu/n)^{-2r} e^{2\pi i(\nu - kn)t} / \sum_k (k-\nu/n)^{-2r}]$$

(3.28)

$$+ n^{-s} \sum_\nu{}' \nu^{-s} ,$$

hence

$$\overset{\circ}{B}_r(1/2n) - S_r^n \overset{\circ}{B}_s(1/2n) = 2(2\pi i n)^{-s} \sum_{\nu \neq 0} (\nu/n)^{-s} e^{\pi i \nu/n}$$

(3.29)

$$\sum_{k \text{ odd}} (k-\nu/n)^{-2r} / \sum_k (k-\nu/n)^{-2r} - 2(2\pi i n)^{-s} \sum_{\nu \text{ odd}} \nu^{-s} .$$

With the use of the function (1.6) and the identity $(2\pi i)^{-s} \sum' \nu^{-s} = -B_s/s!$ for $s = 2, 4, \ldots,$ (3.29) may be written as

$$\overset{\circ}{B}_s(1/2n) - S_r^n \overset{\circ}{B}_s(1/2n) = 2^{1-s} (2\pi i n)^{-s} \sum_{\nu \neq 0} (\nu/n)^{-s} e^{\pi i \nu/n}$$

(3.30)

$$C_{2r}(\nu/2n + \tfrac{1}{2})/C_{2r}(\nu/n) + \delta_s n^{-s} ,$$

where δ_s is defined as in (2.2). Since C_s has period 1,

$$\overset{\circ}{B}_s(1/2n) - S_r^n \overset{\circ}{B}_s(1/2n) = 2^{1-s} (2\pi i n)^{-s} \sum_{\nu=1}^{n-1} e^{\pi i \nu/n} [C_{2r}(\nu/2n + \tfrac{1}{2})$$

(3.31) $$\sum_{k \text{ even}} (\nu/n+k)^{-s} - C_{2r}(\nu/2n) \sum_{k \text{ odd}} (\nu/n+k)^{-s}]/C_{2r}(\nu/n) + \delta_s n^{-s}$$

$$= 2^{1-s} (4\pi i n)^{-s} \sum_{\nu=1}^{n-1} e^{\pi i \nu/n} [C_s(\nu/2n) C_{2r}(\nu/2n + \tfrac{1}{2}) - C_{2r}(\nu/2n)$$

$$C_s(\nu/2n + \tfrac{1}{2})]/C_{2r}(\nu/n) + \delta_s n^{-s} .$$

The function $[C_s(\frac{1}{2}t)\, C_{2r}(\frac{1}{2}t+\frac{1}{2}) - C_{2r}(\frac{1}{2}t)\, C_s(\frac{1}{2}t+\frac{1}{2})]/C_{2r}(t)$ is analytic in $[0,1]$. Therefore, in (3.31) we have once more a trapezoidal rule sum for a periodic analytic function, and one concludes that

$$\overset{\circ}{B}_s(1/2n) - S_r^n \overset{\circ}{B}_s(1/2n) = 2^{1-s}(4\pi i)^{-s} n^{-s+1} \int_0^1 dt\, e^{\pi i t}[C_s(\tfrac{1}{2}t)$$

(3.32)

$$C_{2r}(\tfrac{1}{2}t+\tfrac{1}{2}) - C_{2r}(\tfrac{1}{2}t)C_s(\tfrac{1}{2}t+\tfrac{1}{2})]/C_{2r}(t) + \delta_s n^{-s} + O(n^{-r}) .$$

(3.32) substituted in (3.25) and (3.26) results in (vi).

Next, we take $t = \tau + 1/2n$ in (3.27) and (3.28). Then, instead of (3.30), we obtain

$$\overset{\circ}{B}_s(\tau+1/2n) - S_r^n \overset{\circ}{B}_s(\tau + 1/2n) = 2^{1-s}(4\pi i n)^{-s} \sum_{\nu=1}^{n-1} e^{\pi i \nu/n + 2\pi i \nu \tau}$$

(3.33)

$$[C_s(\nu/2n)C_{2r}(\nu/2n+\tfrac{1}{2}) - C_{2r}(\nu/2n)C_s(\nu/2n+\tfrac{1}{2})]/C_{2r}(t)$$

$$+ (2\pi i n)^{-s} \sum_{\nu \text{ odd}} \nu^{-s} .$$

We proceed as in the derivation of (3.13) from (3.12) and find

$$\overset{\circ}{B}_s(\tau+1/2n) - S_r^n \overset{\circ}{B}_s(\tau+1/2n) = (\frac{1}{m} \sum_{\mu=0}^{m-1} e^{2\pi i \mu \tau}) n^{-s+1} \gamma_{rs}$$

(3.34)

$$+ \delta_s n^{-s} + O(n^{-r}) .$$

Since the factor in parentheses on the right of (3.34) is 0, we have

$$\overset{\circ}{B}_s(\tau+1/2n) - S_r^n \overset{\circ}{B}_s(\tau+1/2n) = \delta_s n^{-s} + O(n^{-r}) , \tag{3.35}$$

and (3.35) substituted in (3.25) and (3.26) proves (vii).

Proof of Theorem 5. The function

$$y = x - \xi^{(s-1)} \overset{\circ}{B}_s \tag{3.36}$$

is in $\overset{\circ}{W}_r$, hence $\|y - S_r^n y\|_2 = O(n^{-r})$ and

$$\|x - S_r^n x\|_2 = |\xi^{(s-1)}| \, \|\overset{\circ}{B}_s - S_r^n \overset{\circ}{B}_s\|_2 + O(n^{-r}) .$$

The ν-th Fourier coefficient of $(2\pi i)^s (\overset{\circ}{B}_s - S_r^n \overset{\circ}{B}_s)$ is, by (3.2) and (3.7)

$$\begin{cases} \nu^{-s} - \nu^{-2r} \sum_k (\nu - kn)^{-s} / \sum_k (\nu - kn)^{-2r} & \text{if} \quad \nu \not\equiv 0 \\ \nu^{-s} & \text{if} \quad \nu \equiv 0, \ \nu \neq 0 \\ -n^{-s} \sum{}' k^{-s} & \text{if} \quad \nu = 0 . \end{cases} \tag{3.38}$$

Therefore, by the Parseval formula

$$(4\pi)^{2s}\|\overset{o}{B}_s - S_r^n\overset{o}{B}_s\|_2^2 = \sum_{\nu\neq 0}[\nu^{-s} - \nu^{-2r}\sum_k(\nu - kn)^{-s}/\sum_k(\nu - kn)^{-2r}]^2$$

$$+ 2n^{-2s}(\sum{}' k^{-s})^2$$

$$= n^{-2s}\sum_{\nu=1}^{n-1}\sum_\ell[(\ell - \nu/n)^{-s} - (\ell - \nu/n)^{-2r}\sum_k(k - \nu/n)^{-s}/\sum_k(k$$

$$-\nu/n)^{-2r}] + O(n^{-2s})$$

$$(3.39) = n^{-2s}\sum_{\nu=1}^{n-1}\{\sum_\ell(\ell - \nu/n)^{-2s} - 2\sum_\ell(\ell - \nu/n)^{-s-2r}\sum_\ell(\ell$$

$$-\nu/n)^{-s}/\sum_\ell(\ell - \nu/n)^{-2r} +$$

$$+ \sum_\ell(\ell - \nu/n)^{-4r}(\sum_\ell(\ell - \nu/n)^{-s})^2/(\sum_\ell(\ell - \nu/n)^{-2r})^2\}$$

$$+ O(n^{-2s})$$

or, in terms of the function C_s,

$$(4\pi)^{2s}\|\overset{o}{B}_s - S_r^n\overset{o}{B}_s\|_2^2 = n^{-2s}\sum_{\nu=1}^{n-1}[C_{2s}(\tfrac{\nu}{n}) - 2C_{s+2r}(\tfrac{\nu}{n})C_s(\tfrac{\nu}{n})/C_{2r}(\tfrac{\nu}{n})$$

(3.40)

$$+ C_{4r}(\tfrac{\nu}{n})C_s^2(\tfrac{\nu}{n})/C_{2r}^2(\tfrac{\nu}{n})] + O(n^{-2s}) \ .$$

The function $C_{2s}(t) - 2C_{s+2r}(t)C_s(t)/C_{2r}(t) + C_{4r}(t)C_s^2(t)/C_{2r}^2$ has period 1 and is analytic for all real t (one verifies easily that its value, at the critical point 0, is $\Sigma'\nu^{-2s} - (\Sigma'\nu^{-s})^2$. Therefore, one deduces from (3.40) by arguments used above,

$$\|\overset{\circ}{B}_s - S_r^n \overset{\circ}{B}_s\|_2 = n^{-s+\frac{1}{2}} (4\pi)^{-s} \{\int_0^1 dt[C_{2r}^2(t) C_{2s}(t)$$

$$(3.41) \quad -2C_{2r+s}(t) C_{2r}(t) C_s(t) + C_{4r}(t) C_s^2(t)]/C_{2r}^2(t)\}^{\frac{1}{2}} + O(n^{-s})$$

$$= \kappa_s n^{-s+\frac{1}{2}} + O(n^{-s}) \quad .$$

(3.41) together with (3.37) prove assertion (viii).

Proof of Theorem 6. The function

$$(3.42) \qquad y = x - \xi^{(s-1)} \overset{\circ}{B}_s$$

is in $\overset{\circ}{\mathfrak{W}}_s$, hence by PSI (Theorem 8.1),

$$(3.43) \qquad \|y^{(p)} - (S_r^n y)^{(p)}\|_2 = O(n^{-s+p}) \quad \text{as } n \to \infty \ .$$

Also,

$$(3.44) \quad \|x - S_r^n x\|_2 = |\xi^{(s-1)}| \, \|\overset{\circ}{B}{}_s^{(p)} - (S_r^n \overset{\circ}{B}_s)^{(p)}\|_2 + O(n^{-s+p}) \ .$$

The ν-th Fourier coefficient of $(2\pi i)^{s-p}(\overset{\circ}{B}{}_s^{(p)} - (S_r^n \overset{\circ}{B}_s)^{(p)})$ is, by (3.2) and (3.7)

$$(3.45) \quad \begin{cases} \nu^{p-s} - \nu^{-2r} \sum_k (\nu - kn)^{p-s} / \sum_k (\nu - kn)^{-2r} & \text{if } \nu \not\equiv 0 \\ \nu^{p-s} & \text{if } \nu \equiv 0, \ \nu \neq 0 \\ 0 & \text{if } \nu = 0 \ . \end{cases}$$

It should be observed that this result also holds in the case $p = s-1$. In this case $\sum_k (\nu - kn)^{p-s}$ should be interpreted as $\lim_{N\to\infty} \sum_{k=-N}^{N} (\nu - kn)^{-1}$. From here on one proceeds as in the proof of Theorem 5 and obtains

$$(3.46) \qquad \|\overset{\circ}{B}{}_s^{(p)} - (S_r^n \overset{\circ}{B}_s)^{(p)}\|_2 = \kappa_{s-p}\, n^{-s+p+\frac{1}{2}} + O(n^{-s+p}) .$$

Proof of Lemma. It is well known (see, for example, [3], Theorem 4) that among all functions z on $[0,1]$ which have an $(r-1)$th absolutely continuous derivative $z^{(r-1)}$ in the open interval $(0,1)$ such that $z^{(r)} \in \mathcal{L}_2(0,1)$ and

$$(3.47) \qquad \begin{aligned} & z(\nu/n) = 0 , && \nu = 0, 1, \ldots, n \\ & z^{(\sigma)}(0+) = z^{(\sigma)}(1-) = \delta_{s\sigma}, && \sigma = 0, 1, \ldots, r-1 \end{aligned}$$

the one that minimizes $\int_0^1 (z^{(r)})^2$ is h_s. Thus,

$$(3.48) \qquad \int_0^1 (h_s^{(r)})^2 \leq \int_0^1 (z^{(r)})^2$$

for any function z as described above. Let $Q(t) = t^s q(t)$ be the polynomial of degree $2r-1$ such that

$$(3.49) \qquad \begin{aligned} & Q^{(\sigma)}(0) = \delta_{s\sigma} \\ & Q^{(\sigma)}(1) = 0 , \quad \sigma = 0, 1, \ldots, r-1 . \end{aligned}$$

Then the function z defined by

(3.50)
$$z(t) = t^s q(nt), \qquad 0 \le t \le 1/n,$$
$$= 0, \qquad 1/n \le t \le 1 - 1/n$$
$$= (-1)^s z(1-t), \qquad 1 - 1/n \le t \le 1$$

satisfies conditions (3.47). Clearly,

(3.51)
$$\int_0^1 (z^{(r)}(t))^2 dt = 2n^{2r-2s} \int_0^{1/n} (Q^{(r)}(nt))^2 dt$$
$$= 2n^{2r-2s-1} \int_0^1 (Q^{(r)})^2 ,$$

hence (3.48) gives

(3.52) $$\{\int (h_s^{(r)})^2\}^{\frac{1}{2}} = O(n^{r-s-\frac{1}{2}}), \quad \text{as} \quad n \to \infty$$

for $s = 1, 2, \ldots, r-1$.

Let I_c be a compact interval in $(0, 1)$, and t an arbitrary point in I_c. From Rolle's Theorem it follows that there is, for $n > 2r$, a point t' with $|t - t'| < r/n$, such that $h_s^{(r-1)}(t') = 0$. Hence,

(3.53)
$$|h_s^{(r-1)}(t)| = |\int_{t'}^{t} h_s^{(r)}(\tau)\, d\tau|$$
$$\le |t - t'|^{\frac{1}{2}} \{\int_{t'}^{t} (h_s^{(r)})^2\}^{\frac{1}{2}}$$

and we conclude, with the aid of (3.52), that

(3.54) $$h_s^{(r-1)}(t) = O(n^{r-s-1}), \quad \text{as} \quad n \to \infty$$

uniformly for $t \in I_C$. Again, there is a point $t'' \in I_C$ with $|t-t''| < r/n$, such that $h_s^{(r-2)}(t'') = 0$, and (3.54) gives

$$|h_s^{(r-2)}(t)| = |\int_{t''}^{t} h_s^{(r-1)}(\tau)\,d\tau|$$

$$\leq |t-t''| \sup_{\tau \in I_C} |h_s^{(r-1)}(\tau)| \tag{3.55}$$

$$= O(n^{r-s-2}), \quad \text{as} \quad n \to \infty .$$

Repeating this argument, one clearly finds

$$h_s(t) = O(n^{-s}), \quad \text{as} \quad n \to \infty$$

uniformly for $t \in I_C$, and this proves the lemma.

REFERENCES

1. Michael Golomb, Approximation by periodic spline interpolants on uniform meshes. J. Appr. Theorem, 1(1968), 26-65.

2. Philip J. Davis, On the numerical integration of periodic analytic functions. Proceedings of Symposium on Numerical Integration, R. E. Langer, ed., Madison 1959, 45-59.

3. M. H. Schultz and R. S. Varga, L-Splines. Numerische Math. 10(1967), 345-369.

Generalized Spline Interpolation and Nonlinear Programming

KLAUS RITTER

1. Introduction

In [1] Ahlberg and Nilson give an extension of spline functions to what they call splines of interpolation. Schoenberg [6] has shown that the splines of interpolation, which he calls g-splines, generalize the Birkhoff interpolation problem [2] in the same way as spline interpolation at simple knots generalizes Lagrange's formula of polynomial interpolation, and spline interpolation with multiple knots generalizes the Hermite interpolation problem.

In many spline interpolation problems, occurring in practice, it might be desirable to relax the interpolation conditions in such a way that it is only required that, at given points x_i, the interpolating spline and some of its (not necessarily consecutive) derivatives attain values in a prescribed interval. Among all splines satisfying the relaxed interpolation conditions one might be interested in the ones which have some additional continuity or smoothness properties.

The purpose of this paper is to study spline interpolation for interpolation problems with relaxed interpolation conditions. It will turn out that it is useful to apply methods which have been developed in the field of mathematical programming.

In the next section a precise definition of the generalized interpolation problem is given. A certain subclass of the set of natural g-splines is specified by some additional continuity properties. The spline functions in this subclass are called extremal g-splines. Section 3) is devoted to the study of a general minimization problem and its dual problem. Based on the results of this section, it is shown in Section 4) that there exists an extremal g-spline which solves the generalized interpolation problem. This extremal g-spline is unique, provided the interpolation problem satisfies a condition which is a generalization of Schoenberg's concept of "m-poised" [6]. Furthermore, minimal properties of the extremal g-spline solution of the generalized interpolation problem as well as numerical methods for its computation are discussed.

2. The generalized interpolation problem

Let

$$x_1 < x_2 < \ldots < x_k$$

be distinct real numbers. For $i = 1, \ldots, k$, let I_i be a prescribed nonvoid subset of $\{0, 1, \ldots, m-1\}$, where m is a given positive integer. Let α_{ij} be prescribed real numbers for $j \in I_i$ and $i = 1, \ldots, k$. Furthermore, suppose that ℓ is the largest number occurring in $I_1 \cup I_2 \cup \ldots \cup I_k$.

The problem of finding functions $f(x) \in C^\ell$ which satisfy the conditions

$$f^{(j)}(x_i) = \alpha_{ij} \quad \text{for } j \in I_i \text{ and } i = 1, \ldots, k \tag{2.1}$$

is called a Hermite-Birkhoff interpolation problem [6], which we abbreviate to HB-problem. Let m be a natural number. Following Schoenberg [6] we say that the HB-problem (2.1) is m-poised provided that

$$\begin{aligned} &P(x) \in \Pi_{m-1} \\ &P^{(j)}(x_i) = 0 \quad \text{if } j \in I_i \text{ and } i = 1, \ldots, k \end{aligned} \tag{2.2}$$

imply that $P(x) \equiv 0$, where Π_{m-1} denotes the class of real polynomials of degree at most $m-1$.

Define the elements e_{ij} of the $(k, m-1)$ matrix E by

$$e_{ij} = \begin{cases} 1 \text{ if } j \in I_i \\ 0 \text{ if } j \notin I_i \end{cases} \qquad i = 1, \ldots, k \ .$$

According to Schoenberg [6] a function $S(x)$ is called a natural g-spline for the knots $x_1, \ldots, x_k$, the matrix E, and order m, provided that it satisfies the following conditions

(2.3) $\quad S(x) \in \Pi_{2m-1}$ in (x_i, x_{i+1}), $i = 1, \ldots, k-1$,

(2.4) $\quad S(x) \in \Pi_{m-1}$ in $(-\infty, x_1)$ and in $(x_k, +\infty)$,

(2.5) $\quad S(x) \in C^{m-1}(-\infty, \infty)$,

(2.6) $\quad$ If $e_{ij} = 0$ then $S^{(2m-j-1)}(x)$ is continuous at $x = x_i$.

The set of all these natural g-splines is denoted by

$$\Omega_m = \Omega_m \ (E, x_1, \ldots, x_k) \ .$$

Schoenberg [6] has shown that there is a uniquely determined natural g-spline $\bar{S}(x) \in \Omega_m$ which solves the HB-problem (2.1), provided that

(2.7)
$$\ell < m \leq \sum_{i,j} e_{ij} = n$$

and the HB-problem is m-poised. Moreover, it has been shown that $\bar{S}(x)$ has a number of interesting optimal properties.

A natural generalization of the HB-problem is obtained, if we replace the conditions (2.1) by

(2.8) $\quad \alpha_{ij} \leq f^{(j)}(x_i) \leq \beta_{ij}$ for $j \in I_i$ and $i = 1, \ldots, k$,

where $\alpha_{ij} \leq \beta_{ij}$ are prescribed real numbers. We shall refer to the problem of finding functions $f(x) \in C^{\ell}$ satisfying (2.8) as a generalized Hermite-Birkhoff interpolation problem, abbreviated to GHB-problem.

Since any solution of the HB-problem (2.1) is also a solution of the GHB-problem (2.8), it is clear that there is an infinite number of $S(x) \in \Omega_m$, which solve (2.8), provided (2.7) holds and $\alpha_{ij} < \beta_{ij}$ for at least one pair (i,j). Therefore, the question arises whether it is possible to describe a subset of Ω_m in which the GHB-problem has a unique solution.

A natural way to decrease the number of solutions of the GHB-problem is to consider only elements of Ω_m which have certain additional continuity properties. Let $S(x) \in \Omega_m$ satisfy (2.8). If for some pair (i,j)

$$\alpha_{ij} < S^{(j)}(x_i) < \beta_{ij} , \tag{2.9}$$

we require that $S^{(2m-1-j)}(x)$ is continuous at $x = x_i$, i.e.

$$S^{(2m-1-j)}(x_i - 0) = S^{(2m-1-j)}(x_i + 0) . \tag{2.10}$$

Furthermore, we require that

$$(-1)^{m-j}(S^{(2m-1-j)}(x_i - 0) - S^{(2m-1-j)}(x_i + 0)) \leq 0 \tag{2.11}$$

$$\text{if } \alpha_{ij} = S^{(j)}(x_i) < \beta_{ij}$$

and

$$(-1)^{m-j}(S^{(2m-1-j)}(x_i - 0) - S^{(2m-1-j)}(x_i + 0)) \geq 0 \tag{2.12}$$

$$\text{if } \alpha_{ij} < S^{(j)}(x_i) = \beta_{ij} .$$

It will be shown that the conditions (2.9) - (2.12) determine a subset of Ω_m in which the GHB-problem has a solution which is unique up to an additive polynomial of degree at most m-1. Furthermore, it will turn out that this solution of the GHB-problem has a number of minimal properties which generalize minimal properties of the solution of the HB-problem [6]. For this reason we shall call any solution $S(x)$ of the GHB-problem an extremal g-spline, provided $S(x) \in \Omega_m$ and it satisfies (2.9) - (2.12).

In order to prove the existence of an extremal solution and to demonstrate its properties we first investigate a minimization problem which is related to the GHB-problem.

3. A minimization problem and its dual problem

Choose the real numbers a and b so that

$$a \leq x_1 < \ldots < x_k \leq b$$

and let $F(a,b)$ denote the space of real functions $f(x)$ on $[a,b]$ such that $f(x) \in C^{m-1}(a,b)$, $f^{(m-1)}(x)$ is absolutely continuous and $f^{(m)}(x) \in L_2(a,b)$.

For every $g(x) \in F(a,b)$ and every $h(x) \in F(a,b)$ we define $D(g,h)$ by

$$D(g,h) = \int_a^b g^{(m)}(x)\, h^{(m)}(x)\, dx \; .$$

With this definition D is a symmetric bilinear functional on $F(a,b) \times F(a,b)$. For every $g(x) \in F(a,b)$ the linear functional Dg on $F(a,b)$ is defined by

$$Dg = D(g, \cdot) \; .$$

Furthermore, the linear functionals ℓ_{ij} on $F(a,b)$ are defined by

$$\ell_{ij} g = g^{(j)}(x_i) \quad \text{for } j \in I_i \quad \text{and} \quad i = 1, \ldots, k \; .$$

Let I_i, α_{ij} and β_{ij} be defined as in Section 2). Then we consider the following minimization problem, which is referred to as (MP):

Find a $g(x) \in F(a,b)$ which minimizes

$$D(g,g)$$

among all $g(x) \in F(a,b)$ satisfying the constraints

$$\alpha_{ij} \leq \ell_{ij} g \leq \beta_{ij} \quad \text{for} \quad j \in I_i \quad \text{and} \quad i = 1, \ldots, k \ .$$

In mathematical programming it is usual to associate with a given minimization problem a maximization problem, the so called dual problem. As we shall see, the investigation of the dual problem gives further information about the solution of the given minimization problem, which is called the primal problem.

In our case the dual problem associated with the given (MP) is defined as follows:

Find $g(x) \in F(a,b)$ and real numbers λ_{ij}, τ_{ij} which maximize

$$(3.1) \quad \sum_{i=1}^{k} \sum_{j \in I_i} [\lambda_{ij}(\alpha_{ij} - \ell_{ij} g) + \tau_{ij}(\ell_{ij} g - \beta_{ij})] + D(g,g)$$

among all $g(x)$, λ_{ij} and τ_{ij} satisfying the constraints

$$(3.2) \quad 2Dg - \sum_{i=1}^{k} \sum_{j \in I_i} (\lambda_{ij} - \tau_{ij})\ell_{ij} = 0; \ \lambda_{ij} \geq 0, \ \tau_{ij} \geq 0 \ .$$

Since for all $g(x)$, λ_{ij} and τ_{ij} which fulfill (3.2)

$$\sum_{i=1}^{k} \sum_{j \in I_i} (\tau_{ij} - \lambda_{ij})\ell_{ij} g = -2D(g,g) \ ,$$

we have the equivalent problem:

Find $g(x) \in F(a,b)$ and real numbers λ_{ij}, τ_{ij} which maximize

$$(3.3) \qquad \sum_{i=1}^{k} \sum_{j \in I_i} (\alpha_{ij}\lambda_{ij} - \beta_{ij}\tau_{ij}) - D(g,g)$$

among all $g(x)$, λ_{ij} and τ_{ij} satisfying (3.2).

We shall refer to each of these two equivalent formulations of the dual problem as (DP).

The following theorem gives necessary and sufficient conditions for $g(x) \in F(a,b)$ to be a solution of (MP).

Theorem (3.1)

A $g(x) \in F(a,b)$ is a solution of (MP) if and only if there are real numbers $\lambda_{ij} \geq 0$ and $\tau_{ij} \geq 0$ such that

1) $2Dg = \sum_{i=1}^{k} \sum_{j \in I_i} (\lambda_{ij} - \tau_{ij})\ell_{ij}$,

2) $\lambda_{ij}[\alpha_{ij} - \ell_{ij}g] = 0, \quad j \in I_i, \quad i = 1, \ldots, k$,

3) $\tau_{ij}[\ell_{ij}g - \beta_{ij}] = 0, \quad j \in I_i, \quad i = 1, \ldots, k$,

4) $\alpha_{ij} \leq \ell_{ij}g \leq \beta_{ij} \quad j \in I_i, \quad i = 1, \ldots, k$.

Proof

a) Suppose $g(x) \in F(a,b)$ is a solution of (MP). For each $i \in \{1, \ldots, k\}$ define the subsets I_i^1 and I_i^2 of I_i in such a way that

$j \in I_i^1$ if and only if $\ell_{ij}g = \alpha_{ij}$ and

$j \in I_i^2$ if and only if $\ell_{ij}g = \beta_{ij}$.

Let $h(x)$ be any element of $F(a,b)$ with

$$(1)\qquad \begin{aligned} &\ell_{ij} h \geq 0 \quad \text{for } j \in I_i^1, \quad i = 1, \ldots, k \quad \text{and} \\ -&\ell_{ij} h \geq 0 \quad \text{for } j \in I_i^2, \quad i = 1, \ldots, k \ . \end{aligned}$$

Since

$$\alpha_{ij} < \ell_{ij} g < \beta_{ij} \quad \text{for } j \notin I_i^1 \cup I_i^2, \quad i = 1, \ldots, k \ ,$$

there exists a positive number t_0 such that

$$g(x) + t\,h(x)$$

satisfies the constraints of (MP) for $0 \leq t \leq t_0$. Therefore, $g(x)$ being a solution of (MP), we have

$$D(g+th, g+th) \geq D(g,g) \quad \text{for } 0 \leq t \leq t_0 \ .$$

From this inequality we obtain

$$2tD(g,h) + t^2 D(h,h) \geq 0 \quad \text{for } 0 \leq t \leq t_0 \ ,$$

which implies $2D(g,h) \geq 0$.

Hence, for every $h(x) \in F(a,b)$, the inequalities (1) imply $2D(g,h) \geq 0$. It follows from a generalization of the Farkas Lemma, given by Fan [3], that

$$2Dg = \sum_{i=1}^{k} \sum_{j \in I_i^1} \lambda_{ij} \ell_{ij} - \sum_{i=1}^{k} \sum_{j \in I_i^2} \tau_{ij} \ell_{ij} \ ,$$

where $\lambda_{ij} \geq 0$ and $\tau_{ij} \geq 0$ are real numbers.

If we choose $\lambda_{ij} = 0$ for $j \notin I_i^1$, $i = 1, \ldots, k$, and $\tau_{ij} = 0$ for $j \notin I_i^2$, $i = 1, \ldots, k$, the first part of the proof is complete.

b) Suppose $g(x) \in F(a,b)$, $\lambda_{ij} \geq 0$ and $\tau_{ij} \geq 0$ satisfy conditions 1) to 4). Let $g_1(x)$ be any element of $F(a,b)$ with

$$\alpha_{ij} \leq \ell_{ij} g_1 \leq \beta_{ij} \quad \text{for } j \in I_i, \quad i = 1, \ldots, k .$$

Put $h(x) = g_1(x) - g(x)$ and define I_i^1 and I_i^2 as in the first part of the proof. Then

$$2Dg = \sum_{i=1}^{k} \sum_{j \in I_i} \lambda_{ij} \ell_{ij} - \sum_{i=1}^{k} \sum_{j \in I_i} \tau_{ij} \ell_{ij} ,$$

$$\ell_{ij} h \geq 0 \quad \text{for } j \in I_i^1, \quad i = 1, \ldots, k \quad \text{and}$$

$$-\ell_{ij} h \geq 0 \quad \text{for } j \in I_i^2, \quad i = 1, \ldots, k .$$

Therefore,

$$2D(g,h) \geq 0$$

and

$$\begin{aligned} D(g_1, g_1) &= D(g+h, g+h) \\ &= D(g,g) + 2D(g,h) + D(h,h) \\ &\geq D(g,g) , \end{aligned}$$

which completes the proof of the theorem.

The above theorem can be used to derive a representation formula for these elements of $F(a,b)$ which are candidates for a solution of (MP). For this purpose we define the truncated power function $(x-\bar{x})_+^j$ by

$$(x-\bar{x})_+^j = \begin{cases} (x-\bar{x})^j & \text{if } x \geq \bar{x} \\ 0 & \text{if } x < \bar{x} . \end{cases}$$

Furthermore, let

$$F_0 = \{f(x) \,|\, f(x) = \frac{1}{2} \sum_{i=1}^{k} \sum_{j \in I_i} \frac{(-1)^{m-j}}{(2m-1-j)!} \sigma_{ij} (x-x_i)_+^{2m-1-j}\} ,$$

where σ_{ij} are real numbers, and

$$F_1 = \{f(x) \in F_0 \,|\, f^{(m)}(x) = 0 \;\; \text{for} \;\; x < x_1 \quad \text{and} \quad x \geq x_k\} .$$

It follows immediately that $F_0 \subset F(a,b)$.

Theorem (3.2)

I) Suppose $g(x) \in F(a,b)$ and λ_{ij}, τ_{ij} satisfy the equation 1) of Theorem (3.1). Let $f(x)$ be the element of F_0 which corresponds to $\sigma_{ij} = \lambda_{ij} - \tau_{ij}$. Then

$$f(x) \in F_1 \quad \text{and}$$

$$g(x) = f(x) + P(x) \quad \text{for all} \;\; x \in [a,b] ,$$

and some $P(x) \in \Pi_{m-1}$.

II) Let $f(x)$ be the element of F_0 which corresponds to $\sigma_{ij} = \lambda_{ij} - \tau_{ij}$. If $f(x) \in F_1$, then $f(x)$, λ_{ij} and τ_{ij} satisfy the equation 1) of Theorem (3.1).

Proof

For any $h(x) \in F(a,b)$ we have

$$\begin{aligned} \ell_{ij} h &= h^{(j)}(x_i) \\ &= h^{(j)}(b) - \int_{x_i}^{b} h^{(j+1)}(x)\,dx \\ &= h^{(j)}(b) - \int_{a}^{b} (x-x_i)_+^0 \, h^{(j+1)}(x)\,dx . \end{aligned}$$

Integrating by parts the last term on the right hand side of the above equation $m-1-j$ times we obtain

$$(1)\qquad \ell_{ij}h = \sum_{\nu=0}^{m-1-j} \frac{(-1)^{\nu}}{\nu!}(b-x_i)^{\nu} h^{(j+\nu)}(b) +$$

$$+ (-1)^{m-j}\int_a^b \frac{(x-x_i)_+^{m-1-j}}{(m-1-j)!} h^{(m)}(x)\,dx \;.$$

For every $j \in \{0, \ldots, m-1\}$, let

$$J_j = \{i \mid j \in I_i\} \;.$$

Then, by (1),

$$(2)\qquad \sum_{i=1}^{k}\sum_{j\in I_i} \sigma_{ij}\ell_{ij}h = \sum_{j=0}^{m-1}\sum_{i\in J_j}\sigma_{ij}\ell_{ij}h =$$

$$= \sum_{\mu=0}^{m-1}\sum_{j=0}^{\mu}\sum_{i\in J_j}\frac{(-1)^{\mu-j}}{(\mu-j)!}\sigma_{ij}(b-x_i)^{\mu-j} h^{(\mu)}(b) +$$

$$+ \sum_{j=0}^{m-1}\sum_{i\in J_j}(-1)^{m-j}\sigma_{ij}\int_a^b \frac{(x-x_i)_+^{m-1-j}}{(m-1-j)!} h^{(m)}(x)\,dx =$$

$$= 2\sum_{\mu=0}^{m-1}(-1)^{m-\mu} f^{(2m-1-\mu)}(b+0)\,h^{(\mu)}(b) + 2\int_a^b f^{(m)}(x)\,h^{(m)}(x)\,dx \,,$$

where $f(x)$ is the element of F_0 which corresponds to σ_{ij}.

Now suppose that $g(x) \in F(a,b)$ and λ_{ij}, τ_{ij} satisfy the equation 1) of Theorem (3.1), i.e.,

(3) $$2D(g,h) = \sum_{i=1}^{k} \sum_{j \in I_i} (\lambda_{ij} - \tau_{ij}) \ell_{ij} h$$

holds for every $h(x) \in F(a,b)$.

If we put $\sigma_{ij} = \lambda_{ij} - \tau_{ij}$, it follows from (2) and (3) that

(4) $f^{(2m-1-\mu)}(b+0) = 0$ for $\mu = 0, \ldots, m-1$, and

(5) $\int_a^b g^{(m)}(x)\, h^{(m)}(x)\, dx = \int_a^b f^{(m)}(x)\, h^{(m)} dx$ for every $h(x) \in F(a,b)$.

Since $f(x) \in \Pi_{2m-1}$ for $x \geq x_k$, it follows from (4) that $f^{(m)}(x) = 0$ for $x \geq x_k$. By definition, $f^{(m)}(x) = 0$ for $x < x_1$, hence $f(x) \in F_1$. Furthermore, it follows from (5) that there exists $P(x) \in \Pi_{m-1}$ such that

$$g(x) = f(x) + P(x) \quad \text{for every} \quad x \in [a,b] .$$

On the other hand, if $f(x) \in F_0$, corresponding to $\sigma_{ij} = \lambda_{ij} - \tau_{ij}$, is an element of F_1, then (4) holds. Hence, it follows from (2) that

$$\sum_{i=1}^{k} \sum_{j \in I_i} \sigma_{ij} \ell_{ij} h = 2 \int_a^b f^{(m)}(x)\, h^{(m)}(x)\, dx \quad \text{for any } h(x) \in F(a,b) ,$$

i.e., $f(x)$ and λ_{ij}, τ_{ij} satisfy the equation 1) of Theorem (3.1).

Lemma (3.3)

I) If $g(x)$ and $h(x)$ are two solutions of (MP), then

$$g(x) - h(x) \in \Pi_{m-1} \quad \text{for } x \in [a,b] .$$

II) If $g(x)$, λ_{ij}, τ_{ij} and $h(x)$, $\bar{\lambda}_{ij}$, $\bar{\tau}_{ij}$ are two solutions of (DP), then

$$g(x) - h(x) \in \Pi_{m-1} \quad \text{for } x \in [a,b] \quad \text{and}$$

$$\lambda_{ij} - \tau_{ij} = \bar{\lambda}_{ij} - \bar{\tau}_{ij} .$$

Proof

I) For $0 \leq t \leq 1$, $g(x) + t(h(x) - g(x))$ satisfies the constraints of (MP). Hence, if we define $\psi(t)$ by

$$\psi(t) = D(g+t(h-g), g+t(h-g)) = D(g,g)+2tD(g,h-g)+t^2 D(h-g,h-g) ,$$

then

$$\psi(0) = \psi(1) = D(g,g) \quad \text{and} \quad \psi(t) \geq D(g,g) \quad \text{for} \quad 0 \leq t \leq 1 .$$

Since $\psi''(t) = 2D(h-g, h-g) \geq 0$, this implies $\psi(t) = \text{const.}$ Thus,

$$D(h-g, h-g) = 0$$

which is equivalent with $h(x) - g(x) \in \Pi_{m-1}$ for $x \in [a,b]$.

II) Again it follows that for $0 \leq t \leq 1$,

$$g(x) + t(h(x) - g(x)), \lambda_{ij} + t(\bar{\lambda}_{ij} - \lambda_{ij}), \tau_{ij} + t(\bar{\tau}_{ij} - \tau_{ij})$$

satisfy the constraints of (DP). Let

$$\phi(t) = \sum_{i=1}^{k} \sum_{j \in I_i} [\alpha_{ij}(\lambda_{ij} + t(\bar{\lambda}_{ij} - \lambda_{ij})) - \beta_{ij}(\tau_{ij} + t(\bar{\tau}_{ij} - \tau_{ij})) -$$
$$- D(g+t(h-g), g+t(h-g)) .$$

Then we have

$$\phi(0) = \phi(1) \quad \text{and} \quad \phi(t) \leq \phi(0) \quad \text{for} \quad 0 \leq t \leq 1 .$$

Since $\phi''(t) = -2D(h-g, h-g) \leq 0$, this implies

$$D(h-g, h-g) = 0 ,$$

which is equivalent to $h(x) - g(x) \in \Pi_{m-1}$ for $x \in [a,b]$. By Theorem (3.2),

$$g(x) - f(x) \in \Pi_{m-1} \text{ for } x \in [a,b] \text{ and}$$

$$h(x) - \bar{f}(x) \in \Pi_{m-1} \text{ for } x \in [a,b] ,$$

where $f(x)$ and $\bar{f}(x)$ are the elements of F_0 which correspond to $\sigma_{ij} = \lambda_{ij} - \tau_{ij}$ and $\bar{\sigma}_{ij} = \bar{\lambda}_{ij} - \bar{\tau}_{ij}$, respectively. Hence, $f(x) - \bar{f}(x) \in \Pi_{m-1}$ which implies $f(x) - \bar{f}(x) \equiv 0$, or $\sigma_{ij} = \bar{\sigma}_{ij}$.

The next lemma gives the first connection between (MP) and (DP) .

Lemma (3.4)

Suppose $g(x) \in F(a,b)$ satisfies the constraints of (MP) and $h(x) \in F(a,b)$ and λ_{ij}, τ_{ij} satisfy the constraints of (DP) . Then

$$(*) \quad \sum_{i=1}^{k} \sum_{j \in I_i} [\lambda_{ij}(\alpha_{ij} - \ell_{ij}h) + \tau_{ij}(\ell_{ij}h - \beta_{ij})] + D(h,h) \leq D(g,g) .$$

If equality holds in (*), then $g(x)$ is a solution of (MP) and $h(x)$, λ_{ij}, τ_{ij} is a solution of (DP) .

Proof

$$\begin{aligned} D(g,g) - D(h,h) &= 2D(h, g-h) + D(g-h, g-h) \\ &\geq \sum_{i=1}^{k} \sum_{j \in I_i} (\lambda_{ij} - \tau_{ij}) \ell_{ij}(g-h) \\ &\geq \sum_{i=1}^{k} \sum_{j \in I_i} [\lambda_{ij}\alpha_{ij} - \tau_{ij}\beta_{ij} - (\lambda_{ij} - \tau_{ij}) \ell_{ij} h] , \end{aligned}$$

where the first inequality follows from $D(g-h, g-h) \geq 0$ and Theorem (3.2) while the second inequality is a consequence of $\lambda_{ij} \geq 0$, $\tau_{ij} \geq 0$ and $\alpha_{ij} \leq \ell_{ij} g \leq \beta_{ij}$.

Suppose equality holds in $(*)$ and let $g_1(x) \in F(a,b)$ satisfy the constraints of (MP). Then, by the first part of the lemma,

$$D(g,g) = \sum_{i=1}^{k} \sum_{j \in I_i} [\lambda_{ij}(\alpha_{ij} - \ell_{ij}h) + \tau_{ij}(\ell_{ij}h - \beta_{ij})] + D(h,h) \leq D(g_1, g_1) .$$

In the same way it can be shown that $h(x)$, λ_{ij}, τ_{ij} is a solution of (DP).

The main result on the relationship between (MP) and (DP) is given by

Theorem (3.5)

I) If $g(x) \in F(a,b)$ is a solution of (MP), then there are λ_{ij} and τ_{ij} such that

$$g(x),\ \lambda_{ij},\ \tau_{ij}$$

is a solution of (DP) and

$$g(x) - f(x) \in \Pi_{m-1} \quad \text{for } x \in [a,b] ,$$

where $f(x) \in F_1$ and corresponds to $\sigma_{ij} = \lambda_{ij} - \tau_{ij}$.

II) Suppose there exists a solution of (MP). If $g(x)$, λ_{ij}, τ_{ij} is a solution of (DP), then there exists $P(x) \in \Pi_{m-1}$ such that

$$g(x) + P(x)$$

is a solution of (MP). Furthermore,

$$g(x) - f(x) \in \Pi_{m-1} \quad \text{for } x \in [a,b] ,$$

where $f(x) \in F_1$ and corresponds to $\sigma_{ij} = \lambda_{ij} - \tau_{ij}$.

Proof

I) It follows from Theorem (3.1) that there are λ_{ij} and τ_{ij} such that $g(x)$, λ_{ij} and τ_{ij} satisfy the constraints of (DP) and

$$\lambda_{ij}(\alpha_{ij}-\ell_{ij}g) = \tau_{ij}(\ell_{ij}g-\beta_{ij}) = 0, \quad j \in I_i, \quad i = 1, \ldots, k \ .$$

Hence, equality holds in the relation (*) in Lemma (3.4) if we insert $g(x)$ and $g(x)$, λ_{ij}, τ_{ij} . Thus it follows from Lemma (3.4) that $g(x)$, λ_{ij}, τ_{ij} is a solution of (DP). Furthermore by the Theorems (3.1) and (3.2), $g(x) - f(x) \in \Pi_{m-1}$ for $x \in [a,b]$, where $f(x) \in F_1$ and corresponds to $\sigma_{ij} = \lambda_{ij} - \tau_{ij}$.

II) Let $h(x)$ be a solution of (MP) . By the first part of the theorem, there are $\bar{\lambda}_{ij}$, $\bar{\tau}_{ij}$ such that $h(x)$, $\bar{\lambda}_{ij}$ and $\bar{\tau}_{ij}$ solve the (DP) . Hence by Lemma (3.3), there exists $P(x) \in \Pi_{m-1}$ such that

$$h(x) = g(x) + P(x) \quad \text{for} \quad x \in [a,b] \ .$$

Therefore, $g(x) + P(x)$ is a solution of (MP) . The remaining statements of the theorem follow from Theorem (3.2) since $g(x)$, λ_{ij} and τ_{ij} satisfy the equation 1) of Theorem (3.1).

4. The solution of the generalized interpolation problem

Having the results of Section 3) at our disposal we can now settle the question of the existence of an extremal g-spline which solves the GHB-problem.

First we establish the connection between the minimization problem, considered in Section 3), and the GHB-problem.

Theorem (4.1)

I) Suppose $g(x) \in F(a,b)$ is a solution of (MP) . Then there exists $f(x) \in F_1$ and $P(x) \in \Pi_{m-1}$ such that

$$S(x) = f(x) + P(x) \in \Omega_m$$

is an extremal solution of the GHB-problem and

$$g(x) = S(x) \quad \text{for} \quad x \in [a,b] \ .$$

II) Suppose $S(x) \in \Omega_m$ is an extremal solution of the GHB-problem. Then $S(x)$ is a solution of (MP) .

Proof

I) By Theorem (3.1) there are $\lambda_{ij} \geq 0$ and $\tau_{ij} \geq 0$ such that

$$2Dg = \sum_{i=1}^{k} \sum_{j \in I_i} (\lambda_{ij} - \tau_{ij}) \ell_{ij} .$$

Let $f(x) \in F_0$ correspond to $\sigma_{ij} = \lambda_{ij} - \tau_{ij}$. By Theorem (3.2), $f(x) \in F_1$ and there exists $P(x) \in \Pi_{m-1}$ such that with $S(x) = f(x) + P(x)$

$$g(x) = S(x) \quad \text{for all } x \in [a,b] .$$

Since $f(x) \in F_1$, $S(x) \in \Omega_m$.
Furthermore, it follows from Theorem (3.1) that

$$\lambda_{ij} = 0 \quad \text{whenever } \alpha_{ij} < \ell_{ij} g = \ell_{ij} S ,$$

$$\tau_{ij} = 0 \quad \text{whenever } \beta_{ij} > \ell_{ij} g = \ell_{ij} S .$$

Hence, the definition of $f(x)$ implies that $S(x)$ satisfies (2.9) - (2.12).

II) It follows from the definition of the g-splines that any $S(x) \in \Omega_m$ can be written in the form

$$S(x) = f(x) + P(x) .$$

where $f(x) \in F_1$ and $P(x) \in \Pi_{m-1}$.
Let σ_{ij}, $j \in I_i$, $i = 1, \ldots, k$, be the numbers which define $f(x)$ as an element of F_0 and put

$$(1) \qquad \begin{array}{ll} \lambda_{ij} = \sigma_{ij}, \ \tau_{ij} = 0 & \text{if } \sigma_{ij} \geq 0 \\ \lambda_{ij} = 0 , \ \tau_{ij} = -\sigma_{ij} & \text{if } \sigma_{ij} < 0 \end{array} \qquad j \in I_i, \ i = 1, \ldots, k .$$

Since $f(x) \in F_1$, it follows from Theorem (3.2) that

$$(2) \qquad 2Dg = 2Df = \sum_{i=1}^{k} \sum_{j \in I_i} (\lambda_{ij} - \tau_{ij}) \ell_{ij} .$$

By assumption, $S(x)$ satisfies (2.9) - (2.12). Therefore,

$$(3) \qquad \begin{aligned} \sigma_{ij} &\leq 0 \quad \text{whenever} \quad \alpha_{ij} < \ell_{ij} S , \\ \sigma_{ij} &\geq 0 \quad \text{whenever} \quad \beta_{ij} > \ell_{ij} S . \end{aligned}$$

Since $\alpha_{ij} \leq \ell_{ij} S \leq \beta_{ij}$, it follows from (1) - (3) that $S(x)$, λ_{ij}, and τ_{ij} satisfy the conditions of Theorem (3.1). Clearly, $S(x)$ is an element of $F(a,b)$. Thus, it follows from Theorem (3.1) that $S(x)$ is a solution of (MP).

Now we shall prove that (MP) has a solution provided that the GHB-problem is m-poised.

The Theorems (3.1) and (3.2) show that in looking for a solution of (MP) we can restrict our attention to the finite dimensional subspace

$$F_1 + \Pi_{m-1} ,$$

i.e. if (MP) has a solution it can be written in the form

$$f(x) + P(x) ,$$

where

$$P(x) = \sum_{\mu=0}^{m-1} \omega_\mu x^\mu \quad \text{and}$$

$$f(x) = \frac{1}{2} \sum_{i=1}^{k} \sum_{j \in I_i} \frac{(-1)^{m-j}}{(2m-1-j)!} (x-x_i)_+^{2m-1-j} \sigma_{ij}$$

with the property that

(4.1) $$f^{(m)}(x) = 0 \quad \text{for} \quad x \geq x_k \quad .$$

On the subspace $F_1 + \Pi_{m-1}$, (MP) reduces to a minimization problem which involves only the finitely many variables σ_{ij} and ω_μ . Before we formulate this special problem we investigate the conditions which must be satisfied by σ_{ij} in order that the corresponding $f(x)$ satisfies (4.1).

By definition,

$$f^{(m)}(x) \in \Pi_{m-1} \quad \text{for} \quad x \geq x_k \quad .$$

Hence, (4.1) is satisfied if and only if

(4.2) $$f^{(2m-1-\mu)}(b+0) = \frac{1}{2} \sum_{j=0}^{\mu} \sum_{i \in J_j} \frac{(-1)^{m-j}}{(\mu-j)!} (b-x_i)^{\mu-j} \sigma_{ij} = 0$$ [1)]

for $\mu = 0, \ldots, m-1$, where $J_j = \{i \mid j \in I_i\}$ for $j = 0, \ldots, m-1$. For $i = 1, \ldots, k$ let the column vector σ_i have the components σ_{ij} for $j \in I_i$. Furthermore, for $\mu = 0, \ldots, m-1$ and $j \in I_i$, define the elements $b^i_{\mu j}$ of the matrices B_i, $i = 1, \ldots, k$, by

$$b^i_{\mu j} = \begin{cases} \dfrac{(-1)^{m-j}}{(\mu-j)!} (b-x_i)^{\mu-j} & \text{if} \quad j \leq \mu \\ 0 & \text{if} \quad j > \mu \quad . \end{cases}$$

Then the element $f(x) \in F_0$, which corresponds to σ_{ij}, satisfies (4.2), and therefore (4.1), if and only if

(4.3) $$B_1\sigma_1 + \ldots + B_k\sigma_k = 0 \quad .$$

If we require that an $f(x) + P(x) \in F_0 + \Pi_{m-1}$ satisfies the constraints of (MP) we obtain further restrictions for the variables σ_{ij} and ω_j :

---

1) Here and below 0^0 is defined to be 1 .

(4.4) $\quad \alpha_{ij} \leq \ell_{ij}P + \ell_{ij}f \leq \beta_{ij}, \quad j \in I_i, \quad i = 1, \ldots, k .$

By the definition of $f(x)$ and $P(x)$ we have

$$\ell_{ij}P = P^{(j)}(x_i) = \sum_{\mu=j}^{m-1} \frac{\mu!}{(\mu-j)!} \omega_\mu x_i^{\mu-j}$$

and

$$\ell_{ij}f = f^{(j)}(x_i) = \sum_{\nu=1}^{i} \sum_{\mu \in I_\nu} \frac{(-1)^{m-\mu}}{2(2m-1-\mu-j)!} \sigma_{\nu\mu}(x_i - x_\nu)^{2m-1-\mu-j} .$$

For $j \in I_i$ and $\mu = 0, \ldots, m-1$ we define the elements $a^i_{\mu j}$ of the matrices A_i, $i = 1, \ldots, k$, by

$$a^i_{j\mu} = \begin{cases} \dfrac{\mu!}{(\mu-j)!} x_i^{\mu-j} & \text{if} \quad j \leq \mu \\ 0 & \text{if} \quad j > \mu . \end{cases}$$

Furthermore, for $j \in I_i$ and $\mu \in I_\nu$ we define the elements $a^{i\nu}_{j\mu}$ of the matrices $A_{i\nu}$, $1 \leq \nu < i \leq k$, by

$$a^{i\nu}_{j\mu} = \frac{(-1)^{m-\mu}}{2(2m-1-\mu-j)!} (x_i - x_\nu)^{2m-1-\mu-j} .$$

If α_i denotes the column vector with the components α_{ij}, β_i denotes the column vector with the components β_{ij} and $\omega' = (\omega_0, \ldots, \omega_{m-1})$, then $f(x) + P(x)$ satisfies (4.4) if and only if

$$(4.5)\quad \begin{array}{lllllll}
\alpha_1 \le A_1\omega & & & & & & \le \beta_1 \\
\alpha_2 \le A_2\omega + A_{21}\sigma_1 + & & & & & & \le \beta_2 \\
\alpha_3 \le A_3\omega + A_{31}\sigma_1 + A_{32}\sigma_2 & & & & & & \le \beta_3 \\
\vdots & & & & & & \vdots \\
\alpha_k \le A_k\omega + A_{k1}\sigma_1 + A_{k2}\sigma_2 + \ldots + A_{kk}\sigma_k & & & & & & \le \beta_k \ ,
\end{array}$$

where all elements of A_{kk} are zero.

Finally, we observe that for every $f(x) \in F_0$

$$D(f,f) = \int_a^b (f^{(m)}(x))^2 dx = \sum_{i,\nu=1} \sum_{\substack{j \in I_i \\ \mu \in I_\nu}} \frac{(-1)^{j+\mu}}{4(m-1-j)!\,(m-1-\mu)!} \times$$

$$\int_a^b (x-x_i)_+^{m-1-j} (x-x_\nu)_+^{m-1-\mu} \sigma_{ij}\sigma_{\nu\mu} dx \ .$$

Defining the elements $c_{j\mu}^{i\nu}$, $j \in I_i$, $\mu \in I_\nu$, of the matrices $C_{i\nu}$, $1 \le \nu \le i \le k$, by

$$(4.6)\quad c_{j\mu}^{i\nu} = \sum_{\ell=0}^{m-\mu-1} \frac{(-1)^{j+\mu+\ell}}{4(m-j+\ell)!\,(m-1-\mu-\ell)!} (b-x_i)^{m-j+\ell} (b-x_\nu)^{m-1-\mu-\ell}$$

and putting

$$C_{i\nu} = C'_{\nu i} \quad \text{for} \quad 1 \le i < \nu \le k \ ,$$

we obtain

$$D(f,f) = \sum_{i,\nu=1}^{k} \sigma_i' C_{i\nu} \sigma_\nu \quad \text{for every} \quad f(x) \in F_0 \ .$$

Using the above definitions we see that, on the subspace $F_1 + \Pi_{m-1}$, (MP) is equivalent to the following quadratic programming problem

(4.7) $$\min\{\sigma'C\sigma \mid \alpha \leq A\omega + M\sigma \leq \beta, \quad B\sigma = 0\} ,$$

where $\sigma'C\sigma$ stands for $\sum_{i,\nu=1}^{k} \sigma_i' C_{i\nu}\sigma_\nu$ and (4.3) and (4.5) are abbreviated by $B\sigma = 0$ and $\alpha \leq A\omega + M\sigma \leq \beta$, respectively. In order to show that (4.7) has a solution we need some properties of this problem which are stated in the following

Lemma (4.2)

1) If $x_k < b$, then C is positive definite.
2) If $x_k = b$, then C is positive semi-definite and

   $\sigma'C\sigma = 0$ and $B\sigma = 0$ imply $\sigma = 0$.

3) If the GHB-problem is m-poised then

$$A\omega + M\sigma = 0 \quad \text{and} \quad B\sigma = 0$$

   imply $\omega = 0$ and $\sigma = 0$.
4) Suppose the GHB-problem is m-poised. Let $(A\omega+M\sigma)_{ij}$ denote the jth component of the vector $A_1\omega$ if $i = 1$ and jth component of the vector $A_i\omega + A_{i1}\sigma_1 + \ldots + A_{i,i-1}$ if $i > 1$. Suppose (ω, σ) satisfies the constraints of (4.7). For any $i_0 \in \{1, \ldots, k\}$ and any $j_0 \in I_{i_0}$ there exists $(\bar{\omega}, \bar{\sigma})$ such that $B\bar{\sigma} = 0$ and

(*) $$(A\bar{\omega} + M\bar{\sigma})_{ij} = \begin{cases} 1 & \text{for } i = i_0 \text{ and } j = j_0 \\ 0 & \text{for } i \neq i_0 \text{ and/or } j \neq j_0 , \end{cases}$$

   and, for any real number t,

$$(\sigma+t\bar{\sigma})'C(\sigma+t\bar{\sigma}) = \sigma'C\sigma + t\sigma_{i_0 j_0} + t^2\bar{\sigma}'C\bar{\sigma} .$$

Proof

1) It follows from the definition that $\sigma'C\sigma \geq 0$. Suppose $\sigma'C\sigma = 0$ and let $f(x)$ be the element of F_0 which corresponds to the components of σ . Then

$$D(f, f) = \int_a^b (f^{(m)}(x))^2 dx = \sigma'C\sigma = 0 \quad .$$

This implies

$$f^{(m)}(x) = 0 \quad \text{for} \quad x \in (x_i, x_{i+1}), \qquad i = 1, \ldots, k-1 \;,$$

and

$$f^{(m)}(x) = 0 \quad \text{for} \quad x \in (x_k, b) \;,$$

since $f^{(m)}(x)$ is continuous in each of these intervals. By the definition of $f(x)$ this is equivalent to $\sigma_{ij} = 0$ for $j \in I_i$ and $i = 1, \ldots, k$. Hence, $\sigma'C\sigma = 0$ implies $\sigma = 0$ and C is positive definite.

2) As under 1) it follows that C is positive semi-definite and that $\bar{\sigma}'C\bar{\sigma} = 0$ implies $\bar{\sigma}_i = 0$ for $i = 1, \ldots, k-1$. Hence,

$$B\bar{\sigma} = 0 \quad \text{is equivalent to} \quad B_k\bar{\sigma}_k = 0 \quad .$$

If $I_k = \{0, 1, \ldots, m\}$, B_k is a lower triangular matrix with nonzero diagonal elements. Hence, the columns of B_k are linearly independent for an arbitrary (nonvoid) I_k . Thus $B_k\bar{\sigma}_k = 0$ implies $\bar{\sigma}_k = 0$; and we have $\bar{\sigma} = 0$.

3) Suppose

$$A\omega + M\sigma = 0 \quad \text{and} \quad B\sigma = 0 \quad .$$

Let $f(x)$ be the element of F_0 and $P(x)$ the element of Π_{m-1}, which corresponds to σ and ω, respectively. Since $B\sigma = 0$, it follows from (4.1) that $f(x) \in F_1$. Hence, by Theorem (3.2),

$$2Df = \sum_{i=1}^{k} \sum_{j \in I_i} \sigma_{ij} \ell_{ij}$$

and

$$2D(f,f) = 2D(f,P+f) = \sum_{i=1}^{k} \sum_{j \in I_i} \sigma_{ij} \ell_{ij}(P+f) = \sigma'(A\omega+M\sigma) \quad ,$$

since $\ell_{ij}(P+f) = (A\omega+M\sigma)_{ij}$. Therefore, it follows from $A\omega + M\sigma = 0$ and $B\sigma = 0$, that

$$\sigma'C\sigma = D(f,f) = 0 \quad ,$$

which by the second part of the lemma implies $\sigma = 0$. Furthermore, $A\omega = 0$ is equivalent to $P^{(j)}(x_i) = 0$ for $j \in I_i$ and $i = 1, \ldots, k$. Thus, $P(x) \equiv 0$, i.e., $\omega = 0$.

4) It follows immediately from the definition that $\binom{A, M}{0, B}$ is a square matrix. By part 3 of the lemma, it is nonsingular. Hence, the existence of $(\bar{\omega}, \bar{\sigma})$ such that $B\bar{\sigma} = 0$ and (*) holds is evident. Let $f(x)$, $\bar{f}(x)$ and $P(x)$, $\bar{P}(x)$ denote the elements of F_0 and Π_{m-1} which correspond to $\sigma, \bar{\sigma}$ and $\omega, \bar{\omega}$ respectively. Since $B\sigma = 0$, $f(x) \in F_1$ and it follows from Theorem (3.2) that

$$2Df = \sum_{i=1}^{k} \sum_{j \in I_i} \sigma_{ij} \ell_{ij} \quad .$$

Hence, for any real t,

$$2D(f, t\bar{P} + t\bar{f}) = \sum_{i=1}^{k} \sum_{j \in I_i} t\sigma_{ij} \ell_{ij}(\bar{P}+\bar{f}) = t\sigma_{i_0 j_0}$$

since $\ell_{ij}\bar{f} = (A\bar{\omega}+M\bar{\sigma})_{ij} = 0$ if $i \neq i_0$ or $j \neq j_0$. Therefore, we have

$$
\begin{aligned}
(\sigma + t\bar{\sigma})'C(\sigma + t\bar{\sigma}) &= D(f+t\bar{f}, f+t\bar{f}) \\
&= D(f,f) + t\sigma_{i_0 j_0} + t^2 D(\bar{f},\bar{f}) \\
&= \sigma' C\sigma + t\sigma_{i_0 j_0} + t^2\bar{\sigma}' C\bar{\sigma} \quad .
\end{aligned}
$$

Theorem (4.3)

If the GHB-problem is m-poised, then the minimization problem (4.7) has a solution.

Proof

By Lemma (4.2), the square matrix $\begin{pmatrix} A, & M \\ 0, & B \end{pmatrix}$ is non-singular. Since $\alpha_{ij} \leq \beta_{ij}$ for $j \in I_i$, $i = 1, \ldots, k$, the constraints of (4.7) are consistent. By Lemma (4.2), $\sigma' C\sigma$ is a positive semi-definite quadratic form. Hence, $\sigma' C\sigma$ attains its minimum on the nonvoid set of vectors (ω, σ) which satisfy the constraints of (4.7).

If a solution of the minimization problem (4.7) is known a solution of (MP) can easily be obtained as is shown by the following

Theorem (4.4)

Suppose (ω, σ) is a solution of the minimization problem (4.7). Let $f(x) \in F_0$ and $P(x) \in \Pi_{m-1}$ correspond to σ and ω, respectively. Then

$$f(x) + P(x)$$

is a solution of (MP).

Proof

Since $B\sigma = 0$, $f(x) \in F_1$. Hence, by Theorem (3.2),

(1) $$2D(f+P) = 2Df = \sum_{i=1}^{k} \sum_{j \in I_i} \sigma_{ij} \ell_{ij} \quad .$$

Furthermore, $\alpha \leq A\omega + M\sigma \leq \beta$ implies

(2) $$\alpha_{ij} \leq \ell_{ij} P + \ell_{ij} f \leq \beta_{ij}, \quad j \in I_i, \quad i = 1, \ldots, k \ .$$

Next we show that

(3) $$\begin{aligned} &\sigma_{ij} \geq 0 \quad \text{whenever} \quad (A\omega + M\sigma)_{ij} < \beta_{ij} \\ &\sigma_{ij} \leq 0 \quad \text{whenever} \quad (A\omega + M\sigma)_{ij} > \alpha_{ij} \ . \end{aligned}$$

Indeed, let $i_0 \in \{1, \ldots, k\}$ and $j_0 \in I_{i_0}$ be arbitrary but fixed. By Lemma (4.2), there exists $(\bar{\omega}, \bar{\sigma})$ with $B\bar{\sigma} = 0$ and

$$(A\bar{\omega} + M\bar{\sigma})_{ij} = \begin{cases} 1 & \text{for } i = i_0 \text{ and } j = j_0 \\ 0 & \text{for } i \neq i_0 \text{ and/or } j \neq j_0 \end{cases}$$

and

(4) $$(\sigma + t\bar{\sigma})'C(\sigma + t\bar{\sigma}) = \sigma'C\sigma + t\sigma_{i_0 j_0} + t^2 \bar{\sigma}'C\bar{\sigma} \ .$$

If $(A\omega + M\sigma)_{i_0 j_0} < \beta_{i_0 j_0}$, then $(\omega + t\bar{\omega}, \sigma + t\bar{\sigma})$ satisfies the constraints of (4.7) for $0 \leq t \leq t_0$ where $t_0 > 0$ is sufficiently small. Hence, (ω, σ) being a solution of (4.7), it follows from (4) that $\sigma_{i_0 j_0} \geq 0$.

In the same way it can be shown that $\sigma_{i_0 j_0} \leq 0$ if $\alpha_{i_0 j_0} < (A\omega + M\sigma)_{i_0 j_0}$.

Now put

(5) $$\begin{aligned} &\lambda_{ij} = \sigma_{ij}, \ \tau_{ij} = 0 \qquad \text{if } \sigma_{ij} \geq 0 \\ &\lambda_{ij} = 0, \quad \tau_{ij} = -\sigma_{ij} \quad \text{if } \sigma_{ij} < 0 \end{aligned} \qquad j \in I_i, \quad i = 1, \ldots, k \ .$$

It follows from (1), (2), (3) and (5) that $f(x) + P(x)$, λ_{ij} and τ_{ij} satisfy the conditions of Theorem (3.1), which are sufficient for $f(x) + P(x)$ to be a solution of (MP) .

Remark 1

Suppose (ω, σ) is a solution of the minimization problem (4.7). Let $f(x) \in F_1$ and $P(x) \in \Pi_{m-1}$ correspond to σ and ω, respectively. By the Theorems (4.1) and (4.4), $S(x) = f(x) + P(x) \in \Omega_m$ is an extremal solution of the GHB-problem. Therefore, it follows from the definition of $f(x)$ that

$$2(S^{(2m-1-j)}(x_i+0) - S^{(2m-1-j)}(x_i-0)) = (-1)^{m-j}\sigma_{ij}, \; j \in I_i, i=1, \ldots, k .$$

If the GHB-problem is m-poised, it follows from the Theorems (4.3) and (4.4) that (MP) has a solution. By Theorem (4.1), there is one-to-one correspondence between solutions of (MP) and extremal g-spline solutions of the GHB-problem. Since, by Lemma (3.3), the solution of (MP) is unique up to an additive $P(x) \in \Pi_{m-1}$ we have proved the following

Theorem (4.5)

Suppose the GHB-problem is m-poised. Then there exists an extremal g-spline $S(x)$ which solves the GHB-problem. $S(x)$ is uniquely determined up to an additive $P(x) \in \Pi_{m-1}$.

In order to guarantee the uniqueness of the extremal solution we have to replace the assumption that the GHB-problem is m-poised by a stronger condition.

Definition (4.6)

The GHB-problem is said to be strongly m-poised, provided that

$$\alpha_{ij} - \beta_{ij} \leq P^{(j)}(x_i) \leq \beta_{ij} - \alpha_{ij}, \qquad j \in I_i, \quad i = 1, \ldots, k$$

imply that $P(x) \equiv 0$ for every $P(x) \in \Pi_{m-1}$.

Using this definition we have the following Corollary of Theorem (4.5):

Corollary (4.7)

If the GHB-problem is strongly m-poised, then it has a uniquely determined extremal solution.

Remark 2

The assumption that the GHB-problem is strongly m-poised is a rather strong condition which by no means is necessary for the uniqueness of the extremal solution. Especially, a GHB-problem is never strongly m-poised, if $\alpha_{i0} < \beta_{i0}$ for $i = 1, \ldots, k$.

A weaker condition can be given, if a solution of the minimization problem (4.7) is known.

Corollary (4.8)

Suppose $(\bar{\omega}, \bar{\sigma})$ is a solution of the minimization problem (4.7).

Suppose that for every $P(x) \in \Pi_{m-1}$,

$$P^{(j)}(x_i) = 0 \qquad \text{for } \bar{\sigma}_{ij} \neq 0 \qquad \text{and}$$

$$\alpha_{ij} - \beta_{ij} \leq P^{(j)}(x_i) \leq \beta_{ij} - \alpha_{ij} \quad \text{for } \bar{\sigma}_{ij} = 0$$

imply that $P(x) \equiv 0$, then the GHB-problem has a uniquely determined extremal solution.

Proof

If the condition of the corollary is satisfied, the GHB-problem is m-poised. By Theorem (4.5), there exists an extremal solution $S(x)$. If $S_1(x)$ is another extremal solution, then $S_1(x) = S(x) + P(x)$, for some $P(x) \in \Pi_{m-1}$. Hence, by Remark 1,

$$2(S^{(2m-1-j)}(x_i+0)-S^{(2m-1-j)}(x_i-0)) =$$

$$= (-1)^{m-j}\sigma_{ij} = 2(S_1^{(2m-1-j)}(x_i+0)-S_1^{(2m-1-j)}(x_i-0)) .$$

Thus it follows from (2.9) - (2.12) that

$$S_1^{(j)}(x_i) = S^{(j)}(x_i) = \alpha_{ij} \quad \text{if} \quad \sigma_{ij} > 0 \qquad \text{and}$$

$$S_1^{(j)}(x_i) = S^{(j)}(x_i) = \beta_{ij} \quad \text{if} \quad \sigma_{ij} < 0 .$$

Since $\alpha_{ij} - \beta_{ij} \le P^{(j)}(x_i) = S_1^{(j)}(x_i) - S^{(j)}(x_i) \le \beta_{ij} - \alpha_{ij}$ for $j \in I_i$ and $i = 1, \ldots, k$, it follows that $P(x) \equiv 0$, i.e. $S_1(x) \equiv S(x)$.

The uniquely determined solution $S_0(x) \in \Omega_m$ of an m-poised HB-problem has two interesting minimal properties (see [1], [5], [6]):

I) $S_0(x)$ minimizes the integral

$$\int_a^b (g^{(m)}(x))^2 dx$$

among all $g(x) \in F(a,b)$ which satisfy the interpolation conditions (2.1).

II) If $h(x) \in F(a,b)$ and $h^{(j)}(x_i) = \alpha_{ij}$ for $j \in I_i$, $i=1, \ldots, k$, then $S_0(x)$ minimizes the integral

$$\int_a^b (h^{(m)}(x) - S^{(m)}(x))^2 dx$$

among all $S(x) \in \Omega_m$.

It follows immediately from Theorem (4.1) that an extremal g-spline solution $S(x)$ of the GHB-problem has a minimal property which corresponds to I) in the sense that $S(x)$ minimizes the integral

$$\int_a^b (g^{(m)}(x))^2 dx$$

among all $g(x) \in F(a,b)$ which satisfy the relaxed interpolation conditions (2.8).

With respect to the second minimal property we have the following

Theorem (4.9)

Let $h(x) \in F(a,b)$ and let $S_0(x)$ be an extremal g-spline which solves the GHB-problem.

If $\ell_{ij}h = \ell_{ij}S_0$, $j \in I_i$, $i = 1, \ldots, k$, then $S_0(x)$ minimizes

$$D(h-S_0, h-S_0) = \int_a^b (h^{(m)}(x) - S_0^{(m)}(x))^2 dx$$

among all $S(x) \in \Omega_m$.

Proof

For any $S(x) \in \Omega_m$ we have

$$D(h-S_0-S, h-S_0-S) = D(h-S_0, h-S_0) - 2D(S, h-S_0) + D(S,S)$$

Furthermore, if $S(x) \in \Omega_m$, there are σ_{ij} such that the element $f(x) \in F_0$ which corresponds to σ_{ij} is in F_1 and

$$S(x) - f(x) \in \Pi_{m-1} .$$

Hence, by Theorem (3.2), $2DS = 2Df = \sum_{i=1}^{k} \sum_{j \in I_i} \sigma_{ij}\ell_{ij}$ and

$$2D(S, h-S_0) = \sum_{i=1}^{k} \sum_{j\in I_i} \sigma_{ij}\ell_{ij}(h-S_0) = 0 .$$

Thus,

$$D(h-S_0-S, h-S_0-S) = D(h-S_0, h-S_0) + D(S,S) \geq D(h-S_0, h-S_0)$$

for every $S(x) \in \Omega_m$.

If $\alpha_{ij} = \beta_{ij}$ for $j \in I_i$, $i = 1, \ldots, k$, then there is a close connection between the second minimal property of $S_0(x)$ and the dual problem (3.2), (3.3) as is shown in the following

Theorem (4.10)

Let $h(x) \in F(a,b)$ and suppose that $\alpha_{ij} = \beta_{ij} = \ell_{ij}h$ for $j \in I_i$ and $i = 1, \ldots, k$. Then the problem of minimizing

$$D(h-S, h-S) \text{ among all } S(x) \in \Omega_m$$

is equivalent to the dual problem (3.2), (3.3) in the sense that

1) $(g(x), \lambda_{ij}, \tau_{ij})$ satisfies the constraints of (DP) if and only if there is some $S(x) \in \Omega_m$ such that

$$g(x) = S(x) \quad \text{for } x \in [a,b] ,$$

2) for all $(S(x), \lambda_{ij}, \tau_{ij})$ satisfying the constraints of (DP)

$$D(h-S, h-S) - D(h,h) = -\sum_{i=1}^{k} \sum_{j\in I_i} \alpha_{ij}(\lambda_{ij} - \tau_{ij}) + D(S,S) ,$$

where the expression on the right hand side is the negative of the function (3.3) to be maximized in (DP) .

Proof

If $(g(x), \lambda_{ij}, \tau_{ij})$ satisfies the constraints of (DP) it follows from Theorem (3.2) that $g(x) \in \Omega_m$. Conversely, if $S(x) \in \Omega_m = F_1 + \Pi_{m-1}$ it follows again from Theorem (3.2) that there are λ_{ij} and τ_{ij} such that $(S(x), \lambda_{ij}, \tau_{ij})$ satisfies the constraints of (DP).

Let $S(x) \in \Omega_m$ and suppose $(S(x), \lambda_{ij}, \tau_{ij})$ satisfies the constraints of (DP). Then

$$D(h-S, h-S) - D(h,h) = -2D(S,h) + D(S,S)$$

$$= -\sum_{i=1}^{k} \sum_{j\in I_i} (\lambda_{ij} - \tau_{ij})\ell_{ij} h + D(S,S)$$

$$= -\sum_{i=1}^{k} \sum_{j\in I_i} (\lambda_{ij} - \tau_{ij})\alpha_{ij} + D(S,S) \quad .$$

Remark 3

In view of the above theorem the second minimal property of $S_0(x)$ is equivalent to the statement that any solution of (MP) is a part of the solution of the associated (DP). This is a well known result of the duality theory in nonlinear programming.

5. Computational methods

The results of the foregoing sections provide also computational methods for determining an extremal solution of the GHB-problem. The first method is based on the minimization problem (4.7), respectively on the smaller problem

$$(5.1) \qquad \min \{\sigma' C\sigma \mid \alpha \le A\omega + M\sigma \le \beta\} \quad .$$

The two problems are connected by the following

Lemma (5.1)

Suppose the GHB-problem is m-poised and $x_k < b$. Then the minimization problem (5.1) has a solution. If (ω^1, σ^1) is a solution of (5.1) then (ω^1, σ^1) is also a solution of the minimization problem (4.7).

Proof

Since the GHB-problem is m-poised it follows as in the proof of Theorem (4.3) that (5.1) has a solution (ω^1, σ^1). By Theorem (4.3), the minimization problem (4.7) has also a solution, say (ω^2, σ^2). Let $f_i(x) \in F_0$ and $P_i(x) \in \Pi_{m-1}$ be the elements which correspond to ω^i and σ^i, respectively. By Theorem (4.4), $f_2(x) + P_2(x)$ is a solution of (MP). Since $f_1(x) + P_1(x)$ satisfies the constraints of (MP) we have

$$\sigma^{2\,\prime} C \sigma^2 = D(f_2, f_2) \le D(f_1, f_1) = \sigma^{1\,\prime} C \sigma^1 .$$

On the other hand, $\sigma^{1\,\prime} C \sigma^1 \le \sigma^{2\,\prime} C \sigma^2$ since (ω^1, σ^1) is a solu- topm of (5.1) and (ω^2, σ^2) satisfies the constraints of (5.1). Hence,

$$\sigma^{2\,\prime} C \sigma^2 = \sigma^{1\,\prime} C \sigma^1 \quad \text{or} \quad D(f_1, f_1) = D(f_2, f_2) .$$

Therefore, $f_1(x) + P_1(x)$ is a solution of (MP). By Theorems (3.1) and (3.2), $f_1(x) \in F_1$. Therefore, $B\sigma^1 = 0$ and (ω^1, σ^1) satisfies the constraints of (4.7). This completes the proof.

Now suppose the GHB-problem is m-poised. Then it follows from Theorem (4.3) that the minimization problem (4.7) has a solution say $(\bar{\omega}, \bar{\sigma})$. Let $f(x) \in F_1$ be the element which corresponds to $\bar{\sigma}$ and $P(x) \in \Pi_{m-1}$ the element which corresponds to $\bar{\omega}$. By the Theorems (4.1) and (4.4)

$$S(x) = f(x) + P(x)$$

is an extremal g-spline which solves the GHB-problem.

Therefore, the evaluation of an extremal solution of the GHB-problem is equivalent to solving the quadratic minimization problem (4.7), respectively (5.1) in the case that $x_k < b$. Since it follows from Lemma (4.2) that $\sigma' C \sigma$ is a convex function there are efficient methods for solving these problems (see e.g. [4]). A second method for determining an extremal solution of the GHB-problem stems from the results on the dual problem.

Suppose $g(x) \in F(a,b)$ and λ_{ij}, τ_{ij} satisfy the constraints of (DP). Let $f(x) \in F_0$ correspond to $\sigma_{ij} = \lambda_{ij} - \tau_{ij}$. By Theorem (3.2),

$$f(x) \in F_1$$

$$g(x) = f(x) + P(x) \quad \text{for all} \quad x \in [a,b] ,$$

where $P(x) \in \Pi_{m-1}$.

On the other hand, if $\lambda_{ij} \geq 0$, $\tau_{ij} \geq 0$ and $f(x) \in F_0$, corresponding to $\sigma_{ij} = \lambda_{ij} - \tau_{ij}$, is an element of F_1, it follows from Theorem (3.2), that $f(x)$, λ_{ij} and τ_{ij} satisfy the constraints of (DP). Using the definitions of Section 4) for α, β, B and C, we can, therefore, write the dual problem in the equivalent form

$$(5.2) \quad \max_{\lambda,\tau} \{\alpha'\lambda - \beta'\tau - (\lambda-\tau)'C(\lambda-\tau) | B(\lambda-\tau) = 0, \ \lambda \geq 0, \ \tau \geq 0\} ,$$

where the column vectors λ and τ have the components λ_{ij} respectively τ_{ij}, $j \in I_i$, $i = 1, \ldots, k$.

Theorem (5.2)

Suppose the GHB-problem is m-poised. Then

1) The maximization problem (5.2) has a solution.
2) If λ^1, τ^1 is a solution of (5.2) then there exists an ω^1 such that

$$\alpha \leq A\omega^1 + M(\lambda^1 - \tau^1) \leq \beta ,$$

where α, β, A and M are defined as in Section 4).

3) Let $f(x) \in F_0$ correspond to $\sigma^1 = \lambda^1 - \tau^1$ and $P(x) \in \Pi_{m-1}$ correspond to ω^1 then

$$S(x) = f(x) + P(x)$$

is an extremal g-spline which solves the GHB-problem.

Proof

1) By the Theorems (4.3) and (4.4), (MP) has a solution. Hence, it follows from Theorem (3.5) that (DP), or equivalently (5.2), has a solution.

2) Let λ^1, τ^1 be a solution of (5.2) and let $f(x) \in F_0$ correspond to $\sigma^1 = \lambda^1 - \tau^1$. Then $f(x)$, λ^1, τ^1 is a solution of (DP). By Theorem (3.5), there exists a $P_1(x) \in \Pi_{m-1}$ such that $f(x) + P(x)$ is a solution of (MP). Hence,

$$\alpha_{ij} \leq \ell_{ij}(P_1 + f) \leq \beta_{ij}, \quad j \in I_i, \quad i = 1, \ldots, k .$$

If we choose ω^1 so that $P_1(x)$ corresponds to ω^1, it follows therefore that

$$\alpha \leq A\omega^1 + M(\lambda^1 - \tau^1) \leq \beta .$$

3) As under 2) it follows that $S(x) = f(x) + P(x)$ is a solution of (MP). Hence, it follows from Theorem (4.1), that $S(x)$ is an extremal g-spline which solves the GHB-problem.

By the above theorem the computation of an extremal solution of an m-poised GHB-problem is equivalent to solving the maximization problem (5.2) and determining an m-dimensional vector ω which satisfies

$$\alpha \leq A\omega + M(\lambda^1 - \tau^1) \leq \beta , \tag{5.3}$$

where λ^1, τ^1 is a solution of (5.2).

Since by Lemma (4.2), $-(\lambda - \tau)'C(\lambda - \tau)$ is a concave function and the constraints of (5.2) consist only of m equations and non-negativity restrictions, (5.2) can be solved very efficiently (see [4]).

In computing an ω^1 which satisfies (5.3) we can take advantage of the fact that, by Theorem (5.2), the resulting $S(x)$ is an extremal g-spline, which by (2.9) - (2.11) implies

$$
\begin{aligned}
(A\omega^1)_{ij} &= \alpha_{ij} - (M\lambda^1 - M\tau^1)_{ij} \quad \text{whenever } \lambda^1_{ij} > 0 \\
(A\omega^1)_{ij} &= \beta_{ij} - (M\lambda^1 - M\tau^1)_{ij} \quad \text{whenever } \tau^1_{ij} > 0 \ .
\end{aligned} \tag{5.4}
$$

Especially, if for m values of i either $\lambda_{i0} > 0$ or $\tau_{i0} > 0$ then the m components of ω^1 are uniquely determined by the corresponding m equations of the form (5.4).

Finally, we remark that for the computation of the extremal solution of the GHB-problem it is not necessary to compute the elements of the matrix C by formula (4.6) since the following lemma holds.

Lemma (5.3)

Let σ, ω, B and M be as defined in Section 4). For every σ with $B\sigma = 0$ we have

$$\sigma' C \sigma = \frac{1}{2}\, \sigma' M \sigma \ .$$

Proof

Let $f(x) \in F_0$ correspond to σ. Since $B\sigma = 0$, $f(x) \in$
Hence, it follows from Theorem (3.2) that

$$2Df = \sum_{i=1}^{k} \sum_{j \in I_i} \sigma_{ij} \ell_{ij} \ .$$

Therefore,

$$2\sigma'C\sigma = 2D(f,f) = \sum_{i=1} \sum_{j \in I_i} \sigma_{ij} \ell_{ij} f = \sigma' M\sigma \quad .$$

In view of this lemma we may replace

$$\sigma'C\sigma \text{ by } \frac{1}{2}\sigma'M\sigma$$

in the minimization problem (4.7) and in the dual problem (5.2).

Appendix

In this section three examples of GHB-problems are given. First we consider the case $m = 1$, i.e., the case of linear spline interpolation. Let $k = 4$, $I_1 = \ldots = I_4 = \{0\}$ and $a = -2$, $b = 10$, $x_1 = 0$, $x_2 = 4$, $x_3 = 6$, $x_4 = 8$. Furthermore, choose $\alpha_{10} = 9$, $\alpha_{20} = 3$, $\alpha_{30} = -6$, $\alpha_{40} = 7$ and $\beta_{10} = 9$, $\beta_{20} = 4$, $\beta_{30} = -5$, $\beta_{40} = 8$. Then we have to construct an extremal g-spline $S(x) \in \Omega_1$, which satisfies the conditions

(1) $$\alpha_{i0} \le S(x_i) \le \beta_{i0}, \qquad i = 1, \ldots, 4 \quad .$$

It is easily seen that this problem is strongly m-poised. Therefore, it follows from Corollary (4.7) that it has a unique extremal solution which can be written in the form

(2) $$S(x) = \sum_{i=1}^{4} \frac{-(x-x_i)_+}{2} \sigma_{i0} + \omega_0 \quad .$$

Since $x_4 < b$, it follows from Lemma (5.1) and the Theorems (4.1) and (4.4) that σ_{10}, σ_{20}, σ_{30}, σ_{40} and ω_0 can be determined by solving the minimization problem

(3) $$\min\{\sigma'C\sigma \,|\, \alpha \le A\omega + M\sigma \le \beta\} ,$$

where

$$\sigma' = (\sigma_{10}, \sigma_{20}, \sigma_{30}, \sigma_{40}), \quad \omega' = (\omega_0) \ ,$$

$$\alpha' = (9, 3, -6, 7), \qquad \beta' = (9, 4, -5, 8) \ \text{and}$$

$$C = \begin{bmatrix} C_{11}, C_{12}, C_{13}, C_{14} \\ C_{21}, C_{22}, C_{23}, C_{24} \\ C_{31}, C_{32}, C_{33}, C_{34} \\ C_{41}, C_{42}, C_{43}, C_{44} \end{bmatrix} \quad A = \begin{bmatrix} A_1 \\ A_2 \\ A_3 \\ A_4 \end{bmatrix} \quad M = \begin{bmatrix} 0 & 0 & 0 & 0 \\ A_{21} & 0 & 0 & 0 \\ A_{31} & A_{32} & 0 & 0 \\ A_{41} & A_{42} & A_{43} & 0 \end{bmatrix}$$

The matrices $C_{i\nu}$, $1 \le \nu \le i \le k$ are determined according to (4.6). For $i > \nu$ we put $C_{i\nu} = C'_{\nu i}$. Thus,

$$C_{11} = (\tfrac{5}{2}), \ C_{21} = (\tfrac{3}{2}), \ C_{31} = (1), C_{41} = (\tfrac{1}{2}) \ ,$$

$$C_{22} = (\tfrac{3}{2}), \ C_{32} = (1), \ C_{42} = (\tfrac{1}{2}), C_{33} = (1) \ ,$$

$$C_{43} = (\tfrac{1}{2}), \ C_{44} = (\tfrac{1}{2}).$$

For $A_1, \ldots, A_4$ we obtain

$$A_1 = (1), \ A_2 = (1), \ A_3 = (1), \ A_4 = (1) \ .$$

Finally,

$$A_{21} = (-2), \ A_{31} = (-3), \ A_{41} = (-4), \ A_{32} = (-1)$$

$$A_{42} = (-2), \ A_{43} = (-1) \ .$$

For these data the problem (3) has the solution

$$\omega_0 = 9, \ \sigma_{10} = 3, \ \sigma_{20} = 5, \ \sigma_{30} = -20, \ \sigma_{40} = 12 \ .$$

Hence, it follows from (2) that

$$S(x) = -\frac{3}{2}x_+ - \frac{5}{2}(x-4)_+ + 10(x-6)_+ - 6(x-8)_+ + 9$$

is the extremal solution of (1).

Next we consider the case of cubic spline interpolation. Let $m = 2$, $k = 3$, $x_1 = 0$, $x_2 = 3$, $x_3 = 9$, $a = -2$, $b = 10$ and $I_1 = \{0,1\}$, $I_2 = \{0\}$, $I_3 = \{1\}$. Furthermore, let $\alpha_{10} = 5$, $\beta_{10} = 6$, $\alpha_{11} = -4$, $\beta_{11} = -4$, $a_{20} = -5$, $\beta_{20} = -5$, $\alpha_{31} = 6$, $\beta_{31} = 7$. Then we have to construct an extremal g-spline $S(x) \in \Omega_2$ which satisfies the conditions

$$(4) \qquad \alpha_{ij} \leq S^{(j)}(x_i) \leq \beta_{ij} \quad \text{for } j \in I_i \quad \text{and} \quad i = 1, 2, 3 \ .$$

Since this problem is strongly m-poised it follows again from Corollary (4.7) that an extremal solution exists which can be written in the form

$$(5) \qquad S(x) = \frac{1}{2} \sum_{i=1}^{3} \sum_{j \in I_i} \frac{(-1)^{2-j}}{(3-j)!} \sigma_{ij} (x-x_i)_+^{3-j} + \omega_0 + \omega_1 x \ .$$

In this example we use the dual problem for the determination of the σ_{ij}'s. Thus we consider the problem (5.2)

$$(6) \qquad \max_{\lambda,\tau} \{\alpha'\lambda - \beta'\tau - (\lambda-\tau)C(\lambda-\tau) \mid B(\lambda-\tau) = 0, \lambda \geq 0, \tau \geq 0\} \ ,$$

where

$$\alpha' = (5, -4, -5, 6), \qquad \beta' = (6, -4, -5, 7) \ ,$$

$$\lambda' = (\lambda_{10}, \lambda_{11}, \lambda_{20}, \lambda_{31}), \quad \tau' = (\tau_{10}, \tau_{11}, \tau_{20}, \tau_{31}) \text{ and}$$

$$C = \begin{bmatrix} C_{11}, C_{12}, C_{13} \\ C_{21}, C_{22}, C_{23} \\ C_{31}, C_{32}, C_{33} \end{bmatrix} \qquad B = (B_1, B_2, B_3)$$

The matrices $C_{i\nu}$, $1 \leq \nu \leq i \leq k$ are determined according to (4.6). For $i < \nu$ we put $C_{i\nu} = C'_{\nu i}$. Thus

$$C_{11} = \begin{bmatrix} \frac{250}{3}, & -\frac{25}{2} \\ -\frac{25}{2}, & \frac{5}{2} \end{bmatrix}, \quad C_{21} = \left(\frac{1127}{24}, -\frac{49}{8}\right), \quad C_{31} = \left(-\frac{19}{8}, \frac{1}{4}\right)$$

$$C_{22} = \left(\frac{343}{12}\right), \qquad C_{32} = \left(-\frac{13}{8}\right), \qquad C_{33} = \left(\frac{1}{4}\right).$$

For the matrices B_i we have

$$B_1 = \begin{pmatrix} 1, & 0 \\ 10, & -1 \end{pmatrix}, \qquad B_2 = \begin{pmatrix} 1 \\ 7 \end{pmatrix}, \qquad B_3 = \begin{pmatrix} 0 \\ -1 \end{pmatrix}.$$

Then the problem (6) has the solution

$$\lambda_{10} = \frac{8}{9}, \quad \lambda_{11} = 0, \quad \lambda_{20} = 0, \quad \lambda_{31} = \frac{8}{3}$$

$$\tau_{10} = 0, \quad \tau_{11} = 0, \quad \tau_{20} = \frac{8}{9}, \quad \tau_{30} = 0.$$

By Theorem (5.2), there are ω_0, ω_1 such that

$$(7) \qquad S(x) = \frac{2}{27} x_+^3 - \frac{2}{27}(x-3)_+^3 - \frac{2}{3}(x-9)_+^2 + \omega_1 x + \omega_0$$

satisfies (4). Inserting (7) into (4) we obtain

$$\begin{aligned} 5 \leq \quad & \omega_0 \leq 6 \\ -4 \leq \quad & \omega_1 \leq -4 \\ -5 \leq 2 + 3\omega_1 + & \omega_0 \leq -5 \\ 6 \leq 10 + \omega_1 & \leq 7 . \end{aligned}$$

For $\omega_0 = 5$ and $\omega_1 = -4$ these inequalities are satisfied. It follows from Theorem (5.2) that with $\omega_0 = 5$, $\omega_1 = -4$, (7) is an extremal solution to (4). This can be verified by inserting (7) into (2.9), ..., (2.12).

Finally, we consider a problem which is m-poised but not strongly m-poised and has an infinite number of extremal solutions.

Let $m = 3$, $k = 5$, $x_i = i$, $i = 1, \ldots, 5$, $a = 0$, $b = 6$ and $I_1 = I_5 = \{0\}$, $I_2 = I_3 = I_4 = \{1\}$. Furthermore, let $\alpha_{10} = -2$, $\beta_{10} = -1$, $\alpha_{21} = 3$, $\beta_{21} = 4$, $\alpha_{31} = 6$, $\beta_{31} = 7$, $\alpha_{41} = 1$, $\beta_{41} = 2$, $\alpha_{50} = 10$, $\beta_{50} = 11$. Then we have to construct an extremal g-spline $S(x) \in \Omega_3$ which satisfies the conditions

$$(8) \qquad \alpha_{ij} \le S^{(j)}(x_i) \le \beta_{ij} \quad \text{for } j \in I_i \text{ and } i = 1, \ldots, 5 .$$

Since with $\omega_2 = \omega_1 = 0$ and $-1 \le \omega_0 \le 1$, $P(x) = \omega_2 x^2 + \omega_1 x + \omega_0 \in \Pi_2$ and

$$\alpha_{ij} - \beta_{ij} \le P^{(j)}(x_i) \le \beta_{ij} - \alpha_{ij}, \quad j \in I_i \text{ and } i = 1, \ldots, 5 ,$$

the problem is not strongly m-poised. However, it is easily seen that it is m-poised. Hence it follows from Theorem (4.5) that it has an extremal solution which can be written in the form

$$(9) \qquad S(x) = \frac{1}{2} \sum_{i=1} \sum_{j \in I_i} \frac{(-1)^{3-j}}{(5-j)!} \sigma_{ij} (x-x_i)_+^{5-j} + \sum_{j-0}^{2} \omega_j x^j .$$

By the Theorems (4.1) and (4.4) the σ_{ij} and ω_j can be determined by solving the minimization problem

$$\min\{\sigma' C \sigma \mid \alpha \le A\omega + M\sigma \le \beta,\ B\sigma = 0\} .$$

Since $\sigma' C \sigma = \frac{1}{2} \sigma' M \sigma$ for all σ with $B\sigma = 0$ the problem

$$(10) \qquad \min\{\sigma' M \sigma \mid \alpha \le A\omega + M\sigma \le \beta,\ B\sigma = 0\}$$

has the same solution. Here is

$$\sigma' = (\sigma_{10}, \sigma_{21}, \sigma_{31}, \sigma_{41}, \sigma_{50}), \quad \omega' = (\omega_0, \omega_1, \omega_2)$$

$$\alpha' = (-2, 3, 6, 1, 10), \qquad \beta' = (-1, 4, 7, 2, 11)$$

and

$$A = \begin{bmatrix} A_1 \\ A_2 \\ A_3 \\ A_4 \\ A_5 \end{bmatrix} = \begin{bmatrix} 1 & 1 & 1 \\ 0 & 1 & 4 \\ 0 & 1 & 6 \\ 0 & 1 & 8 \\ 1 & 5 & 25 \end{bmatrix}, \qquad B = \begin{bmatrix} -1 & 0 & 0 & 0 & -1 \\ -5 & 1 & 1 & 1 & -1 \\ \frac{-25}{2} & 4 & 3 & 2 & -\frac{1}{2} \end{bmatrix}$$

$$M = \begin{bmatrix} 0 & 0 & 0 & 0 & 0 \\ A_{21} & 0 & 0 & 0 & 0 \\ A_{31} & A_{32} & 0 & 0 & 0 \\ A_{41} & A_{42} & A_{43} & 0 & 0 \\ A_{51} & A_{52} & A_{53} & A_{54} & 0 \end{bmatrix} = \begin{bmatrix} 0 & 0 & 0 & 0 & 0 \\ -\frac{1}{48} & 0 & 0 & 0 & 0 \\ -\frac{1}{3} & \frac{1}{12} & 0 & 0 & 0 \\ -\frac{27}{16} & \frac{2}{3} & \frac{1}{12} & 0 & 0 \\ -\frac{64}{15} & \frac{27}{16} & \frac{1}{3} & \frac{1}{48} & 0 \end{bmatrix}$$

For these data we obtain as a solution to (10)

$$\sigma_{10} = \sigma_{50} = \omega_0 = 0, \quad \sigma_{21} = \sigma_{41} = -18 ,$$

$$\sigma_{31} = 36, \quad \omega_1 = -3, \quad \omega_2 = \frac{7}{4} .$$

Hence, it follows from (9) that

$$S(x) = -\frac{3}{8}(x-2)_+^4 + \frac{3}{4}(x-3)_+^4 - \frac{3}{8}(x-4)_+^4 - 3x + \frac{7}{4}x^2 \tag{11}$$

is an extremal solution of (8). It is easy to verify that for any $\omega_0 \in [0, \frac{1}{4}]$

$$S(x) + \omega_0$$

is also an extremal solution to (10), where $S(x)$ is given by (11).

REFERENCES

[1] J. H. Ahlberg and E. N. Nilson: The approximation of linear functionals, SIAM J. on Num. Analysis, 3 (1966), 173-182.

[2] G. D. Birkhoff: General mean value and remainder theorems with applications to mechanical differentiation and integration, Trans. Amer. Math. Soc. 7 (1906), 107-136.

[3] K. Fan: On systems of linear inequalities, Linear Inequalities and Related Systems, Annals of Mathematical Studies, Princeton University Press, Princeton 1956, 99-156.

[4] H. P. Künzi and W. Krelle: Nonlinear Programming, Blaisdell Publishing Company, Waltham 1966.

[5] I. J. Schoenberg: On interpolation by spline functions and its minimal properties, Oberwolfach Symposium: On Approximation Theory, edited by P. L. Butzer, Internat. Series of Numerical Math., 5(1964), 109-129.

[6] ____________: On the Ahlberg-Nilson extension of spline interpolation: The g-splines and their optimal properties, Mathematics Research Center Technical Summary Report 716, University of Wisconsin, 1966.

Splines via Optimal Control

O.L. MANGASARIAN AND L.L. SCHUMAKER

§1. Introduction. The purpose of this paper is to employ results from the theory of optimal control to investigate certain constrained variational problems whose solutions may be regarded as generalized splines. We intend to investigate a sufficiently general minimization problem to encompass many of the existing varieties of generalized splines such as Lg-splines [8], splines satisfying inequalities at a finite number of points (see [2, 8, 14]), and splines interpolating to infinite sets of data [5, 8]. Mixtures of discrete and continuous inequality constraints will be allowed.

It seems reasonable to refer to the solutions of our minimum problems as splines in view of the fact that most of the earlier authors who introduced new kinds of splines constructively took special pains to establish minimal properties (see e.g. the references in [8]), while most of the recent papers discussing generalized splines simply define them via extremal problems (cf. [1, 2, 4, 5, 8, 9, 14, 15]). The primary tools in these latter papers are Hilbert space techniques. Our minimization problem will be posed in the Banach space $\mathcal{L}_p$, $1 < p < \infty$, where methods analogous to those in Hilbert spaces are not immediately evident.

Although control theory may rightfully be regarded as a complicated and at times deep subject, we wish to emphasize that our main concern will be with a certain sufficiency theorem from control theory (see §A.2) which is particularly easy to use, and whose proof is not formidable. In most instances, this theorem appears also to provide the correct necessary conditions, and hence is extremely useful in identifying the properties of the splines as well as in verifying their optimality We shall also illustrate how optimal control results dealing with necessity can be used to characterize the splines.

Throughout this paper we shall adhere to the policy of relying primarily on existing optimal control results. At times it would be useful to have modifications or extensions of presently available control theorems, but here we shall have to resist the temptation to meddle in control theory. If what we discuss serves to stimulate interest in new kinds of spline functions or in developing extensions of optimal control theory, our purpose will have been fulfilled.

§2. Examples. To motivate the reader to examine the rather general constrained minimization problem introduced in §3, we begin by quoting some simple examples whose solutions clearly exhibit spline-like behavior. For our present purposes we shall be content with pointing out the properties of these "splines", and shall leave the verification of their extremality to §10.

Example 2.1. For $p > 1$ the function

$$(2.1)\quad s_p(t) = \begin{cases} 1 - \left(\frac{2p-1}{p}\right)(t+1) & -2 \le t < -1 \\ 1 - \left(\frac{2p-1}{p}\right)(t+1) + \left(\frac{p-1}{p}\right)(t+1)^{\frac{2p-1}{p-1}} & -1 \le t < 0 \\ 1 - \left(\frac{2p-1}{p}\right)(1-t) + \left(\frac{p-1}{p}\right)(1-t)^{\frac{2p-1}{p-1}} & 0 \le t < 1 \\ 1 - \left(\frac{2p-1}{p}\right)(1-t) & 1 \le t \le 2 \end{cases}$$

uniquely minimizes $\int_{-2}^{2} |f''(t)|^p dt$ subject to the constraint $1 \le f(-1) \le 3/2$, $-1/4 \le f(0) \le 0$, $1 \le f(1) \le 3/2$. This functic is illustrated in Figure 2.1. We note that $s_p \in C^2[-2,2]$ and that its third derivative has jumps at $-1, 0, 1$. The value of $J_p = \int_{-2}^{2} |f''|^p \, dt$ is $2(\frac{2p-1}{p-1})^{p-1}$.

For $p = 2$, $s_2(t)$ is the natural interpolating cubic spline with inequality data (cf. [8], [14]), and consists of cubics in $[-1,0]$ and $[0,1]$ and of linear segments in $[-2, -1]$, $[1,2]$. It is interesting to notice that as $p \to \infty$, $s_p(t)$ converges uniformly to

$$s(t) = \begin{cases} 1 - 2(t+1) & -2 \leq t < -1 \\ t^2 & -1 \leq t < 1 \\ 1 - 2(1-t) & 1 \leq t \leq 2 \end{cases},$$

while $(J_p)^{1/p} \to 2$.

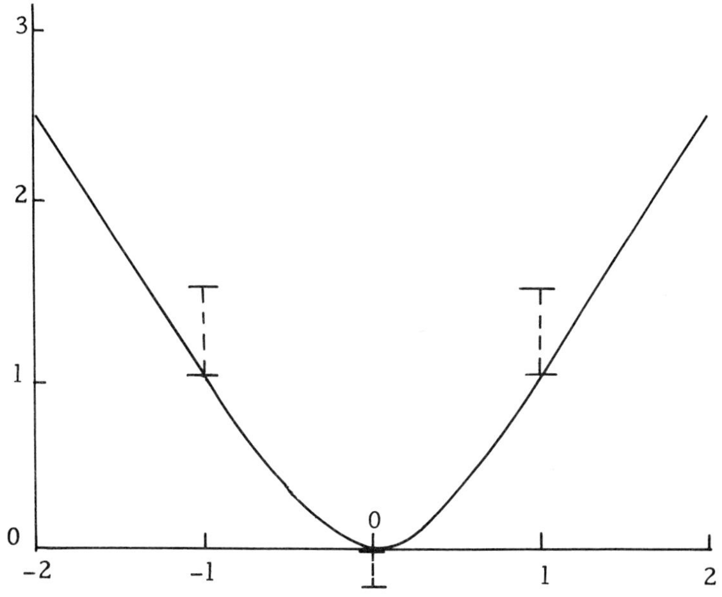

Figure 2.1. The spline (2.1)

Example 2.2. The function

$$(2.2)\qquad s(t) = \begin{cases} t - t^2 & 0 \le t < 1 - \sqrt{\gamma} \\ (2\sqrt{\gamma} - 1)t + (1 + \gamma - 2\sqrt{\gamma}) & 1 - \sqrt{\gamma} \le t \le 1 \end{cases}$$

uniquely minimizes $\int_0^1 (f')^2 dt$ subject to the constraints $t - t^2 \le f(t) \le t$ and $f(1) = \gamma$, $0 \le \gamma \le 1$. This spline is of continuity class $C^1[0,1]$, and is illustrated in Figure 2.2. When $\gamma = 0$ or 1, $s(t)$ follows the lower or upper boundari to γ, respectively.

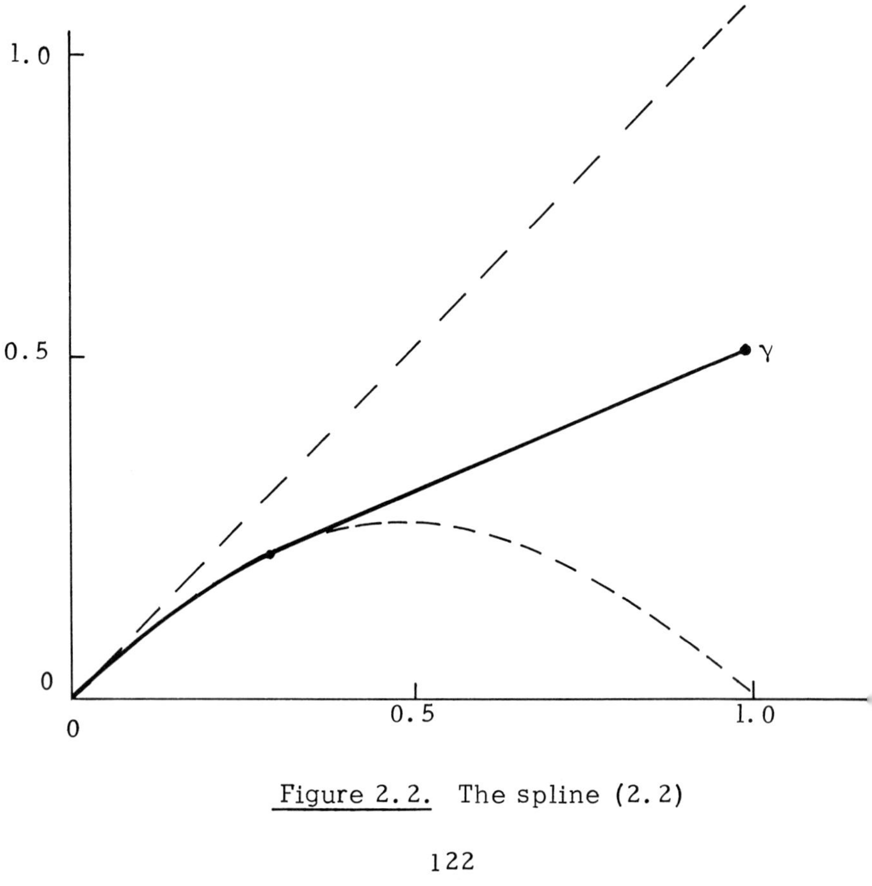

Figure 2.2. The spline (2.2)

Example 2.3. The function

$$(2.3)\quad s(t) = \begin{cases} \gamma + \left(\frac{6-5\gamma-\delta}{4}\right)(t+1) + \left(\frac{\gamma+\delta-2}{4}\right)(t+1)^3, & -1 \le t < 0 \\ \delta + \left(\frac{5\delta+\gamma-6}{4}\right)(t-1) - \left(\frac{\gamma+\delta-2}{4}\right)(t-1)^3, & 0 \le t \le 1 \end{cases}$$

uniquely minimizes $\int_{-1}^{1} (f'')^2 dt$ subject to the constraints $\alpha(t) \le f(t) \le \beta(t)$ and $f(-1) = \gamma$, $f(1) = \delta$, where

$$\alpha(t) = \begin{cases} -t & -1 \le t < 0 \\ t & 0 \le t \le 1 \end{cases} \qquad \beta(t) = \begin{cases} 1-t & -1 \le t < 0 \\ 1+t & 0 \le t \le 1 \end{cases}$$

and γ, δ are real numbers satisfying $1 \le \gamma \le 2$, $1 \le \delta \le 2$. We point out that $s \in C^2[-1,1]$ and $s''(-1) = s''(1) = 0$, $s(0) = 1$. At the point 0 the third derivative s''' has a jump. Figure 2.3 is a graph of (2.3) for typical values of γ, δ. When $\gamma = \delta = 1$, $s(t) \equiv 1$. When $\gamma = \delta > 1$, $s'(0) = 0$ and s is symmetric.

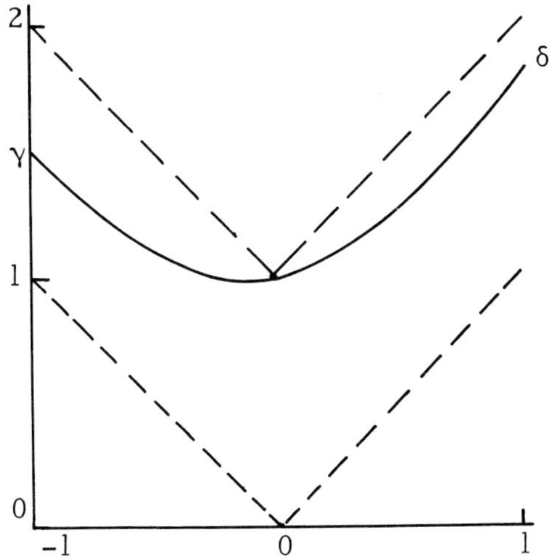

Figure 2.3. The spline (2.3)

Example 2.4. The function

$$(2.4)\quad s(t) = \begin{cases} 0 & 0 \le t \\ \eta^4 \left(\frac{t-\xi}{\eta-\xi}\right)^3 & \xi \le t \\ \eta^4 + 4\eta^3 (t-\eta) + \frac{16}{3}\eta^2 (t-\eta)^2 - \frac{16}{9}\eta^2 \frac{(t-\eta)^3}{(1-\eta)}, & \eta \le t \end{cases}$$

where $\xi = \eta/4$ and η is a root of $5\eta^4 - 28\eta^3 + 32\eta^2 - 9\gamma$ in $[0,1]$, uniquely minimizes the expression $\int_0^1 (f'')^2 dt$ subject tc the constraints $0 \le f(t) \le t^4$ and $f(1) = \gamma$, $0 \le \gamma \le 1$. The spline s is in class $C^2[0,1]$ and satisfies $s''(1) = 0$. It is depicted in Figure 2.4 for a representative choice of γ.

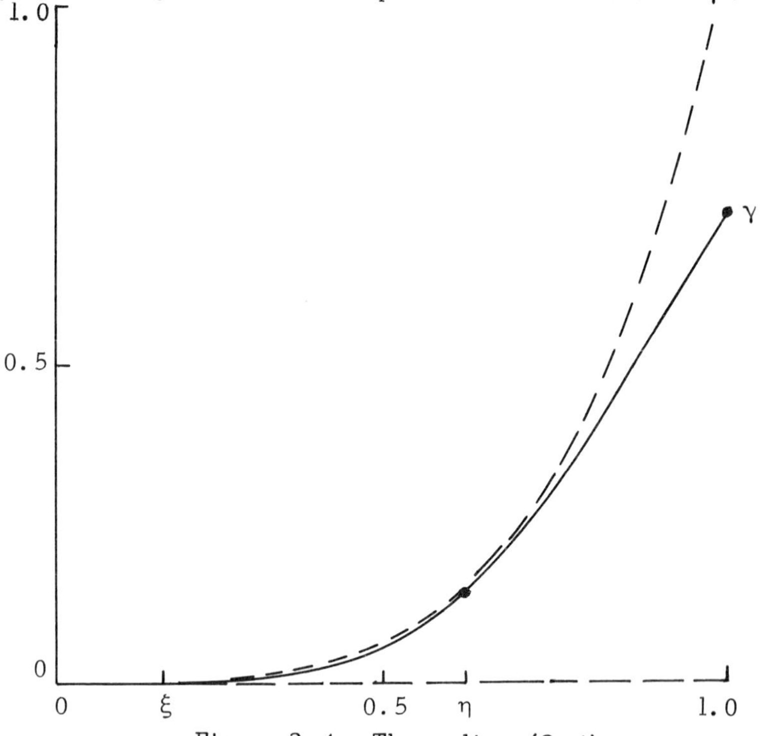

Figure 2.4. The spline (2.4)

Example 2.5. The function

$$(2.5)\quad s(t) = \begin{cases} (-1/4 + e/2 - e^2/4)t^3 + (5/4 - 3e/2 + e^2/4)t + e^t, & 0 \le t < 1, \\ (-1/4 + e/2 - e^2/4)(-t^3 + 6t^2) + (11/4 - 9e/2 + 7e^2/4)t & \\ \qquad + (-1/2 + e - e^2/2) + e^t, & 1 \le t \le 2 \end{cases}$$

uniquely minimizes $\int_0^2 (f'' - e^t)^2 dt$ subject to the constraints $f(0) = f(2) = 1$ and $1 \le f(1) \le 5/4$. At 0 and 2 we have $s''(0) = 1$, $s''(2) = e^2$, while globally $s \in C^2[0,2]$. At 1, $s'''(1+) - s'''(1-) > 0$. Figure 2.5 is an illustration of s.

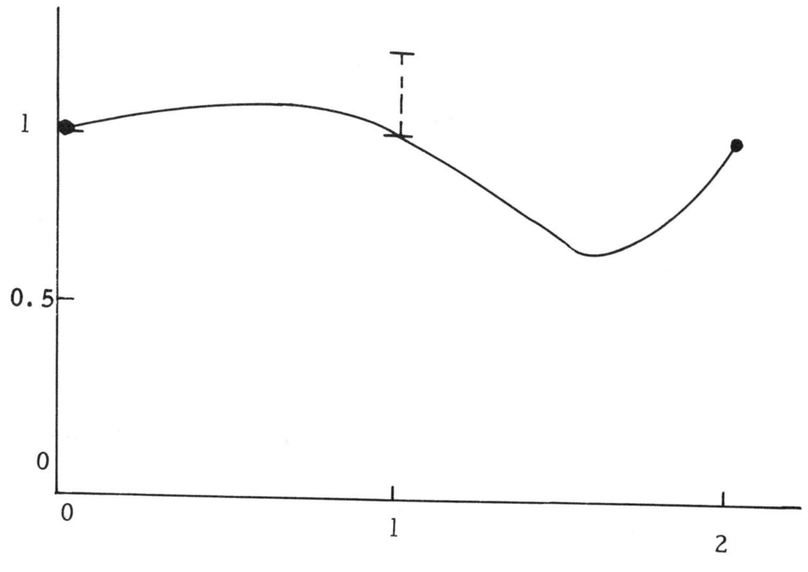

Figure 2.5. The spline (2.5)

§3. The variational problem. In this section we define a constrained minimization problem of sufficient generality to include the examples cited in §2 as well as most of the previous notions of generalized splines (see [8] and references therein).

Let $1 < p < \infty$ and $-\infty < a < b < \infty$. Suppose L is a linear differential operator of the form

$$(3.1) \quad L = \sum_{j=0}^{m} a_j(t)\left(\frac{d}{dt}\right)^j, \quad a_m(t) \neq 0 \text{ on } [a,b], \quad a_j \in C^j[a,b] .$$

Let

$$(3.2) \quad \mathcal{H}_p^m = \{f : f \in C^{m-1}[a,b],\ f^{(m-1)} \text{ absolutely continuous},\ Lf \in \mathcal{L}_p(a,b)\} .$$

$\mathcal{H}_p^m$ is a Banach space under a variety of norms, for example

$$\|f\|_{\mathcal{H}_p^m} = \sum_{j=1}^{m} |f(t_j)| + \|Lf\|_{\mathcal{L}_p} \quad \text{for some } a \leq t_1 < \ldots < t_m \leq b .$$

Clearly L is a bounded linear operator from $\mathcal{H}_p^m$ onto $\mathcal{L}_p$, whose null space we denote by N. For $i = 1, 2, \ldots, k$ let $\mathfrak{M}_i$ be linear differential operators of the form

$$(3.3) \quad \mathfrak{M}_i = \sum_{j=0}^{m-1} b_{ij}(t)\left(\frac{d}{dt}\right)^j, \quad b_{ij}(t) \in C[a,b] .$$

Suppose $\Lambda = \{\lambda_i\}_1^n$ is a set of bounded linear functionals defined on $\mathcal{H}_p^m$, and let $\gamma = (\gamma_1, \ldots, \gamma_n)$ and $\delta = (\delta_1, \ldots, \delta_n)$ be vectors in $\mathbb{R}_n$ with $\gamma_i \leq \delta_i$, $i = 1, 2, \ldots, n$. Let $\alpha_i(t)$ and $\beta_i(t)$ be prescribed functions in $\mathfrak{M}_i \mathcal{H}_p^m$ with $\alpha_i(t) \leq \beta_i(t)$ $i = 1, 2, \ldots, k$. We now define the following sets of functions

$$(3.4) \quad U_1 = \{f \in \mathcal{H}_p^m : \alpha_i(t) \leq \mathfrak{M}_i f(t) \leq \beta_i(t),\ i = 1, 2, \ldots, k\} ,$$

(3.5) $U_2 = \{f \in \mathcal{H}_p^m : \gamma_i \leq \lambda_i f \leq \delta_i, \quad i = 1, 2, \ldots, n\}$,

(3.6) $U = U_1 \cap U_2$.

Finally, let $g(t) \in \mathcal{L}_p$ be prescribed.

Throughout this paper we shall be concerned with the constrained variational problem:

(3.7) $$\underset{f \in U}{\text{minimize}} \|Lf - g\|_{\mathcal{L}_p} .$$

§4. Existence and uniqueness of minima. The following theorem settles the questions of existence and uniqueness for solutions of (3.7).

Theorem 4.1. If $U \neq \emptyset$, then there exists a solution of (3.7). Any two solutions of (3.7) differ by an element in N, and if $s \in U$ is a solution of (3.7), then it is unique if and only if $N \cap U(s) = (0)$, where $U(s) = \{f - s : f \in U\}$.

The proof of this theorem is relegated to the Appendix (§A.1) although it is neither particularly difficult nor overly lengthy.

§5. A general optimal control problem. The optimal control literature is voluminous, and considerable variation in notation and generality can be found in the various references. Using the notation of [10], we quote a rather general control problem which subsumes most of those considered in the literature.

Let $-\infty < a < b < \infty$ be prescribed. We seek vector functions $u(t)$ and $x(t)$ which will minimize the functional

(5.1) $$J(u,x) = \int_a^b \phi(t, x(t), u(t))\, dt + \theta(x(a), x(b))$$

subject to the differential equations

(5.2)† $$\dot{x} = g(t,x,u) \quad ,$$

the constraints

(5.3) $$h(t,x(t),u(t)) \leq 0 \quad ,$$

the initial conditions

(5.4) $$p(x(a)) \leq 0 \quad ,$$

and the terminal conditions

(5.5) $$q(x(b)) \leq 0 \quad .$$

Here g, h, p and q are also vector functions. It is customary to refer to x as the state vector and to u as the control vector. Throughout this paper we shall assume that u is an element of an appropriate class of admissible controls, whose precise definition we can neglect for the moment.

In recent years, considerable effort has been devoted to establishing the existence of optimal controls, and to deriving sets of necessary or sufficient conditions for them. In most cases this involves imposing additional differentiability and/or convexity properties on some of the above vector functions. The presently available existence results involve more restrictions than we can afford here, so we shall ignore them (having already disposed of the existence and uniqueness questions directly for our variational problem in §§A.1 and 4). In the appendix we quote two recent theorems dealing with necessity and sufficiency due to Neustadt [11] and Mangasarian [10], respectively. These results will be our principal tools in investigating problem (3.7).

§6. The variational problem of §3 as a control problem. Throughout the remaining sections we assume that the set Λ consists of linear functionals defined as follows: Let $a = t_0 < t_1 < \ldots < t_{\mu-1} < t_\mu = b$ and integers $0 \leq m_i \leq m-1$ be given

† Throughout this paper we shall use interchangeably the notations $\dot{x}$ and x' to denote differentiation with respect to t.

with $\sum_{i=0}^{\mu} m_i = n$. We define the n linear functionals of Λ by

$$(6.1) \qquad \lambda_{ij} f = \sum_{\nu=0}^{m-1} c_{ij\nu} f^{(\nu)}(t_i) \qquad \begin{cases} j = 1, 2, \ldots, m_i \\ i = 0, 1, \ldots, \mu , \end{cases}$$

where $c_{ij} = (c_{ij0}, \ldots, c_{ij,m-1})$ are prescribed vectors of real numbers. Without loss of generality we may assume that for each $i = 0, 1, \ldots, \mu$, the set of vectors $\{c_{ij}\}_{j=1}^{m_i}$ is linearly independent. In accordance with this renumbering of the λ's, it is convenient to renumber the data of (3.5) as δ_{ij} and γ_{ij}. A set Λ of linear functionals of this type is said to generate Extended-Hermite-Birkhoff interpolation (see [8]).

We now translate our basic variational problem (3.7) into the framework of §5. We begin by setting $x_1 = f$ and defining functions $x_2(t), \ldots, x_m(t)$ and $u(t)$ via the differential equations

$$(6.2) \begin{cases} \dot{x}_i = x_{i+1} & i = 1, 2, \ldots, m-1 \\ \dot{x}_m = (u - \sum_{i=0}^{m-1} a_i x_{i+1})/a_m . \end{cases}$$

This prescription yields $x_{i+1} = x_1^{(i)}$, $i = 0, 1, \ldots, m-1$ and $u = Lx_1$. The constraints describing the class U_1 can now be written as

$$(6.3) \begin{cases} \sum_{j=0}^{m-1} b_{ij}(t) x_{j+1}(t) - \beta_i(t) \leq 0 \\ & i = 1, 2, \ldots, k . \\ -\sum_{j=0}^{m-1} b_{ij}(t) x_{j+1}(t) + \alpha_i(t) \leq 0 . \end{cases}$$

The constraints which delineate the flat U_2 may be translated as

$$(6.4)\quad \begin{cases} \sum_{\nu=0}^{m-1} c_{ij\nu}\, x_{\nu+1}(t_i) - \delta_{ij} \le 0 \\ -\sum_{\nu=0}^{m-1} c_{ij\nu}\, x_{\nu+1}(t_i) + \gamma_{ij} \le 0 \end{cases} \qquad \begin{matrix} j = 1, 2, \ldots, m_i \\ i = 0, 1, \ldots, \mu \ . \end{matrix}$$

Problem (3.7) is thus equivalent to

$$(6.5)\qquad \text{minimize} \sum_{i=0}^{\mu-1} \int_{t_i}^{t_{i+1}} |u|^p dt$$

over all measurable functions $u(t)$ such that (6.2-4) hold. This is clearly an optimal control problem which is essentially of the form quoted in §5. Here, however, we have a sort of composite problem defined on a sequence of intervals $[t_0, t_1]$, $\ldots, [t_{\mu-1}, t_\mu]$, so that a certain amount of care must be exercised in applying the results cited in the Appendix.

For later reference, we identify the functions appearing in (5.1-5) and (A.12) for problem (6.5). We take $\phi(u) = |u|^p$ and $\theta = 0$. The vector g is defined as the right hand side of (6.2) while h is

$$\begin{cases} h_{2i-1}(t,x) = \sum_{j=0}^{m-1} b_{ij}(t)x_{j+1}(t) - \beta_i(t) \\ h_{2i}(t,x) = -\sum_{j=0}^{m-1} b_{ij}(t)x_{j+1}(t) + \alpha_i(t) \end{cases} \qquad i = 1, 2, \ldots, k \ .$$

The vector p is defined with the $2m_0$ components

$$(6.6) \quad \begin{cases} p_{2j-1}(x(t_0)) = \sum_{\nu=0}^{m-1} c_{0j\nu} x_{\nu+1}(t_0) - \delta_{0j} \\ p_{2j}(x(t_0)) = -\sum_{\nu=0}^{m-1} c_{0j\nu} x_{\nu+1}(t_0) + \gamma_{0j} \end{cases} \quad j = 1, 2, \ldots, m_0 .$$

Similarly, q is defined with $2m_\mu$ components

$$(6.7) \quad \begin{cases} q_{2j-1}(x(t_\mu)) = \sum_{\nu=0}^{m-1} c_{\mu j\nu} x_{\nu+1}(t_\mu) - \delta_{\mu j} \\ q_{2j}(x(t_\mu)) = -\sum_{\nu=0}^{m-1} c_{\mu j\nu} x_{\nu+1}(t_\mu) + \gamma_{\mu j} \end{cases} \quad j = 1, 2, \ldots, m_\mu .$$

Finally, to take account of the constraints of (6.4) corresponding to t_i's in the interior of (a, b), we need to introduce constraints via σ's as in (A.12). We must introduce vector σ's for each of the $t_1, t_2, \ldots, t_{\mu-1}$ in (a, b). For $i = 1, 2, \ldots, \mu-1$ define σ_i to be a $2m_i$-vector with components

$$(6.8) \quad \begin{cases} \sigma_{i,2j-1}(x(t_i)) = \sum_{\nu=0}^{m-1} c_{ij\nu} x_{\nu+1}(t_i) - \delta_{ij} \\ \sigma_{i,2j}(x(t_i)) = -\sum_{\nu=0}^{m-1} c_{ij\nu} x_{\nu+1}(t_i) + \gamma_{ij} \end{cases} \quad j = 1, 2, \ldots, m_i .$$

§7. Sufficient conditions for a solution of (3.7). In this section we employ the sufficiency theorem of §A.2 to derive conditions which assure that a function $s \in U$ solves (3.7).

Recalling the equivalence of (3.7) with the control problem set forth in §6, we now seek sufficient conditions for a solution of (6.5). It is easily seen that the problem described by (6.2-5) satisfies the hypotheses of Theorem A.1. In this connection we point out that for $p > 1$ the objective function $\phi(u) = |u|^p$ is convex and is $[p]$-times continuously differentiable with respect to u.

By a straightforward application of Theorem A.1 we have

<u>Theorem 7.1.</u> Suppose $1 < p < \infty$ is either an integer or an arbitrary real number satisfying $p \geq m+1$. Let $s \in U$, $u = Ls$, $x_1 = s$, and $\{x_i\}_2^m$ be defined as in (6.2). Set $v_m(t) = -a_m(t) \dfrac{d\phi(u)}{du}$ and for $i = 1, 2, \ldots, m-1$

$$v_i(t) = -\dot{v}_{i+1}(t) + a_i(t)v_m(t)/a_m(t) \quad . \tag{7.1}$$

Then $s(t)$ is a solution of (3.7), or equivalently (6.5), provided there exist non-negative functions $\{\underline{w}_i(t)\}_1^k$, $\{\overline{w}_i(t)\}_1^k$, $\{\underline{z}_i(t)\}_1^k$, $\{\overline{z}_i(t)\}_1^k$ and non-negative scalars $\{\underline{r}_{ij}\}_{0,1}^{\mu, m_i}$, $\{\overline{r}_{ij}\}_{0,1}^{\mu, m_i}$ such that

$$(7.2) \quad \begin{cases} \dot{v}_1 + \sum_{j=1}^{k} b_{j0}(\overline{w}_j - \underline{w}_j) - v_m a_0/a_m = 0 \\[2ex] \dot{v}_{i+1} + \sum_{j=1}^{k} b_{ji}(\overline{w}_j - \underline{w}_j) - v_m a_i/a_m + v_i = 0, \quad i = 1, 2, \ldots, m \end{cases}$$

$$(7.3) \quad \begin{cases} \overline{w}_i(t) = 0 \quad \text{whenever} \quad \beta_i(t) > \mathfrak{m}_i s(t) \\[2ex] \underline{w}_i(t) = 0 \quad \text{whenever} \quad \alpha_i(t) < \mathfrak{m}_i s(t) \end{cases}$$

$$(7.4) \quad \begin{cases} v(a) = \sum_{j=1}^{m_0} (\underline{r}_{0j} - \overline{r}_{0j}) c_{0j} \\[2ex] \sum_{j=1}^{m_0} (\underline{r}_{0j}(\gamma_{0j} - \lambda_{0j}s) + \overline{r}_{0j}(\lambda_{0j}s - \delta_{0j})) = 0 \end{cases}$$

(7.5)
$$\begin{cases} v(b) = \sum_{j=1}^{m_\mu} (\underline{r}_{\mu j} - \bar{r}_{\mu j}) c_{\mu j} \\ \sum_{j=1}^{m_\mu} (\underline{r}_{\mu j}(\gamma_{\mu j} - \lambda_{\mu j} s) + \bar{r}_{\mu j}(\lambda_{\mu j} s - \delta_{\mu j})) = 0 \ . \end{cases}$$

(7.6) The vector $v(t)$ is continuous on $[a, b]$ except for a finite number of points (including possibly the $\{t_i\}_1^{\mu-1}$) at which the following jump conditions are permitted:

(7.7)[†]
$$\begin{cases} [v(t_i)] = \sum_{j=1}^{m_i} (\underline{r}_{ij} - \bar{r}_{ij}) c_{ij} \qquad i = 1, 2, \ldots, \mu-1 \\ \sum_{j=1}^{m_i} (\underline{r}_{ij}(\gamma_{ij} - \lambda_{ij} s) + \bar{r}_{ij}(\lambda_{ij} s - \delta_{ij})) = 0 \end{cases}$$

(7.8)
$$\begin{cases} [v(\hat{t})] = \sum_{i=1}^{k} (\underline{z}_i(\hat{t}) - \bar{z}_i(\hat{t})) b_i \\ \sum_{i=1}^{k} (\underline{z}_i(\hat{t})(\alpha_i(\hat{t}) - \mathfrak{m}_i s(\hat{t})) + \bar{z}_i(\hat{t})(\mathfrak{m}_i s(\hat{t}) - \beta_i(\hat{t})) = 0 \ , \end{cases}$$

where $b_i = (b_{i,0}, \ldots, b_{i,m-1})$. In (7.4-7) we have used the notation λ_{ij} as defined in (6.1), (see also (6.4) and (6.6-8)). We also recall that the symbol $\mathfrak{m}_i$ was defined in (3.3) (compare with (6.3)).

To assist in interpreting this theorem, we proceed to relate some of the above conditions to direct properties of $s(t)$, and also consider some special cases which occur frequently.

† The notation $[v(\hat{t})]$ denotes $v(\hat{t}+) - v(\hat{t}-)$ whenever $a < \hat{t} < b$ Later we shall have use for the notation $[v(a)] = v(a+)$, $[v(b)] = v(b-)$.

First, we note that the equations (7.1) are equivalent to

$$v_i(t) = \sum_{j=0}^{m-i} (-1)^j (a_{i+j} v_m/a_m)^{(j)}, \qquad i = 1, 2, \ldots, m \; . \tag{7.9}$$

For $i = m$ (7.9) is an identity. To complete the verification of (7.9), suppose it has been established for $i+1$. Then by (7.1)

$$v_i = v_m a_i/a_m - \Big(\sum_{j=0}^{m-1-i} (-1)^j (a_{i+j+1} v_m/a_m)^{(j)} \Big)'$$

$$= v_m a_i/a_m + \sum_{j=1}^{m-i} (-1)^j (a_{i+j} v_m/a_m)^{(j)}$$

which is (7.9) for i. Thus by induction (7.9) holds for all $i = 1, 2, \ldots, m$.

Set $I_1 = \{t \in [a,b] : \alpha_i(t) < \mathfrak{m}_i s(t) < \beta_i(t), \; i = 1, 2, \ldots,$ [illegible] $I_2 = \{t \in [a,b] : t \neq t_i, \; i = 0, 1, \ldots, \mu\}$, and $I = I_1 \cap I_2$. Then for $t \in I$ the equations for $i = 1, 2, \ldots, m-1$ of (7.2) are automatically satisfied in view of (7.1) and (7.3). In addition the first equation of (7.2) is equivalent for $t \in I$ to

$$L^*(v_m/a_m) = \sum_{j=0}^{m} (-1)^j (a_j v_m/a_m)^{(j)} = 0 \; . \tag{7.10}$$

Equations (7.9), (7.10) are directly related to properties of s since $v_m(t) = -a_m(t) \frac{d\phi}{du}(Ls)$.

If $m_0 = 0$ (i.e. none of the linear functionals in Λ involve evaluation of s or its derivatives at $t = a$), and if $\hat{t}_1 = \min(t_1, \inf\{t : \alpha_i(t) < \mathfrak{m}_i s(t) < \beta_i(t), \; i = 1, 2, \ldots, k\})$, then (7.2), (7.4) can hold for these t if and only if

$$-\frac{d\phi}{du}(Ls) = \frac{v_m(t)}{a_m(t)} = 0 \qquad a \le t < \hat{t}_1 \; . \tag{7.11}$$

Indeed by (7.10) $L^*(v_m/a_m) = 0$ for $a \le t < \hat{t}_1$, while (7.4) coupled with (7.9) leads to $v_m(a) = v'_m(a) = \ldots = v_m^{(m-1)}(a) = 0$, whence we conclude that $v_m(t) \equiv 0$. Similarly, if $m_\mu = 0$ and $\hat{t}_{\mu-1} = \max(t_{\mu-1}, \sup\{t: \alpha_i(t) < \mathfrak{m}_i s(t) < \beta_i(t),\ i = 1, 2, \ldots, k\})$, then (7.11) also must hold for $t_{\mu-1} < t \le b$.

In the special case that Λ corresponds to what is called Hermite-Birkhoff interpolation, (see [8], i.e. (cf. (6.1))

$$\lambda_{ij} f = f^{(\ell_{ij})}(t_i) \quad \text{where } 0 \le \ell_{ij} \le m-1, \tag{7.12}$$

condition (7.7) is considerably simplified. In this situation the vector c_{ij} has all zero components except for the ℓ_{ij} th which is 1. Then (7.7) is equivalent to

$$\text{(7.13)} \quad \begin{cases} [v_{\ell_{ij}+1}(t_i)] = 0 & \text{if } \gamma_{ij} < s^{(\ell_{ij})}(t_i) < \delta_{ij} \\ [v_{\ell_{ij}+1}(t_i)] \ge 0 & \text{if } \gamma_{ij} = s^{(\ell_{ij})}(t_i) < \delta_{ij} \\ [v_{\ell_{ij}+1}(t_i)] \le 0 & \text{if } \gamma_{ij} = s^{(\ell_{ij})}(t_i) = \delta_{ij} \\ [v_{\ell_{ij}+1}(t_i)] \text{ is arbitrary} & \text{if } \gamma_{ij} = s^{(\ell_{ij})}(t_i) = \delta_{ij} . \end{cases}$$

In this case the relations (7.4) and (7.5) are also equivalent to (7.13) for $i = 0$ and μ, respectively, if we agree that $[v(a)] = v(a+)$ and $[v(b)] = v(b-)$.

Finally, if the operators $\mathfrak{m}_i f$ have the special form $\mathfrak{m}_i f = f^{(i-1)}$, then (7.8) is equivalent to

$$(7.14)\quad \begin{cases} [v_i(\hat{t})] = 0 \ \text{if} \ \alpha_i(\hat{t}) < s^{(i-1)}(\hat{t}) < \beta_i(\hat{t}) \\ [v_i(\hat{t})] \geq 0 \ \text{if} \ \alpha_i(\hat{t}) = s^{(i-1)}(\hat{t}) < \beta_i(\hat{t}) \\ [v_i(\hat{t})] \leq 0 \ \text{if} \ \alpha_i(\hat{t}) < s^{(i-1)}(\hat{t}) = \beta_i(\hat{t}) \\ [v_i(\hat{t})] \ \text{arbitrary if} \ \alpha_i(\hat{t}) = s^{(i-1)}(\hat{t}) = \beta_i(\hat{t}) . \end{cases}$$

In addition, in this case equations (7.2) become

$$(7.15)\quad \begin{cases} \dot{v}_1 + \bar{w}_1 - \underline{w}_1 - v_m a_0/a_m = 0 \\ \dot{v}_{i+1} + v_i + \bar{w}_{i+1} - \underline{w}_{i+1} - v_m a_i/a_m = 0, \qquad i = 1, 2, \ldots, m-1 \end{cases}$$

§8. Polynomial g-splines. As indicated in §7, Theorem 7.1 admits of considerable simplification when (3.7) is made less general. For convenience we devote this section to combining the remarks of §7 into a corollary of Theorem 7.1 which is still sufficiently general to include most of the examples in §2.

For this section let $p = 2$, $L = (\frac{d}{dt})^m = L^*$, let Λ generate Hermite-Birkhoff interpolation (see (7.12)) and in (3.4) let $k = 1$, $\mathfrak{M}_1 f = f$. In other words, we consider the problem

$$(8.1)\qquad \underset{f \in U}{\text{minimize}} \int_a^b (f^{(m)})^2 \, dt$$

$$(8.2)\quad U = \{f \in H_2^m : \alpha(t) \leq f(t) \leq \beta(t) \ \text{and} \ \gamma_{ij} \leq f^{(\ell_{ij})}(t_i) \leq \delta_{ij} ,$$
$$j = 1, 2, \ldots, m_i, \ i = 0, 1, \ldots, \mu\} .$$

Theorem 8.1. The function $s(t) \in U$ is a solution of (8.1) provided

$$(8.3)\quad s^{(2m)}(t) = 0 \ \text{when} \ \alpha(t) < s(t) < \beta(t), \ \text{and} \ t \neq t_i ,$$
$$i = 0, 1, \ldots, \mu .$$

$$(8.4)\qquad s^{(m)}(t) = 0, \quad \begin{cases} a \le t < \hat{t}_1 \text{ if } m_0 = 0 \quad \text{and} \\ \hat{t}_1 = \min(t_1, \inf\{t : \alpha(t) < s(t) < \beta(t)\}) \\ \hat{t}_{\mu-1} < t \le b \text{ if } m_\mu = 0 \text{ and} \\ \hat{t}_{\mu-1} = \max(t_{\mu-1}, \sup\{t : \alpha(t) < s(t) < \beta(t)\}) \end{cases}$$

(8.5) $[s^{(2m-\ell-1)}(t\ )] = 0$, for $0 \le \ell \le m-1$, if $s^{(\ell)}$ is not constrained at t or if the constraint is not active e.g. if $\ell = $ some ℓ_{ij} and

$$\gamma_{ij} < s^{(\ell_{ij})}(t\ ) < \delta_{ij}\ .$$

$$(8.6)\qquad (-1)^{m-\ell_{ij}}[s^{(2m-\ell_{ij}-1)}(t_i)] \ge 0 \text{ if } s^{(\ell_{ij})}(t_i) < \delta_{ij}$$

$$(8.7)\qquad (-1)^{m-\ell_{ij}}[s^{2m-\ell_{ij}-1)}(t_i)] \le 0 \text{ if } s^{(\ell_{ij})}(t_i) > \gamma_{ij}$$

$$(8.8)\qquad \begin{cases} (-1)^m\ [s^{(2m-1)}(\hat{t})] \ge 0 \text{ if } s(\hat{t}) = \alpha(\hat{t}) \\ (-1)^m\ [s^{(2m-1)}(\hat{t})] \le 0 \text{ if } s(\hat{t}) = \beta(\hat{t}) \end{cases}$$

$$(8.9)\qquad \begin{cases} (-1)^m s^{(2m)}(t) \ge 0 \text{ whenever } s(t) < \beta(t) \text{ on an interval of positive length} \\ (-1)^m s^{(2m)}(t) \le 0 \text{ whenever } s(t) > \alpha(t) \text{ on an interval of positive length.} \end{cases}$$

<u>Proof:</u> We note that $u = Ls = s^{(m)}$. Following Theorem 7.1 we set

$$v_m(t) = -d\phi/du = -2u, \quad \text{and} \quad v_i(t) = -\dot{v}_{i+1}(t) \quad ,$$

$i = 1, 2, \ldots, m-1$. Now condition (7.10) is implied by (8.3) since $L^*(v_m/a_m) = v_m^{(m)} = 0$ holds whenever $s^{(2m)}(t) = 0$. Property (8.4) clearly implies (7.11). Next we notice that (8.5-7) implies (7.13), (7.6) since here (cf. (7.9))

$$v_i(t) = (-1)^{m-i} v_m^{(m-i)} = 2(-1)^{m-i-1} s^{(2m-i)}, \quad i = 1, 2, \ldots, m .$$

Property (8.8) yields (7.14) easily, while clearly (7.15) can be satisfied if and only if (8.9) holds. (The choice of $\underline{w}_1, \bar{w}_1$ is obvious.)

When no continuous constraint of the form $\alpha(t) \le f(t) \le \beta(t)$ is imposed, conditions (8.8) and (8.9) may be removed. Then Theorem 8.1 corresponds to g-splines with inequality constraints on Hermite-Birkhoff data (see [8], [14]).

§9. Necessary conditions for a solution of (3.7). It is always convenient to have a complete characterization of a solution to a problem such as (3.7), that is, a set of conditions which are both necessary and sufficient. In this section we shall report on some progress in this direction. In line with our stated philosophy, we shall concern ourselves here only with some special cases of the general problem (3.7). The key difficulty with using presently known control theorems is the fact already mentioned in §6 that (3.7) is a sort of composite problem involving transfer through a sequence of states at prescribed times. It is likely that the necessary conditions for such control problems can be treated with suitable modifications of existing techniques, but such a program is outside the scope of this paper.

To illustrate how necessary conditions can be derived from optimal control, we consider the following variational problem of type (3.7):

$$(9.1) \quad \begin{cases} \underset{f \in U}{\text{minimize}} \; \|Lf\|_{\mathcal{L}_2} \\ \\ U = \{f \in \mathcal{H}_p^m : \alpha(t) \le f(t) \le \beta(t), \; f(a) = \gamma_0, \; f(b) = \gamma_1\} \end{cases}$$

where $\alpha(t), \beta(t) \in \mathcal{H}_p^m$ and γ_0, γ_1 are prescribed real numbers, $(\alpha(t) < \beta(t), \alpha(a) < \gamma_0 < \beta(a), \alpha(b) < \gamma_1 < \beta(b).)$

Theorem 9.1. Let $s \in U$ be a solution of (9.1). Then

$$L^* Ls(t) = 0 \text{ whenever } \alpha(t) < s(t) < \beta(t) \quad , \tag{9.2}$$

(9.3) The functions $v_i(t) = -\sum_{j=0}^{m-i} (-1)^j (a_{i+j} Ls/a_m)^{(j)}$ are absolutely continuous on $[a,b]$ with the exception of possible jumps in $v_1(t)$ at points $\hat{t}$ where $s(t)$ equals $\alpha(t)$ or $\beta(t)$, in which case

$$\begin{cases} [v_1(\hat{t})] \geq 0 & \text{if} \quad s(\hat{t}) = \alpha(\hat{t}) \\ [v_1(\hat{t})] \leq 0 & \text{if} \quad s(\hat{t}) = \beta(\hat{t}) \\ [v_1(\hat{t})] \text{ arbitrary if } & \alpha(\hat{t}) < s(\hat{t}) < \beta(\hat{t}) \quad , \end{cases} \tag{9.4}$$

$$v_i(a) = v_i(b) = 0, \qquad i = 2, 3, \ldots, m \quad . \tag{9.5}$$

Proof: We intend to invoke Neustadt's theorem which is stated as Theorem A.2. First, we translate (9.1) into Neustadt's framework. Let $\dot{x}_0(t) = u^2$, $x_1(t) = s(t)$, $\{x_i(t)\}_2^m$ and $u(t)$ be as in (6.2). The constraints of the U of (9.1) may then be written as

$$\begin{cases} h(t,x) = (x_1(t) - \beta(t))(x_1(t) - \alpha(t)) \leq 0 \\ p_1(x(a),x(b)) = x_1(a) - \gamma_0 = 0 \\ p_2(x(a),x(b)) = x_1(b) - \gamma_1 = 0 \quad . \end{cases} \tag{9.6}$$

Setting $p_0(x(a), x(b)) = x_0(b) - x_0(a)$, it is clear that s minimizes $p_0(x(a), x(b))$ over all measurable u such that (9.6) holds.

The function $\rho(\xi, t)$ of Theorem A.2 in this case is

$$\rho(\xi, t) = (2\xi_1 - \alpha - \beta)\xi_2 - (\dot{\alpha}(\xi_1 - \beta) + \dot{\beta}(\xi_1 - \alpha)) \quad .$$

We cannot apply Neustadt's theorem directly since it gives necessary conditions only for minimization problems where the set of admissible controls u consists of essentially bounded measurable functions. To overcome this difficulty we shall have to rely on the sufficiency theorem of §7 as well as the uniqueness (a.e.) of the control (see Theorem 4.1).

We begin by posing a further constrained version of (9.1). In particular, let $U_K = \{f \in U: \|Lf\|_\infty \leq K\}$, where U is as in (9.1) and $\|\cdot\|_\infty$ denotes the essential supremum norm. It is an easy exercise to verify that U_K is non-empty for sufficiently large K. We sketch the proof: Since $\alpha(t) < \beta(t)$ and $\alpha(a) < \gamma_0 < \beta(a)$, $\alpha(b) < \gamma_1 < \beta(b)$, there exists a function $\tilde{f} \in U$ with $\alpha(t) < \tilde{f}(t) < \beta(t)$, for all $a \leq t \leq b$. For arbitrary n the corresponding Bernstein polynomial $B_n(\tilde{f};t)$ agrees with $\tilde{f}$ at a and b, while for sufficiently large n it is arbitrarily close to $\tilde{f}$ in the ∞-norm. Hence for sufficiently large n we have a polynomial B in U. Taking $K = \|LB\|_\infty$, we have $B \in U_K$.

Next we notice that U_K is a closed subset of U. Thus by a simple modification of the proof of Theorem 4.1, we conclude that there exists $s_K \in U_K$ which minimizes $\int_a^b (Lf)^2 dt$ subject to $f \in U_K$. For the time being, let $s(t) = s_K(t)$, $x_1(t) = s(t)$, and $\{x_i\}_2^m$ and u be defined as in (6.2) Applying Theorem A.2 we conclude there exists $\omega_0, \omega_1, \omega_2$ and a scalar function $\lambda(t)$ satisfying (A.17-20), and absolutely continuous functions $\{\psi_i(t)\}_0^m$ satisfying

$$(9.7) \quad \begin{cases} \dot{\psi}_0(t) = 0 \\ \dot{\psi}_1(t) = \lambda(t)(2x_2 - \dot{\alpha} - \dot{\beta}) + \psi_m a_0/a_m \\ \dot{\psi}_2(t) = -\psi_1(t) + \lambda(t)(2x_1 - \alpha - \beta) + \psi_m a_1/a_m \\ \dot{\psi}_i(t) = -\psi_{i-1}(t) + \psi_m a_{i-1}/a_m, \quad i = 3, 4, \ldots, m \ . \end{cases}$$

In addition, (A.23) implies

$$\psi_0 u^2 + (\psi_1 - \lambda(2x_1 - \alpha - \beta))x_2 + \psi_2 x_3 + \psi_3 x_4 + \cdots + \psi_{m-1}x_m + \psi_m\left(-\sum_{i=0}^{m-1} a_i x_{i+1} + u\right)/a_m$$

is maximal with respect to u, i.e.

(9.8) $$2u\psi_0 + \psi_m/a_m = 0 . \qquad \text{(a.e.)}$$

Finally, the transversality conditions (A.22) assure that

(9.9) $$\begin{cases} \psi_0(a) = \omega_0 & \psi_0(b) = \omega_0 \\ \psi_1(a) = \lambda(a)((2x_1 - \alpha - \beta)(a)) - \omega_1 & \psi_1(b) = \omega_2 \\ \psi_i(a) = 0 & \psi_i(b) = 0 \quad i = 2, 3, \ldots, m . \end{cases}$$

Our first observation is that by (9.7) and (9.9), $\psi_0(t) \equiv \omega_0$. By (A.20), $\omega_0 \leq 0$, but we claim more, namely that $\omega_0 < 0$. Indeed, if $\omega_0 = 0$, then $\psi_0 \equiv 0$, whence by (9.7-8) $\psi_m \equiv \ldots \equiv \psi_2 \equiv 0$. But then (9.7) implies $\psi_1(t) = \lambda(t)(2x_1 - \alpha - \beta)$. By the absolute continuity of $\psi_1(t)$ we conclude λ is also absolutely continuous and $\dot{\lambda}$ exists (a.e.). Then $\dot{\psi}_1 = \lambda(2x_2 - \dot{\alpha} - \dot{\beta}) + \dot{\lambda}(2x_1 - \alpha - \beta)$, which by (9.7) yields $\dot{\lambda}(2x_1 - \alpha - \beta) = 0$. If $x_1 \neq (\alpha + \beta)/2$, then $\dot{\lambda} = 0$. On the other hand, when $x_1 = (\alpha + \beta)/2$ then certainly $\alpha < x_1 < \beta$ and by (A.17), $\dot{\lambda} = 0$ here also. Thus λ = constant, which by (A.18) must be 0 . Hence $\psi_1 \equiv 0$ and equations (9.9) then require $\omega_1 = \omega_2 = 0$. Finally, this contradicts (A.19) and hence the assumption $\omega_0 = 0$ is untenable.

To derive properties (9.3-4) of s we use (9.8) in (9.7). The <u>mth</u> equation of (9.7) yields $\psi_{m-1} = \psi_m a_{m-1}/a_m - \dot{\psi}_m$.

Arguing inductively as in §7 we obtain $\psi_i = \sum_{j=0}^{m-i} (-1)^j (a_{i+j}\psi_m/a_m)^{(j)}$

for $i = 2, 3, \ldots, m$. In addition, we obtain $\psi_1 = \lambda(2x_1 - \alpha - \beta) + \sum_{j=0}^{m-1} (-1)^j (a_{1+j}\psi_m/a_m)^{(j)}$. Thus by (9.8)

$$(9.10) \quad \begin{cases} v_i(t) = +\dfrac{\psi_i}{2\omega_0}, \quad i = 2, \ldots, m \text{ and} \\ v_1(t) = \dfrac{-\lambda(2x_1 - \alpha - \beta) + \psi_1}{2\omega_0}. \end{cases}$$

The absolute continuity of the ψ's yields the continuity of the v_i as asserted in (9.3). Furthermore, $2\omega_0[v_1] = -[\lambda](2x_1 - \alpha - \beta) + [\psi_1]$, and since λ is decreasing, $\omega_0 < 0$, and ψ_1 is continuous, we conclude that $[v_1]$ is proportional to $-(2x_1 - \alpha - \beta)$. This yields (9.4).

To establish (9.2), we notice that the expression for ψ_1 obtained above can be differentiated whenever $\alpha(t) < s(t) < \beta(t)$ since then $\lambda(t)$ is constant. Differentiating, we obtain $\dot{\psi}_1 = \lambda(2x_2 - \dot{\alpha} - \dot{\beta}) + \sum_{j=0}^{m-1} (-1)^j (a_{1+j}\psi_m/a_m)^{(j+1)}$, which after using the second equation of (9.7) becomes $\sum_{j=0}^{m} (-1)^j (a_j\psi_m/a_m)^{(j)} = 0$. Substituting from (9.8) leads to (9.2). The conditions (9.5) follow directly from (9.9–10).

We have shown that there exists a function $s(t) = s_K(t)$ satisfying (9.2–5). Referring to Theorem 7.1, we find that $s_K(t)$ also provides a minimum for the variational problem defined in (9.1), without additional constraints. Now, if $\tilde{s}(t)$ is any other solution of (9.1), then by the uniqueness assertion in Theorem 4.1, $L\tilde{s} = Ls_K$ (a.e.), and thus $L\tilde{s}$ is also essentially bounded on $[a, b]$. In other words, the above argument is valid for any solution of (9.1) and the proof of Theorem 9.1 is complete.

When λ is differentiable, we remark that the functions appearing in Theorem 9.1 may be identified with those in Theorem 7.1 via

$$w(t) = -d\lambda/dt, \qquad v(t) = -\psi(t) + \lambda(t)\nabla_x h(x,t) \ .$$

Remark 9.2. Theorem 9.1 also holds for the variational problem described by (9.1) with the end constraint $f(a) = \gamma_0$ deleted. In addition, we have

(9.11) $\quad Ls = 0 \quad \text{for} \quad a \leq t < \hat{t}_1 = \min(t_1, \inf\{t : \alpha(t) < s(t) < \beta(t)\})$.

Similarly, if the constraint $f(b) = \gamma_1$ is deleted from (9.1) then s also satisfies

(9.12) $\quad Ls = 0 \quad \text{for} \quad b \geq t > \hat{t}_2 = \max(t_2, \sup\{t : \alpha(t) < s(t) < \beta(t)\})$.

To establish (9.11) we first notice that by (9.2), $L^*(\psi_m/a_m) = 0$ for $a \leq t < \hat{t}_1$. Moreover, by (9.7) and an easy inductive argument we obtain

$$\psi_m^{(i)} = (-1)^i\Big(\psi_{m-i} - \sum_{j=0}^{i-1}(-1)^j(\psi_m a_{m-i+j}/a_m)^{(j)}\Big), \quad i = 1, 2, \ldots, m-2$$

while

$$\psi_m^{(m-1)} = (-1)^{m-1}\Big(\psi_1 - \lambda(2x_1 - \alpha - \beta) - \sum_{j=0}^{m-2}(-1)^j(\psi_m a_{j+1}/a_m)^{(j)}\Big) \ .$$

Using (9.9) we infer that $\psi_m^{(i)}(a) = 0, \ i = 0, 1, \ldots, m-1$. (In this case ω_1 does not appear in (9.9).) Since ψ_m/a_m is a solution of an m-th order homogeneous linear differential equation, these m boundary conditions imply $\psi_m(t) = -2u\omega_0 \equiv 0$ for $a \leq t < \hat{t}_1$. The proof of (9.12) is similar.

§10. Discussion of the examples of §2. Our main tools for the ensuing discussion are Theorems 7.1, 8.1 and the uniqueness assertion of Theorem 4.1. We outline how the solutions in the examples of §2 were constructed, and verify that they provide unique minimums.

Example 2.1. Here $L = L^* = (\frac{d}{dt})^2$, $u = s''$, and

$$v_2(t) = \begin{cases} -pu^{p-1} & u \geq 0 \\ p(-u)^{p-1} & u < 0, \end{cases} \qquad v_1(t) = \begin{cases} p(p-1)u^{p-2}u' & u \geq 0 \\ p(p-1)(-u)^{p-2}u', & u < 0 . \end{cases}$$

By condition (7.10) we require $(s'')^{p-1}$ to reduce to a polynomial of degree 1 in each of the intervals $[-1, 0]$ and $[0, 1]$. In view of (7.11) in the intervals $[-2, -1]$, $[1, 2]$ we need $s''(t) = 0$. Condition (7.6) requires s'' to be globally continuous, while (7.13) allows jumps in $v_1(t)$ at $t = -1, 0, 1$. These conditions suffice to construct the spline (2.1). To complete the verification of optimality we need only check the signs of the jump conditions at $-1, 0, 1$. We have

$$[v_1(-1)] \cong [s'''(-1)] = \infty \geq 0$$

$$[v_1(0)] \cong [s'''(0)] = \frac{-2(2p-1)}{(p-1)^2} \leq 0$$

$$[v_1(1)] \cong [s'''(1)] = \infty \geq 0$$

(the relation $\cong$ means that the two quantities have the same sign).

The uniqueness assertion of Example 2.1 is easy to establish. For $L = (d/dt)^2$ the null space N consists of polynomials of degree 1, and hence any two solutions must differ by such a polynomial. On the other hand, $g \in U(s)$ for this example if and only if $0 \leq g(-1) \leq 1/2$, $-1/4 \leq g(0) \leq 0$, and $0 \leq g(1) \leq 1/2$. Since no linear polynomial except the null polynomial can satisfy these three conditions we conclude that $N \cap U(s) = (0)$ and (2.1) provides the unique minimum.

Several of the remaining examples are most easily discussed via Theorem 8.1.

Example 2.2. Here $m = 1$ and Λ defines only the constraint $\gamma = f(1)$. Theorem 8.1 points the way to the construction of

(2.2). By (8.9), s can follow either the upper or lower boundary since $\beta'' = 0$ and $\alpha'' = -2 \leq 0$. By (8.3), s must be linear when not following the boundaries. At the point $1 - \sqrt{\gamma}$, (8.8) requires $[s'] \leq 0$. But since $s \in \mathcal{H}_2^1$ must be continuous, it is easily seen that s cannot get off the lower boundary with $[s'] \leq 0$ and still satisfy the constraints; hence we require $s \in C^1[0,1]$. The conditions suffice for the construction of (2.2), which then automatically fulfills the hypotheses of Theorem 8.1.

In this example N = {constant} and $g \in U(s)$ only if $g(0) = 0$ and $g(1) = 0$. Thus $N \cap U(s) = (0)$ and uniqueness follows from Theorem 4.1.

Example 2.3. Here $m = 2$. By (8.3), s must reduce to cubic polynomials except when s follows one of the boundaries $\alpha(t)$ or $\beta(t)$. Condition (8.9) could be satisfied with s following either the upper or lower boundaries except that it cannot follow through the point 0. By (8.8), $[s'''] \geq 0$ when on α, and $[s'''] \leq 0$ when on β. To satisfy (8.5) we need $s''(-1) = s''(1) = 0$ and $[s''(0)] = 0$. These requirements suggest seeking s of the form (2.3) (it is fairly easy to rule out the possibility of following the boundaries with all of the above conditions). To verify that (2.3) indeed minimizes $\int_{-1}^{1} (f'')^2 dt$, we check the hypotheses of Theorem 8.1. For the prescribed γ, δ it is easily checked that s meets the constraints. To check the jump at 0 we note that $[s'''(0)] = -3(\gamma + \delta - 2) \leq 0$ as desired. Finally, since $g \in U(s)$ is equivalent to $g(-1) = g(1) = 0$, while N consists of linear polynomials, it follows that $N \cap U(s) = (0)$ and (2.3) is a unique extremal.

Example 2.4. Here again $m = 2$ and s must be cubic when it is not following the boundary. By (8.9), s cannot follow the upper boundary since $s^{(4)}(t) = \beta^{(4)}(t) = 4! > 0$ contrary to (8.9). Since $s^{(4)}(t) = \alpha^{(4)}(t) \equiv 0$, s may equal $\alpha(t)$ on intervals. To satisfy (8.5) we need $[s''(1)] = s''(1) = 0$, while (8.8) requires $s'' \in C^2$ and $[s'''] \geq 0$, $[s'''] \leq 0$ when getting on or off of $\alpha(t)$ or $\beta(t)$, respectively. These prescriptions lead us to seek $s(t)$ of the form (2.4). We claim (2.4) meets the hypotheses of Theorem 8.1. In particular, we note that $[s'''(\xi)] = 6\eta^4/(\eta-\xi)^3 > 0$ while $[s'''(\eta)] = \frac{6\eta^4}{(\eta-\xi)^2}\left(\frac{1}{\eta-\xi} - \frac{1}{1-\eta}\right) < 0$ as desired. We also notice that the equation $\Delta(\eta) = 5\eta^4 - 28\eta^3 + 32\eta^2 - 9\gamma = 0$ always has a solution in $[0,1]$ since $\Delta(0) = -9\gamma$ while $\Delta(1) = 9 - 9\gamma$. In addition, (2.4) meets the constraints (to check this in $[\xi,\eta]$ consider $\beta - s$ which is readily seen to be non-negative). The uniqueness of (2.4) is a consequence of the fact that no non-trivial linear polynomial can vanish at 0 and 1 simultaneously.

Example 2.5. We take $\phi(u) = (u - e^t)^2$, $u = Lf = f''$, $v_2 = -2(u - e^t)$ and $v_1 = -\dot{v}_2$. By (7.10), for t in the intervals $[0,1]$ and $[1,2]$, $u - e^t$ must be linear. To satisfy (7.4), (see (7.13)), we require $0 = v_2(0) = v_2(2)$, i.e., $s''(0) = 1$, $s''(2) = e^2$. For (7.6) we need v_2 continuous, i.e. $s \in C^2[0,2]$. At 1, s''' is allowed a jump by (7.13). Clearly (2.5) satisfies these conditions and also meets the constraints.

The uniqueness of (2.5) is a consequence of the fact that $U(s) \cap N = \{$polynomials p of degree 1 such that $p(0) = p(1) = p(2) = 0\} = (0)$.

§11. Further remarks

(11.1) We have illustrated several problems with both discrete and continuous constraints, but have limited ourselves to examples where only one continuous constraint is specified. It is very easy to formulate mixed problems with simultaneous continuous inequality constraints, but the solutions usually become considerably more complicated.

(11.2) The existence theorem of §A.2 could be used equally as well on problems defined as in §3 where L and the $\mathfrak{m}$'s of (3.1), (3.3) are allowed to be non-linear (but convex) functions of the x_i's.

(11.3). In view of the behavior as $p \to \infty$ of Example 2.1, it would be interesting to consider constrained variational problems involving the uniform norm.

(11.4) As an examination of the discussion in §10 of the examples of §2 will indicate, the necessity and sufficiency theorems quoted here do not specify where or how many times the splines get on or off of the boundaries. These points are determined in the examples as roots of certain polynomial equations. It appears that this question is relevant to the construction of algorithms for computing the splines (cf. [8] and [14] where algorithms are given for some spline problems).

(11.5). Optimal control methods have been utilized to study a certain constrained approximation problem (see [12] and [13]), not completely unlike some considered here.

(11.6). Mangasarian (unpublished) has developed sufficiency theorems for optimal control problems in several time variables. It is conceivable that certain constrained variational problems of this type may have relevance for developing multi-dimensional splines.

(11.7). Although the emphasis here has been on variational problems with spline-like solutions, it seems likely that, conversely, presently known results on splines may be useful in developing algorithms for solving certain practical optimal control problems.

(11.8). We have chosen to use the necessity theorem of Neustadt (see §A.3) because of certain convenient features, but several others in the literature could be used. Although we have omitted them, it is possible to derive necessary conditions for certain problems where the linear functionals Λ involve evaluation of f or its derivatives at points in the interior of (a, b). This proceeds by prescribing a continuous inequality constraint and passing to the limit. Such theorems would be more conveniently established if composite control theory were to be developed.

APPENDIX

§A.1. Proof of Theorem 4.1. Since the minimum problem (3.7) is posed in $\mathcal{L}_p$, existence will follow immediately upon showing that LU is a closed, convex, nonempty set in $\mathcal{L}_p$. The assumption that $U \neq \phi$ assures that LU is also nonempty. Since convexity is obvious, it remains only to establish that LU is closed.

To show that LU is closed, let φ_i be a sequence of elements in LU converging in $\mathcal{L}_p$ to a function φ. We must show that $\varphi \in LU$ also. The key to proving this is a careful construction of the norm for $\mathcal{H}_p^m$. Since L is of order m, its null space N_L is spanned by precisely m functions. Each of the operators $\mathfrak{M}_i$ has a null space $N_{\mathfrak{M}_i}$ of dimension at most $m-1$. We shall have use for the notation

$$U_1(0) = \{f \in \mathcal{H}_p^m : \lambda_i f = 0, \quad i = 1,2,\ldots,n\} .$$

For $\ell = 0,1,\ldots,k+1$ let $\{e_i^{(\ell)}\}_1^{d_\ell}$ be sets of linearly independent functions on $[a,b]$ such that for $j = 0,1,\ldots,k$

$$\bigcup_{\ell=k-j+1}^{k+1} \{e_i^{(\ell)}\}_1^{d_\ell} \quad \text{spans} \quad N_L \cap \bigcap_{i=1}^{k-j} N_{\mathfrak{M}_i} \cap U_1(0) ,$$

and

$$\bigcup_{\ell=0}^{k+1} \{e_i^{(\ell)}\}_1^{d_\ell} \quad \text{spans} \quad N_L .$$

By construction we notice that $\mathfrak{M}_\ell$ is $1-1$ on the span of $\{e_i^{(\ell)}\}_1^{d_\ell}$, $\ell = 1,2,\ldots,k+1$, where for convenience we define $\mathfrak{M}_{k+1} \equiv I$, the identity operator. Thus by the linear independence there exist sets $\{t_i^{(\ell)}\}_1^{d_\ell}$ with $a \leq t_1^{(\ell)} < \ldots < t_{d_\ell}^{(\ell)} \leq b$ such that

$$\det(\mathfrak{m}_\ell\, e_i^{(\ell)}(t_j^{(\ell)})) \neq 0, \quad \ell = 1, 2, \ldots, k+1 \ .$$

We note that $d_0 \leq n$, and consider the matrix $\{\lambda_i e_j^{(0)}\}_{i=1, j=1}^{n \quad d_0}$ Its rank is d_0; for otherwise, some linear combination of the $\{e_i^{(0)}\}_1^{d_0}$ will lie in $U_1(0)$, which is impossible by the construction of the e's. We can thus choose $\{\tilde{\lambda}_i\}_1^{d_0}$ from $\{\lambda_i\}_1^n$ so that

$$\det\{\tilde{\lambda}_i e_j^{(0)}\}_{i,j=1}^{d_0} \neq 0 \ .$$

We are now ready to define the norm for $\mathcal{H}_p^m$; namely

$$\|f\|_{\mathcal{H}_p^m} = \sum_{\ell=1}^{k+1} \sum_{j=1}^{d_\ell} |\mathfrak{m}_\ell f(t_j^{(\ell)})| + \sum_{i=1}^{d_0} |\tilde{\lambda}_i f| + \|Lf\|_{\mathcal{L}_p} \ .$$

The homogeneity and the triangle inequality are evident. If $\|f\|_{\mathcal{H}_p^m} = 0$, then $f \in N_L$, whence

$$f = \sum_{\ell=0}^{k+1} \sum_{j=1}^{d_\ell} \alpha_{\ell j} e_j^{(\ell)}(t) \ .$$

Now $\tilde{\lambda}_i f = \sum_{j=1}^{d_0} \alpha_{0j} \tilde{\lambda}_i e_j^{(0)} = 0$ for $i = 1, \ldots, d_0$, which implies $\alpha_{0j} = 0$, $j = 1, 2, \ldots, d_0$. Next applying $\mathfrak{m}_1$ we have

$$\mathfrak{m}_1 f(t_j^{(1)}) = \sum_{j=1}^{d_1} \alpha_{1j} \mathfrak{m}_1 e_j^{(1)}(t_j^{(1)}) = 0, \qquad j = 1, \ldots, d_1 \ ,$$

which implies $\alpha_{1j} = 0$, $j = 1, 2, \ldots, d_1$. Repeating this process step by step for $\mathfrak{m}_2, \ldots, \mathfrak{m}_{k+1}$ we deduce that $f \equiv 0$. Thus we have verified that the above expression is a norm for $\mathcal{H}_p^m$. That

$\mathcal{H}_p^m$ is complete in this norm is also easily checked.

Returning to our original objective, let $f_i \in U$ be pre-images of φ_i, i.e., $Lf_i = \varphi_i$. We can choose the f_i so that

$$f_i(t_j^{(k+1)}) = 0, \quad j = 1, 2, \ldots, d_{k+1} .$$

(For example, this can be accomplished by adding a linear combination of the $\{e_j^{(k+1)}\}_1^{d_{k+1}}$ which will not affect Lf_i or the fact that $f_i \in U$ since $e_j^{(k+1)} \in N_L \cap \bigcap_{i=1}^{k} N_{\mathfrak{M}_i} \cap U_1(0)$.)

The fact that $f_i \in U$ implies that for all i the values $\{\mathfrak{M}_\ell f_i(t_j^{(\ell)})\}_{j=1}^{d_\ell}$ are uniformly bounded for $\ell = 1, 2, \ldots, k$ as are the values $\{\tilde{\lambda}_j f_i\}_{j=1}^{d_0}$. Hence we may extract a subsequence such that each of these numbers converges to a finite value. Call the new subsequence f_i again, and notice that φ_i still converges on this new sequence in $\mathcal{L}_p$. Now by the definition of the $\mathcal{H}_p^m$-norm, the sequence f_i is Cauchy since φ_i is and so are the $\mathfrak{M}_\ell f_i(t_j^{(\ell)})$ and $\tilde{\lambda}_j f_i$. The fact that $\mathcal{H}_p^m$ is a Banach space implies that f_i converges to some $f \in \mathcal{H}_p^m$. We claim $f \in U$, since $\mathcal{H}_p^m$ convergence implies uniform convergence of f and its first $m-1$ derivatives so that U is closed. Finally we notice that $Lf = \varphi$, i.e. $\varphi \in LU$, since $\|Lf - \varphi\|_{\mathcal{L}_p} = \|Lf - Lf_i\|_{\mathcal{L}_p} + \|Lf_i - \varphi\|_{\mathcal{L}_p} \leq \|L\| \, \|f - f_i\|_{\mathcal{H}_p^m} + \|Lf_i - \varphi\|_{\mathcal{L}_p} \to 0$ as $i \to \infty$. This completes the proof that LU is closed.

The fact that $\mathcal{L}_p$ is uniformly convex for $1 < p < \infty$ guarantees that the element of LU closest to g in the p-norm is unique in $\mathcal{L}_p$, i.e., any two solutions of (3.7) differ by an element in N. The last assertion of the theorem is quite easy to prove. Indeed, if $N \cap U(s) \neq (0)$, then we may add any non-trivial element of $N \cap U(s)$ to s to obtain a second minimizing function in U. Conversely, if s_1 and s_2 are two solutions of (3.7), then $\Delta = s_1 - s_2 \in N \cap U(s_2)$, which implies $s_2 \equiv s_1$ whenever this set consists of the zero element only.

§A.2. A sufficiency theorem for optimal control. We quote here an extended version of Theorem 1 of Mangasarian [10]. We refer to §5 for the statement of the general optimal control problem to which the theorem applies.

Theorem A.1. Let $\phi(t,x,u)$ and each component of $g(t,x,u)$ and $h(t,x,u)$ be differentiable and convex in the variables (x,u) for $t \in [a,b]$, let each component of $p(x(a))$ and $q(x(b))$ be differentiable and convex in $x(a)$ and $x(b)$, respectively, and let $\theta(x(a),x(b))$ be differentiable and convex in $(x(a),x(b))$. If there exist vectors $\bar{u}(t), \bar{x}(t), \bar{v}(t), \bar{w}(t)$, $\bar{r}$, and $\bar{s}$ satisfying (5.1-5) with $\bar{x}(t), \bar{v}(t)$ continuous and $\bar{w}(t)$ integrable and such that

(A.1) $\nabla_x \phi(t,\bar{x},\bar{u}) + \nabla_x \bar{v}g(t,\bar{x},\bar{u}) + \nabla_x \bar{w}h(t,\bar{x},\bar{u}) + \dot{\bar{v}}(t) = 0$,

(A.2) $\nabla_u \phi(t,\bar{x},\bar{u}) + \nabla_u \bar{v}g(t,\bar{x},\bar{u}) + \nabla_u \bar{w}h(t,\bar{x},\bar{u}) = 0$,

(A.3) $\nabla_{x(a)} \theta(\bar{x}(a),\bar{x}(b)) + \nabla_{x(a)} \bar{r}p(\bar{x}(a)) + \bar{v}(a) = 0$,

(A.4) $\nabla_{x(b)} \theta(\bar{x}(a),\bar{x}(b)) + \nabla_{x(b)} \bar{s}q(\bar{x}(b)) - \bar{v}(b) = 0$,

(A.5) $\bar{r}p(\bar{x}(a)) = 0, \quad \bar{r} \geq 0$,

(A.6) $\bar{s}q(\bar{x}(b)) = 0, \quad \bar{s} \geq 0$,

(A.7) $\bar{w}(t)h(t,\bar{x}(t),\bar{u}(t)) = 0, \quad \bar{w}(t) \geq 0$,

(A.8) $\bar{v}_i(t) \geq 0$ when g_i is nonlinear in x or u ,

then $\bar{u}(t), \bar{x}(t)$ will minimize the functional (5.1) subject to the conditions (5.2-5). Moreover,

(A.9) If h is a function of t and x alone, then the adjoint variable $\bar{v}(t)$ is permitted to have a finite number of finite jump discontinuities at any point $\hat{t}$ in $[a,b]$ provided that the jump in $\bar{v}$ at $\hat{t}$ satisfies

$$[\bar{v}(\hat{t})] = -\nabla_x \xi h(\hat{t},\bar{x}(\hat{t}))$$

$$\xi h(\hat{t},\bar{x}(\hat{t})) = 0, \quad \xi \geq 0$$

where ξ is a vector and the notation [ ] is defined in the footnote to (7.7).

(A.10) If $x_i(a)$ does not appear in θ and p, then $\bar{v}_i(a) = 0$
If $x_i(b)$ does not appear in θ and q, then $\bar{v}_i(b) = 0$

(A.11) If $x_i(a)$ is fixed, then delete the i-th equation from (A.3) and $\bar{r}_i$ from (A.5).
If $x_i(b)$ is fixed, then delete the i-th equation from (A.4) and $\bar{s}_i$ from (A.6).

(A.12) If at some $\hat{t} \in (a,b)$ we impose an additional convex inequality constraint on (5.1-5), say $\sigma(x(\hat{t})) \leq 0$ where σ has ν components, then the adjoint variable $\bar{\bar{v}}(t)$ is permitted to have a jump at $\hat{t}$ satisfying

$$[\bar{v}(\hat{t})] = -\nabla_x \eta\sigma(\bar{x}(\hat{t}))$$

$$\eta\sigma(\bar{x}(\hat{t})) = 0, \quad \eta \geq 0 ,$$

where $\eta \in \mathbb{R}^{\nu}$.

For convenience, we note that $\bar{v}(t)$ is usually called the adjoint variable, (A.1) the adjoint equation, (A.2) the maximum (or minimum principle), and (A.3-6) the transversality conditions.

§A.3. <u>A necessity theorem for optimal control</u>. The following theorem is due to Neustadt [11], and was obtained as a special case of very general variational results. Following Neustadt, we now repose problem (3.7). We are forced to give up some generality, however.

Let there be given a compact interval $I = [a,b]$, an

open set $G \subset R^n$, an arbitrary set $V \subset R^r$, a continuous function $g(\xi, v, t)$ from $G \times V \times I$ into R^n which is of class C^1 with respect to ξ, a scalar-valued function $h(\xi, t)$ which is defined on $R^n \times I$ and of class C^2 on $G \times I$, and scalar-valued functions $p_i(\xi_1, \xi_2)$, $i = 0, 1, \ldots, \tilde{m}$, which are defined on $R^n \times R^n$ and of class C^1 on $G \times G$. Let $\mathcal{U}$ denote the set of all measurable, essentially bounded functions u from I into V. Then the problem consists in finding a function x from I into G such that

(A.13) x is absolutely continuous on I,

(A.14) there is a function $u \in \mathcal{U}$ such that

$$\dot{x}(t) = g(x(t), u(t), t) \quad \text{for almost all } t \in I,$$

(A.15) $p_i(x(a), x(b)) = 0$ for $i = 1, \ldots, \tilde{m}$,

(A.16) $h(x(t), t) \leq 0$ for all $t \in I$,

and such that $p_0(x(a), x(b))$ achieves a minimum (subject to the preceding four constraints).

<u>Theorem A. 2.</u> Let $\bar{x}(t), \bar{u}(t)$ be a solution of the above problem. Then there exist numbers $\omega_0, \omega_1, \ldots, \omega_{\tilde{m}}$ and a scalar function λ defined on I such that

(A.17) λ is non-increasing on I, is right-continuous in (a, b), and is constant on subintervals of I on which $h(\bar{x}(t), t) < 0$,

(A.18) $$\lambda(b) = 0,$$

(A.19) $$\sum_{i=0}^{\tilde{m}} |\omega_i| + |\lambda(a)| > 0,$$

(A.20) $$\omega_0 \leq 0.$$

Moreover, there exists an absolutely continuous n-vector function $\psi(t)$ defined on I such that

(A.21) $\dot{\psi} = -\psi(t)\nabla_\xi g(\bar{x}(t),\bar{u}(t),t) + \lambda(t)\nabla_\xi \rho(\bar{x}(t),t)$ (a.e.) $t \in I$

where for $\xi \in G$, $t \in I$,

$$\rho(\xi,t) = (\nabla_\xi h(\xi,t))g(\xi,\bar{u}(t),t) + \dot{h}(\xi,t) \quad ,$$

(A.22)
$$\begin{cases} \psi(a) = -\sum_{i=0}^{\tilde{m}} \omega_i \nabla_{\xi_1} p_i(\bar{x}(a),\bar{x}(b)) + \lambda(a)\nabla_\xi h(\bar{x}(a),a) \\ \psi(b) = \sum_{i=0}^{\tilde{m}} \omega_i \nabla_{\xi_2} p_i(\bar{x}(a),\bar{x}(b)) \quad , \end{cases}$$

(A.23) $(\psi(t) - \lambda(t)\nabla_\xi h(\bar{x}(t),t))g(\bar{x}(t),\bar{u}(t),t)$

$$= \max_{u \in \mathcal{U}} (\psi(t) - \lambda(t)\nabla_\xi h(\bar{x}(t),t))g(\bar{x}(t),u,t) \quad \text{(a.e.)} \quad t \in I \; .$$

Here (A.21) is the adjoint equation, (A.22) the transversality conditions, and (A.23) the maximum principle.

Acknowledgement: We wish to thank J. W. Jerome for his intere[st] and helpful remarks.

REFERENCES

1. Anselone, P. M. and P. J. Laurent, A general method for the construction of interpolating or smoothing spline functions, MRC Technical Summary Report #834, Madison, Wisconsin (1967).

2. Atteia, Fonctions "spline" definies sur un ensemble convexe, Numerische Mathematik 12, 192-210, (1968).

3. Goldberg, S., Unbounded linear operators, McGraw-Hill, New York, 1966.

4. Golomb, M., Splines, n-widths, and optimal approximati[on], MRC Technical Summary Report #784, Madison, Wisconsin (

5. Golomb, M. and J. W. Jerome, Ordinary differential equations with boundary conditions on arbitrary sets, to appear.

6. Grimmell, W. C., The existence of piecewise continuous fuel optimal controls, SIAM J. Con. 5 (1967), 515-519.

7. Halkin, H., A generalization of LaSalle's bang-bang principle, SIAM J. Con. 2 (1965), 199-202.

8. Jerome, J. W. and L. L. Schumaker, On Lg-splines, J. Approx. Th, 2 (1969), 29-49.

9. Jerome, J. W. and L. L. Schumaker, Characterizations of functions with higher order derivatives in $\mathcal{L}_p$, to appear, Trans. A.M.S.

10. Mangasarian, O. L., Sufficient conditions for the optimal control of nonlinear systems, J. SIAM Control, 4 (1966), 139-152.

11. Neustadt, L. W., A general theory of extremals, J. of Computer and System Sciences, 3 (1969), 57-92.

12. Pontryagin, L. S., et al, The Mathematical Theory of Optimal Processes, Interscience, New York, 1962.

13. Reid, W. T., A simple optimal control problem involving approximation by monotone functions, J. Optimization Th., 2 (1968), 365-377.

14. Ritter, K., Generalized spline interpolation and nonlinear programming, this volume.

15. Sard, A., Optimal Approximation, J. Fun. Anal. 1, No. 2 (1967), 222-244.

O. L. MANGASARIAN AND L. L. SCHUMAKER

O. L. Mangasarian, Computer Sciences Department and Mathematics Research Center, University of Wisconsin, supported by U.S. Army DA-31-124-ARO-D-462 and NSF GJ-362 and L. L. Schumaker, University of Texas, partially supported by USAFOSR-69-1812.

On the Approximation by γ-Polynomials

CARL de BOOR

1. Introduction. As was first pointed out by Hobby and Rice [5], many nonlinear approximation problems — such as approximation by exponential sums or by splines with variable knots — admit of the following abstract description: One has given a real normed linear space X and a map

$$\gamma: T \to X$$

from some subset T of the real line $\mathbb{R}$ to X. One then defines a γ-polynomial of order n to be any element of X of the form

$$p = p(\alpha, \tau) = \sum_{i=1}^{n} \alpha_i \gamma(\tau_i) \quad ,$$

where $\alpha = (\alpha_1, \ldots, \alpha_n)$ is a real vector and $\tau = (\tau_1, \ldots, \tau_n) \subset T^n$ with $\tau_1 < \tau_2 < \ldots < \tau_n$. Let

$$\mathbb{P}_{\gamma, n}$$

denote the set of all γ-polynomials of order n. Then the approximation problem consists in finding, for given $f \in X$, a $p^* \in \mathbb{P}_{\gamma, n}$ such that

$$\|f - p^*\| = \operatorname{dist}(f, \mathbb{P}_{\gamma, n}) = \inf_{p \in \mathbb{P}_{\gamma, n}} \|f - p\| \quad .$$

As usual, p^* is called a best approximation (b.a.) to f in (or, by elements of) $\mathbb{P}_{\gamma,n}$.

To give some examples, let $X = L_p[0,1]$ and set $\gamma(t) = G(\cdot,t)$, where $G(s,t)$ is defined on $[0,1] \times T$. With G Green's function for a k-th order ordinary linear initial value problem on $(0,1]$ and $T = [0,1)$, one has approximation by generalized splines. With $G(s,t) = e^{st}$ and $T = \mathbb{R}$, one has approximation by exponential sums. With $G(s,t) = (1+st)^{-1}$ and $T = (-1,\infty)$, one has an approximation problem which shares many features with approximation by rational functions. The last two examples lend themselves easily to an extension of T to the complex plane, and reveal their essential properties only after such an extension has been made [5]. Other examples may be found in [7].

A seemingly different example occurs in Numerical Analysis. Here X is the topological dual Y^* of a normed linear space Y of functions defined on T, and, for $t \in T$, $\gamma(t)$ is the linear functional of point evaluation at t, i.e.,

$$\text{for all } y \in Y, \quad \gamma(t)y = y(t) .$$

Best approximation by γ-polynomials of order n to $f \in X$ amounts to the construction of a best approximate rule of the form $\sum_{i=1}^{n} a_i y(t_i)$ for the evaluation of the linear functional f at y.

But it is not difficult to see that many approximation problems by γ-polynomials of fixed order can be considered to be special cases of the last example. For, with X, γ and T given, let Y be the linear space of functions on T whose general element y is given by

$$y(t) = y_\lambda(t) = \lambda\gamma(t), \text{ all } t \in T,$$

for some $\lambda \in X^*$. If the linear span of $\{\gamma(t) \mid t \in T\}$ is dense in X, then the map

$$\varphi : X^* \to Y : \lambda \mapsto y_\lambda$$

is one-one, hence Y is normed by

$$\|y_\lambda\| \underset{\text{def}}{=} \|\lambda\|, \quad \text{all } y_\lambda \in Y ,$$

and X is mapped linearly and isometrically into Y^* by $(\varphi^*)^{-1}$. In particular, $\gamma(t)$ is mapped by $(\varphi^*)^{-1}$ to the linear functional of point evaluation at t, all $t \in T$.

Hobby and Rice [5] studied the problems of existence, uniqueness, and characterization of best approximations by γ-polynomials when X is an L_p-space on [0,1], for some p with $1 \leq p \leq \infty$. It is one purpose of this paper to give, in a more abstract setting, simpler proofs for some of their results.

2. <u>Existence of best approximations</u>. In most of the examples mentioned in the introduction, $\mathbb{P}_{\gamma,n}$ fails to be an existence set of for $n > 1$, since it fails to be closed. The reason for this is quite clear. If $t_1 \neq t_2$ are points in T, then $\mathbb{P}_{\gamma,n}$ contains the first divided difference

$$\gamma(t_1,t_2) = (t_2-t_1)^{-1} (\gamma(t_2) - \gamma(t_1))$$

of γ at the points t_1, t_2. Hence, if γ is strongly differentiable at t_1, then $\gamma^{(1)}(t_1) = \lim_{t_2 \to t_1} \gamma(t_1,t_2)$ is in the closure of $\mathbb{P}_{\gamma,n}$, but usually fails to be in $\mathbb{P}_{\gamma,n}$.

To get an existence set, one must at least adjoin to $\mathbb{P}_{\gamma,n}$ all strong limits of the form

$$\lim_{t_0,\ldots,t_k \to t} \gamma(t_0,\ldots,t_k) ,$$

where $\gamma(t_0,\ldots,t_k)$ denotes the k-th divided difference of γ at $t_0,\ldots,t_k$.

We denote by $C_X^{(k)}(T)$ the linear space of all functions on T to X which are k times continuously strongly differentiable on T. If $g \in C_X[a,b]$, where [a,b] is a finite interval, then

$$\omega_{[a,b]}(g, h) = \sup\{\|g(s) - g(t)\| \mid s,t \in [a,b], |s-t| \leq h\}$$

denotes the modulus of continuity of g on $[a,b]$.

(1) Lemma. Let $k > 0$, $[a,b]$ a finite interval, and $g \in C_X^{(k)}[a,b]$. If

$$a \le t_0 < t_1 < \dots < t_k \le b$$

and $\hat{t} \in [t_0, t_k]$, then

$$\|k!\, g(t_0, \dots, t_k) - g^{(k)}(\hat{t})\| \le \omega_{[a,b]}(g^{(k)}, t_k - t_0) .$$

Hence

$$\lim_{t_0,\dots,t_k \to \hat{t}} g(t_0,\dots,t_k) = g^{(k)}(\hat{t})/k! .$$

Proof. By Taylor's theorem with integral remainder (cf., e.g., Graves [3]), one has

$$g(t) = \sum_{i=0}^{k-1} [(t-a)^i/i!]g^{(i)}(a) + \int_a^b M_k(t;s)g^{(k)}(s)\,ds ,$$

where

$$M_k(t;s) \equiv (t-s)_+^{k-1}/(k-1)! .$$

Hence

$$g(t_0,\dots,t_k) = \int_a^b M_k(t_0,\dots,t_k;s)g^{(k)}(s)\,ds .$$

Since

$$M_k(t_0,\dots,t_k;s) \ge 0 \text{ with equality iff } s \notin (t_0,t_k) ,$$

and

$$\int_a^b M_k(t_0,\ldots,t_k;s) = \int_{t_0}^{t_k} M_k(t_0,\ldots,t_k;s)\,ds = 1/k! \quad ,$$

(cf., e.g., [1]), one has

$$\|k!g(t_0,\ldots,t_k) - g^{(k)}(\hat{t})\| = \|\int_{t_0}^{t_k} k!\, M_k(t_0,\ldots,t_k;s)[g^{(k)}(s)$$

$$- g^{(k)}(\hat{t})]\,ds \leq \int_{t_0}^{t_k} k!\; M_k(t_0,\ldots,t_k;s)\|g^{(k)}(s) - g^{(k)}(\hat{t})\|\,ds$$

$$\leq \max_{t_0 \leq s \leq t_k} \|g^{(k)}(s) - g^{(k)}(\hat{t})\| k! \int_{t_0}^{t_k} M_k(t_0,\ldots,t_k;s)\,ds$$

$$\leq \omega_{[a,b]}(g^{(k)},\ t_k - t_0)\ ; \qquad \text{q.e.d.}$$

If $\gamma \in C_X^{(k-1)}(T)$, then we define a k-<u>extended</u> γ-<u>polynomial of order</u> n to be any element of X of the form

$$p = p(\alpha,\tau) = \sum_{i=1}^{r} \sum_{j=0}^{m_i} \alpha_{ij}\, \gamma^{(j)}(t_i) \quad , \tag{2}$$

with

$$m_i + 1 \leq k,\quad t_i \in T,\quad i = 1,\ldots,r,$$

$$t_1 < t_2 < \ldots < t_r,\ \sum_{i=1}^{r} (m_i + 1) = n\ .$$

Further, we take the τ-vector for p in (2) to be the vector

$$\tau = (t_1, \ldots, t_1, t_2, \ldots, t_2, t_3, \ldots, t_r) ,$$

with t_i appearing $m_i + 1$ times, $i = 1, \ldots, r$. Denote by

$$\mathbb{P}^k_{\gamma, n}$$

the set of all k-extended γ-polynomials of order n.

Remark. This rather narrow definition of extended γ-polynomial suffices for this paper. But one may want to enlarge it at times to include also weak limits of $\gamma(t_0, \ldots, t_k)$ as $t_0, \ldots, t_k \to t$. Again, the strong continuity of $\gamma^{(k-1)}$ is not essential, nor does it seem necessary to demand that $\gamma^{(k-1)}$ exist on all of T.

(3) Theorem. If (i) $T = [a,b]$ is a finite interval, (ii) $\gamma \in C_X^{(n-1)}(T)$ and (iii) the set

$$\{\gamma^{(j)}(t_i) \mid j = 0, \ldots, m_i;\ i = 1, \ldots, r\}$$

is linearly independent whenever $a \leq t_1 < t_2 < \ldots < t_r \leq b$ and $\sum_{i=1}^{r} (m_i + 1) = n$, then $\mathbb{P}^n_{\gamma, n}$ is the strong closure, $\overline{\mathbb{P}}_{\gamma, n}$, of $\mathbb{P}_{\gamma, n}$. Further, $\mathbb{P}_{\gamma, n}$ is boundedly compact in $\mathbb{P}^n_{\gamma, n}$; hence $\mathbb{P}^n_{\gamma, n}$ is an existence set.

Proof. Let $\{p_m\}_{m=1}^{\infty}$ be a bounded sequence in $\mathbb{P}_{\gamma, n}$. Since T is compact, we may assume (after going to a subsequence, if necessary) that the sequence $\{\tau^{(m)}\}$ of corresponding τ-vectors converges to some $\tau \in T^n$. Hence, we can write p_m as

$$(4) \qquad p_m = \sum_{i=1}^{r} \sum_{j=0}^{m_i} \alpha_{ij}^{(m)} j!\gamma(t_{i0}^{(m)}, \ldots, t_{ij}^{(m)}), \qquad m = 1, 2, \ldots ,$$

where

$$\lim_{m \to \infty} t_{ij}^{(m)} = t_i, \quad j = 0, \ldots, m_i;\ i = 1, \ldots, r ,$$

with $a \le t_1 < \ldots < t_r \le b$ and $\sum_{i=1}^{r} (m_i + 1) = n$. Since $\gamma \in C_X^{(n-1)}$ and $m_i + 1 \le n$, all i, it follows from Lemma (1) that

$$(5) \quad \lim_{m\to\infty} j!\, \gamma(t_{i0}^{(m)}, \ldots, t_{ij}^{(m)}) = \gamma^{(j)}(t_k), \quad j = 0, \ldots, m_i ;$$

$$i = 1, \ldots, r ,$$

in norm. By assumption (iii), the set $\{\gamma^{(j)}(t_i) \mid j = 0, \ldots, m_i;\ i = 1, \ldots, r\}$ is linearly independent, hence there exists $K > 0$ and m_0 such that $m \ge m_0$ implies

$$\|p_m\| \ge K \max_{i,j} |\alpha_{ij}^{(m)}| .$$

Since $\{p_m\}$ is, by assumption, bounded, this implies that each of the n sequences $\{\alpha_{ij}^{(m)}\}_{m=1}^{\infty}$ is bounded. Hence, after going to a subsequence if necessary, we may assume that

$$\lim_{m\to\infty} \alpha_{ij}^{(m)} = \alpha_{ij}, \quad j = 0, \ldots, m_i;\ i = 1, \ldots, r .$$

But then

$$\lim_{m\to\infty} p_m = \sum_{i=1}^{r} \sum_{j=0}^{m_i} \alpha_{ij}\, \gamma^{(j)}(t_i) \in \mathbb{P}_{\gamma,n}^{n} .$$

This proves that $\mathbb{P}_{\gamma,n}$ is boundedly compact in $\mathbb{P}_{\gamma,n}^{n}$, hence, that $\overline{\mathbb{P}}_{\gamma,n} \subset \mathbb{P}_{\gamma,n}^{n}$. As to the converse containment, observe that, by Lemma (1), for $j \le n-1$ and $t \in T$,

$$\|\gamma^{(j)}(t) - j!\, \gamma(t, t+h, \ldots, t+jh)\| \le \omega_{[a,b]}(\gamma^{(j)}, jh) ,$$

so that certainly $\mathbb{P}_{\gamma,n}^{n} \subset \overline{\mathbb{P}}_{\gamma,n}$. q. e. d.

Theorem (3) by itself has little applicability, since in practice either assumption (i) or assumption (ii) fails. E.g., in the case of approximation by exponential sums in $L_p[0,1]$, T is the whole real line and (i) fails, while assumption (ii) is certainly unjustified when approximating by splines of fixed order with a large enough number of variable knots.

But one can extend the argument for Theorem (3) to include these and other examples in the following way. Suppose that $f \in X$ is to be approximated by elements in $\mathbb{P}_{\gamma,n}$, and let $\{p_m\}$ be a minimizing sequence for f in $\mathbb{P}_{\gamma,n}$. Now write

$$p_m = q_m + r_m, \quad m = 1, 2, \ldots \ ,$$

where the "nice" part, q_m, involves only those $\tau_i^{(m)}$ which converge to certain $t_i \in T$ and have, in the limit, no more than k coincident, where γ is known to be in $C_X^{(k-1)}(T)$. It can often be shown that the remainder sequence $\{r_m\}$ becomes eventually "orthogonal" to every $x \in X$ in the sense that

$$\text{(6)} \qquad \text{for all } x \in X, \ \underline{\lim}_{m \to \infty} \|r_m + x\| \geq \|x\| \ .$$

This can be shown to imply, together with the boundedness of $\{p_m\}$, that $\{q_m\}$ is bounded, hence Theorem (3) then implies that some subsequence of $\{q_m\}$ converges to some $q \in \mathbb{P}^k_{\gamma,n}$. Further, (6) then implies that this subsequence is a minimizing sequence for f in $\mathbb{P}_{\gamma,n}$, since $\{p_m\}$ is, thus showing q to be a b.a. to f in $\mathbb{P}^k_{\gamma,n}$.

As a preliminary for a rigorous argument along the lines just indicated, we investigate in the next section the "limit" concept suggested by (6).

3. Weak convergence concepts and existence sets. Let S be a subset of the normed linear space X. To show that $f \in X$ has a b.a. in S, one usually proceeds as follows. One picks a minimizing sequence $\{p_m\}$ in S for f, and then

attempts to show that some subsequence $\{p_{j(m)}\}$ of $\{p_m\}$ converges to some element p of S in some sense. The weaker the convergence concept used, the easier it should be to establish the compactness of $\{p_m\}$ in S. On the other hand, the convergence concept should be strong enough to imply that

$$(1) \qquad \|f - p\| \leq \lim_{m\to\infty} \|f - p_{j(m)}\| = \operatorname{dist}(f, S) \quad ,$$

which then would finish the argument showing p to be a b.a. to f in S.

To give an example, one might use the following convergence concept: The sequence $\{x_m\}$ in X "converges" to x iff

$$\text{for all } y \in X, \quad \overline{\lim_{m\to\infty}} \|x_m - y\| \geq \|x - y\| \quad .$$

Clearly, if some subsequence of the minimizing sequence $\{p_m\}$ for f in S "converges" to some $p \in S$, then p is a b.a. to f in S. But since such "convergence" is not even preserved when going to a subsequence, we prefer the following slightly stronger notion.

(2) Definition. The sequence $\{x_m\}$ in the normed linear space X comes close to $x \in X$ iff

$$\text{for all } y \in X, \quad \underline{\lim}_{m\to\infty} \|x_m - y\| \geq \|x - y\| \;.$$

A subset S of X is nearly compact iff every sequence in S has a subsequence which "comes close to" some element of S.

With this, the preceding discussion establishes

(3) Theorem. Let S be a subset of the normed linear space X. If bounded subsets of S are nearly compact, then S is an existence set.

It seems worthwhile to point out by an example that bounded existence sets need not be nearly compact. For this, let $X = c_0 = \{f: \mathbb{N} \to \mathbb{R} \mid \lim_{n\to\infty} f(n) = 0\}$ be the Banach space of real null sequences with norm

$$\|f\| = \sup_{n \in \mathbb{N}} |f(n)| ,$$

and set

$$S = \{f_n \mid n = 1, 2, \ldots\} ,$$

where

$$f_n(m) = \begin{cases} 1, & m \leq n \\ 0, & m > n, \end{cases} \qquad n, m = 1, 2, \ldots \quad .$$

Then S is an existence set: If $f \in X$, then there is n_0 such that $n \geq n_0$ implies $|f(n)| < 1/2$. But then, for all $n \geq n_0$,

$$|(f - f_n)(m)| = |(f - f_{n-1})(m)|, \quad \text{all } m \neq n ,$$

while

$$|(f - f_n)(n)| > |f(n)| = |(f - f_{n-1})(n)| .$$

Hence

$$\text{for all } n \geq n_0, \quad \|f - f_n\| \geq \|f - f_{n-1}\| ,$$

therefore,

$$\operatorname{dist}(f, S) = \min_{n \leq n_0} \|f - f_n\| .$$

Further, S is bounded. But, S is not nearly compact. For, if $f \in X$, then there exists n_0 such that $|f(n_0)| < 1$. Set

$$g(n) = 2\delta_{n, n_0}, \qquad n = 1, 2, \ldots \quad .$$

Then $g \in X$ and

$$\|f-g\| \geq |(f-g)(n_0)| > 1 = \overline{\lim_{n\to\infty}} \|f_n - g\| .$$

Hence, every subsequence of the sequence $\{f_n\}$ in S "comes close to" no $f \in X$, let alone an $f \in S$.

In the remainder of this section, we make some simple remarks, and prove two technical lemmata concerning sequences which "come close to" some element, and, finally, prove an existence theorem.

(4) Remarks. (i) If $\{x_m\}$ "comes close to" x, then so does every subsequence of $\{x_m\}$.

(ii) But, a sequence may "come close to" more than one element. Thus, the sequence $\{x_m\}$ in $L_1[0,1]$ given by

$$x_m(t) = \begin{cases} m, & 0 \leq t \leq 1/m , \\ 0, & 1/m < t \leq 1 , \quad m = 1,2,\dots , \end{cases}$$

"comes close to" every $y \in L_1[0,1]$ with $\|y\|_1 \leq 1$.

(iii) If $\{x_m\}$ "comes close to" x and $\{y_m\}$ converges in norm to y, then $\{x_m + y_m\}$ "comes close to" $x+y$.

(iv) If $\{x_m\}$ "comes close to" x and is bounded, and the sequence $\{\alpha_m\}$ of scalars converges to α, then $\{\alpha_m x_m\}$ "comes close to" αx.

(v) If $\{x_m\}$ converges in norm to x, then $\{x_m\}$ "comes close to" x and to no other element of X. Hence, if $\{x_m\}$ "comes close to" x, then all strongly convergent subsequences of $\{x_m\}$ converge to x.

(vi) If $\{x_m\}$ converges weakly to x, then $\{x_m\}$ "comes close to" x. This is just a restatement of the fact that the norm is lower semicontinuous with respect to weak sequential convergence.

A slight but important generalization of (4) (vi) concerns convergence with respect to a family of seminorms.

(5) Definition. Let X be a linear space, and Φ a

family of seminorms on X. The sequence $\{x_m\}$ in X converges Φ to $x \in X$ iff

$$\text{for all } \varphi \in \Phi, \quad \lim_{m\to\infty} \varphi(x-x_m) = 0 .$$

(6) Lemma. Let X be a normed linear space, let Φ be a family of seminorms on X with the property that

$$(7) \qquad \text{for all } x \in X, \quad \sup_{\varphi \in \Phi} \varphi x = \|x\| .$$

If the sequence $\{x_m\}$ in X converges Φ to x, then $\{x_m\}$ "comes close to" x.

Proof. Since vector addition is continuous with respect to Φ-convergence, it is sufficient to prove that

$$\underline{\lim}_{m\to\infty} \|x_m\| \geq \|x\| ,$$

whenever $\{x_m\}$ converges Φ to x. For this, observe that $\varphi \in \Phi$ and $\lim_{m\to\infty} \varphi(x-x_m) = 0$ implies

$$\varphi x = \lim_{m\to\infty} \varphi x_m .$$

By (7), $\varphi x_m \leq \|x_m\|$, therefore,

$$\varphi x = \lim_{m\to\infty} \varphi x_m = \underline{\lim}_{m\to\infty} \|x_m\| ,$$

hence, again by (7),

$$\|x\| = \sup_{\varphi \in \Phi} \varphi x \leq \underline{\lim}_{m\to\infty} \|x_m\| ;$$

q. e. d.

(8) Lemma. Let X be a normed linear space, and let $\{x_m\}$ be a bounded sequence in X which "comes close to"

a certain $x \in X$; if $x = 0$, assume in addition that $\underline{\lim} \|x_m\| > 0$. Further, let $\{y_m\}$ be a sequence in X converging in norm to some $y \in X$ which does not depend linearly on x. If the sequence $\{\alpha_m y_m + \beta_m x_m\}$ is bounded, then so is the sequence $\{\alpha_m y_m\}$.

Proof. Since $\{y_m\}$ converges in norm, it is bounded. Hence, if $\{\alpha_m y_m\}$ is not bounded, then, (after going to a subsequence if necessary) one has

$$\lim_{m\to\infty} |\alpha_m| = \infty .$$

By assumption, there exists C such that

$$\text{for all } m, \quad \|\alpha_m x_m + \beta_m y_m\| \le C .$$

Hence,

$$\|y_m + \frac{\beta_m}{\alpha_m} x_m\| \le C/|\alpha_m| \underset{m\to\infty}{\longrightarrow} 0 ,$$

showing that $\{\frac{\beta_m}{\alpha_m} x_m\}$ converges in norm to $-y$. In particular, $\{\frac{\beta_m}{\alpha_m} x_m\}$ is bounded, hence, as $\underline{\lim} \|x_m\| > 0$ by assumption, $\{\beta_m/\alpha_m\}$ is bounded, therefore, - after going to a subsequence, if necessary, - we may assume that

$$\lim_{m\to\infty} \beta_m/\alpha_m = \alpha .$$

But then, by (4) (iv) above, $\{(\beta_m/\alpha_m)x_m\}$ "comes close to" αx, hence, with (4) (v) above,

$$-y = \alpha x ,$$

contradicting the assumption that y does not depend linearly on x. q. e. d.

(9) Definition. If $\{x_m\}$ is a sequence in the normed linear space X, then the normalization of $\{x_m\}$ is the sequence $\{\hat{x}_m\}$, given by

$$\hat{x}_m = \begin{cases} x_m / \|x_m\|, & x_m \neq 0 \\ 0 & , x_m = 0 \end{cases}, \qquad m = 1, 2, \ldots .$$

(10) Theorem. Let X be a normed linear space, and let S, $\hat{S}$ be nonempty subsets of X, closed under scalar multiplication, with the properties: (i) $\hat{S} \subset \bar{S}$, the (norm) closure of S; (ii) every sequence in S can be written as the sum of two sequences, $\{q_m\}$ and $\{r_m\}$, such that the normalization $\{\hat{q}_m\}$ of $\{q_m\}$ is compact in $\hat{S}$, and some subsequence of the normalization $\{\hat{r}_m\}$ of $\{r_m\}$ "comes close to" zero. Then, $\hat{S}$ is an existence set.

Proof. Let $f \in X$, and let $\{p_m\}$ be a minimizing sequence for f in $\bar{S}$. We may assume that $\{p_m\}$ is in S. After going to a subsequence, if necessary, we may assume that $\{p_m\}$ is the sum of two sequences $\{q_m\}$ and $\{r_m\}$, such that the normalization $\{\hat{q}_m\}$ of $\{q_m\}$ converges in norm to some $\hat{q} \in \hat{S}$, while the normalization $\{\hat{r}_m\}$ of $\{r_m\}$ "comes close to" zero. Also,

$$\{p_m\} = \{\|q_m\|\hat{q}_m + \|r_m\|\hat{r}_m\}$$

is bounded.

We begin by proving that some subsequence of $\{q_m\}$ converges to an element of $\hat{S}$. Since $\{\hat{q}_m\}$ converges to $\hat{q} \in \hat{S}$, and $\hat{S}$ is closed under scalar multiplication, it is sufficient to show that some subsequence of $\{q_m\}$ is bounded. This, in turn, is trivial in case $\underline{\lim}\,\|q_m\| = 0$. It is also trivial in case $\underline{\lim}\,\|r_m\| = 0$, since $\{q_m + r_m\}$ is bounded by assumption. Otherwise, $\|\hat{q}\| = \lim \|\hat{q}_m\| = 1$, hence $\hat{q} \neq 0$, and $\underline{\lim}\,\|\hat{r}_m\| = 1 > 0$. But then, the boundedness of $\{q_m\}$ follows from (8).

With this, we may assume, after going to a subsequence, if necessary, that $\{q_m\}$ converges in norm to some $q \in \hat{S}$.

Next, we show that

$$\text{for all } x \in X, \ \overline{\lim} \|r_m - x\| \geq \|x\| .$$

This is trivially true in case $\lim \|r_m\| = \infty$. Otherwise, some subsequence $\{\|r_{j(m)}\|\}$ of $\{\|r_m\|\}$ converges to some scalar α. But then, by (4) (i) and (4) (iv), $\{r_{j(m)}\} = \{\|r_{j(m)}\|\hat{r}_{j(m)}\}$ "comes close to" $\alpha \cdot 0 = 0$, since the bounded sequence $\{\hat{r}_m\}$ "comes close to" 0, by assumption. Hence,

$$\text{for all } x \in X, \ \overline{\lim_{m\to\infty}} \|r_m - x\| \geq \lim_{m\to\infty} \|r_{j(m)} - x\| \geq \|x\| .$$

But this implies that $q \in \hat{S} \subset \bar{S}$ is a b.a. to f in $\bar{S}$. For, one has

$$\begin{aligned} \operatorname{dist}(f, \bar{S}) = \lim \|p_m - f\| &\geq \overline{\lim} (\|r_m + q - f\| - \|q_m - q\|) \\ &= \overline{\lim} \|r_m + q - f\| \\ &\geq \|q - f\| . \end{aligned}$$

Specifically, with $f \in \bar{S}$, it follows that $f = q$ for some $q \in \hat{S}$, hence, as $\hat{S} \subset \bar{S}$, $\hat{S} = \bar{S}$ follows. q.e.d.

To give a simple example, consider best approximation by exponential sums in $X = C[0,1]$. Here

$$\gamma(t) = e^{ts}, \quad t \in T = (-\infty, \infty) .$$

Since T is not compact, Theorem 2. (3) is not directly applicable. But one verifies that $\mathbb{P}^n_{\gamma,n}$ is in this case an existence set, and $\overline{\mathbb{P}}_{\gamma,n} = \mathbb{P}^n_{\gamma,n}$, by verifying that the assumptions of Theorem (10) are satisfied:

Set $S = \mathbb{P}_{\gamma,n}$, $\hat{S} = \mathbb{P}^n_{\gamma,n}$. Since γ is infinitely often strongly differentiable, $\hat{S} \subset \bar{S}$, by Lemma 2. (1). Further, $\{\gamma^{(j)}(t_i) \mid j = 0,\dots,m_i;\ i = 1,\dots,k\}$ is a linearly independent set whenever $t_1 < t_2 < \dots < t_k$, and for arbitrary integers $m_1,\dots,m_k$. If now $\{p_m\}$ is a bounded sequence in $S = \mathbb{P}_{\gamma,n}$, then after going to a subsequence if necessary, we can write

$$p_m = q_m + r_m, \quad q_m = \sum_{i=1}^{r} a_i^{(m)} \gamma(t_i^{(m)}), \quad r_m = \sum_{i=r+1}^{n} a_i^{(m)} \gamma(t_i^{(m)}),$$

$$m = 1, 2, \ldots,$$

where

$$\lim_{m\to\infty} t_i^{(m)} = \begin{cases} t_i \in T, & i \le r \\ \pm\infty, & i > r. \end{cases}$$

By Theorem 2.(3), some subsequence of the normalization $\{\hat{q}_m\}$ of $\{q_m\}$ converges strongly to some $q \in \hat{S} = \mathbb{P}^n_{\gamma,n}$. Further, it can be shown [8] that some subsequence of the normalization $\{\hat{r}_m\}$ of $\{r_m\}$ converges pointwise to zero on $(0,1)$. Since, for $f \in C[0,1]$, $\|f\|_\infty = \sup_{s \in (0,1)} |f(s)|$, this implies, as in (6), that such a subsequence "comes close to" zero.

Remark. Theorem (10) is needed to complete the proof of Theorem 4 in [5].

4. Existence of best generalized spline approximants. Let M be the linear differential operator defined by

(1) $$(Mx)(s) = x^{(k)}(s) + \sum_{i=0}^{k-1} a_i(s) x^{(i)}(s), \quad s \in [0,1],$$

with $a_i \in C^{(i)}[0,1]$, $i = 0, \ldots, k-1$. Let $G(s,t)$ be Green's function for the initial value problem

$$(Mx)(s) = g(s), \quad s \in (0,1]$$

$$x^{(j)}(0) = 0, \quad j = 0, \ldots, k-1,$$

and consider the curve

$$\gamma(t) = G(\cdot, t), \quad t \in T = [0,1),$$

in $X = L_p[0,1]$.

For $p < \infty$, γ is in $C_X^{(k-1)}(T)$, but not in $C_X^{(k)}(T)$, with

$$(3) \qquad \gamma^{(j)}(t) = (\partial/\partial t)^j \, G(\cdot, t), \quad j = 0, \ldots, k-1,$$

independently of p. Hence, $\mathbb{P}^k_{\gamma,n}$ does not change with p, and is a subset of $L_\infty[0,1]$. To emphasize this fact, we will denote this set of functions by $S^e_{M,n}$ in the sequel.

For $p = \infty$, γ is merely in $C_X^{(k-2)}(T)$, since $(\partial/\partial t)^{k-1} G(\cdot, t)$ has a jump discontinuity at t, hence can not be the uniform limit of the continuous function $\gamma^{(k-2)}(t, t+h)$ as $h \to 0$, except for $t = 0$. This fact produces complications in an existence proof, which will be dealt with elsewhere. Here, we will be satisfied with showing that $S^e_{M,n}$ is an existence set in L_∞.

We note that $\gamma^{(j)}(t)$ vanishes identically on $[0, t)$, and is k times continuously differentiable on $[t, 1]$. Hence

$$\{\gamma^{(j)}(t_i) \mid j = 0, \ldots, m_i; \; i = 1, \ldots, r\}$$

is a linearly independent set whenever $0 \leq t_1 < \ldots < t_r < 1$ and $m_i + 1 \leq k$, $i = 1, \ldots, r$. Each element $\sum_i \sum_j \alpha_{ij} \gamma^{(j)}(t_i)$ of $S^e_{M,n}$ reduces to a function in the k-dimensional null space, $\ker M$, of M on each of the intervals $(0, t_1), (t_1, t_2), \ldots, (t_r, 1)$. Hence, the elements of $S^e_{M,n}$ are generalized splines [4], and are (deficient) L-splines [9] only if $M = L^* L$ for some differential operator L of order $k/2$.

The following observations will be of use later on. Since $\ker M$ is finite-dimensional, all norms on $\ker M$ are equivalent, and bounded sets in $\ker M$ are compact in $\ker M$. Also, no element of $\ker M$ other than zero vanishes identically on any subinterval of $[0,1]$ of positive length. This implies

(4) Lemma. Let $0 \leq a < b \leq 1$, and let $\{u_m\}$ be a sequence in $\ker M$ which is bounded in $L_1[a,b]$. Then $\{u_m\}$ is uniformly bounded in $[0,1]$, hence some subsequence of $\{u_m\}$ converges to some $u \in \ker M$ uniformly on $[0,1]$.

(5) Definition. With $0 \le t_1 < t_2 < \ldots < t_r \le 1$, and $1 \le p \le \infty$, let $\Phi_p(t_1, \ldots, t_r)$ denote the collection $\{\varphi_\partial \mid \partial > 0\}$ of seminorms on $L_p(0,1)$, with φ_∂ defined by

$$\varphi_\partial f = \|f \cdot \chi_\partial\|_p \quad \text{all } f \in L_p ,$$

where

$$\chi_\partial(s) = \begin{cases} 1, & \text{for all } i, \quad |s - t_i| \ge \partial \\ 0, & \text{otherwise.} \end{cases}$$

(6) One observes that

$$\sup\{\varphi f \mid \varphi \in \Phi_p(t_1, \ldots, t_r)\} = \lim_{\partial \to 0} \|f \cdot \chi_\partial\|_p = \|f\|_p ,$$

for all $f \in L_p[0,1]$. Hence, if $\{f_m\} \subset L_p[0,1]$ converges $\Phi_p(t_1, \ldots, t_r)$ to some $f \in L_p[0,1]$, (cf. 3.(5), 3.(6)), then $\{f_m\}$ "comes close to" f in $L_q[0,1]$, all $q \le p$.

For $p < \infty$, and $f \in L_p[0,1]$, one has

(7)
$$\lim_{\partial \to \infty} \|f \cdot (1 - \chi_\partial)\|_p = 0 .$$

This implies

(8) Lemma. Let $p < \infty$, $\{f_m\} \subset L_p[0,1]$ converging $\Phi_p(t_1, \ldots, t_r)$ to some $f \in L_p[0,1]$. If $\lim_{m \to \infty} \|f_m\|_p = \|f\|_p$, then $\{f_m\}$ converges L_p to f.

Proof. One has

$$\|f_m\|_p^p = \|f_m \chi_\partial\|_p^p + \|f_m(1 - \chi_\partial)\|_p^p$$

$$\ge \big|\, \|f\chi_\partial\|_p - \|(f_m - f)\chi_\partial\|_p \big|^p + \big|\, \|(f_m - f)(1 - \chi_\partial)\|_p - \|f(1 - \chi_\partial)\|_p \big|^p .$$

Let $\mu > 0$ be given. Then, by (7), there exists $\partial > 0$ such that

$$\|f \cdot (1 - \chi_\partial)\|_p < \eta \ ,$$

hence

$$\|f\chi_\partial\|_p \geq \|f\|_p - \eta \ .$$

For this ∂, there exists $\bar{m}$ such that $m \geq \bar{m}$ implies

$$\|(f_m - f)\chi_\partial\|_p \leq \eta \ ,$$

hence

$$\|(f_m - f)(1 - \chi_\partial)\|_p \geq \|f_m - f\|_p - \eta \ .$$

But then, for all $m \geq \bar{m}$,

$$\|f_m\|_p^p \geq |\ \|f\|_p - 2\eta|^p + |\ \|f - f_m\|_p - 2\eta|^p \ ,$$

therefore,

$$\|f\|_p^p = \lim_{m \to \infty} \|f_m\|_p^p \geq |\ \|f\|_p - 2\eta|^p + |\overline{\lim_{m \to \infty}}\ \|f - f_m\|_p - 2\eta|^p \ .$$

Since η is arbitrary, $\overline{\lim_{m \to \infty}}\ \|f - f_m\|_p = 0$ follows i q. e. d.

(9) Lemma. Let $0 \leq t_1 < t_2 < \ldots < t_r < 1$, let $\varepsilon > 0$ be small enough so that $t_r - 1 < t_r - \varepsilon$, $t_r + \varepsilon \leq 1$, and let $\{p_m\}$ be a sequence in $S^e_{M,n}$ which is bounded in $L_p[0, t_r + \varepsilon]$ for some $1 \leq p \leq \infty$, and whose corresponding sequence $\{\tau^{(m)}\}$ of τ-vectors converges to some vector

$$\tau = (t_1, \ldots, t_1, t_2, \ldots, t_2, t_3, \ldots, t_r) \ .$$

Then some subsequence of $\{p_m\}$ converges $\Phi_\infty(t_1, \ldots, t_r)$ to an element of $S^e_{M,n}$.

Proof by induction on r, it being vacuously true for $r = 0$. Assume $r > 0$ and assume the correctness of the statement for $r-1$. Let h be the number of components of τ which equal t_r. Then, after going to a subsequence if necessary, we may assume that

$$|\tau_i^{(m)} - t_r| \leq \varepsilon/2, \quad i = n-h+1, \ldots, n; \; m = 1, 2, \ldots .$$

For $m = 1, 2, \ldots,$ write

$$p_m = q_m + u_m ,$$

where u_m involves only the last h terms of p_m. Then

$$p_m(s) = q_m(s), \quad \text{all } s \in [0, t_r - \varepsilon], \quad m = 1, 2, \ldots ,$$

since $\gamma^{(j)}(t)$ vanishes on $[0, t)$. It follows that $\{q_m\}$ is bounded in $L_p[0, t_r - \varepsilon]$, and is in $S^e_{M, n-h}$, hence, by induction hypothesis, we may assume (after going to a subsequence if necessary) that $\{q_m\}$ converges $\Phi_\infty(t_1, \ldots, t_{r-1})$ to some element q of $S^e_{M, n-h}$. This implies that for some constant c, and all large enough m,

$$|q_m(s)| \leq c, \quad \text{all } s \in [t_r - \varepsilon, 1] ,$$

hence, $\{u_m\}$ is bounded in $L_p[0, t_r + \varepsilon]$.

For $m = 1, 2, \ldots,$ let u_m^+ be the element of ker M for which

$$u_m^+(s) = u_m(s), \text{ all } s \in [t_r + \varepsilon/2, 1] .$$

By Lemma (4) and the boundedness of $\{u_m\}$ in $L_p[0, t_r + \varepsilon]$, we may assume, after going to a subsequence if necessary, that $\{u_m^+\}$ converges uniformly on $[0, 1]$ to an element $u^+ \in \ker M$. Set

$$u(s) = \begin{cases} 0, & s < t_r \\ u^+(s), & s \geq t_r \end{cases}.$$

Then $u \in S^e_{M,k}$, and $\{u_m\}$ converges $\Phi_\infty(t_r)$ to u. Hence, $\{p_m\}$ converges $\Phi_\infty(t_1,\ldots,t_r)$ to $q+u$, and $q+u \in S^e_{M,n-h+k}$. If $h \geq k$, we are done. Otherwise, use Theorem 2.(3) together with the fact that γ is $(k-1)$-times continuously differentiable in L_p and $\{\gamma^{(j)}(t_r)\}_{j=0}^{k-1}$ is linearly independent to conclude from the boundedness of $\{u_m\}$ in $L_p[0, t_r + \varepsilon]$ that some subsequence of $\{u_m\}$ converges L_p to an element $\hat{u}$ of $\mathbb{P}^h_{\gamma,h}$. Since $\{u_m\}$ converges $\Phi_\infty(t_r)$ to u, it follows then by 3.(4)(v), Lemma 3.(6), and by (6) above, that $u = \hat{u}$, hence $p = q+u \in S^e_{M,n}$; q.e.d.

(10) Theorem. Let $S^e_{M,n} = \mathbb{P}^k_{\gamma,n}$ be the set of k-extended γ-polynomials of order n in $L_1[0,1]$, with γ given by (2). Then $S^e_{M,n}$ is an existence set in $L_p[0,1]$, $1 \leq p \leq \infty$. For $p < \infty$, $S^e_{M,n}$ is approximatively compact, and is the strong closure of $\mathbb{P}_{\gamma,n}$.

Proof. Let $f \in L_p[0,1]$, and let $\{p_m\}$ be a minimizing sequence for f in $S^e_{M,n}$. If $\{\tau^{(m)}\}$ is the corresponding sequence of τ-vectors, then, after going to a subsequence if necessary, we may assume that $\{\tau^{(m)}\}$ converges to some $\tau \in [0,1]^n$. Let $t_1 < t_2 < \ldots < t_r$ be the distinct ones among the components of τ. Then $0 \leq t_1 < t_2 < \ldots < t_r \leq 1$. Since $\{p_m\}$ is bounded in $L_p[0,1]$, we may assume (after going to a subsequence if necessary) that $\{p_m\}$ converges $\Phi_\infty(t_1,\ldots,t_r)$ to some element $\hat{p} \in S^e_{M,n}$: For, if $t_r < 1$, this follows directly from Lemma (9). If $t_r = 1$, then those terms of p_m which involve $\tau_i^{(m)}$ with $\lim_{m\to\infty} \tau_i^{(m)} = t_r$ converge trivially $\Phi_\infty(t_r)$ to zero, hence using Lemma (9) for the sequence of remaining terms, onereaches the same conclusion in this case.

By (6), it then follows that $\{p^{(m)}\}$ "comes close to" $\hat{p}$ in $L_p[0,1]$, hence

$$\operatorname{dist}(f, S^e_{M,n}) = \lim \|p_m - f\|_p = \underline{\lim}\, \|p_m - f\|_p \geq \|\hat{p} - f\|_p .$$

This shows $S^e_{M,n}$ to be an existence set. But, it also follows that

$$\lim \|p_m - f\|_p = \|\hat{p} - f\|_p .$$

Hence, if $p < \infty$, then, by Lemma (8), $\{p_m - f\}$ converges in norm to $\hat{p} - f$, therefore, $\{p_m\}$ converges in norm to $\hat{p}$, showing $S^e_{M,n}$ to be approximatively compact; q. e. d.

(11) Corollary. If $1 < p < \infty$, then some $f \in L_p[0,1]$ has more than one b.a. in $S^e_{M,n}$.

Proof. By [2; Theorem 3], an approximatively compact subset S of $L_p[0,1]$, $1 < p < \infty$, is a uniqueness set iff S is convex. Since $S^e_{M,n}$ is closed under scalar multiplication, convexity of $S^e_{M,n}$ would imply that $S^e_{M,n}$ is a linear subspace of $L_p[0,1]$, contradicting the fact that $\{\gamma^{(j)}(t_i) \mid j = 0, \ldots, m_i;\ i = 1, \ldots, r\}$ is linearly independent for $0 \le t_1 < t_2 < t_r < 1$ and $m_i + 1 \le k$, $i = 1, \ldots, r$, with arbitrary r.

Remark. If $\{u_i \mid i = 1, \ldots, k\}$ is a basis for ker M, then

$$G(s,t) = \begin{cases} h(s,t), & s \ge t \\ 0, & s < t \end{cases},$$

where

$$h(s,t) = \sum_{i=1}^{k} u_i(s) v_i(t) ,$$

$\{v_i \mid i = 1, \ldots, k\}$ being the set of adjunct functions for $\{u_i \mid i = 1, \ldots, k\}$. This means [6; p. 669] that

$$\text{for all } t \in [0,1], \text{ for } j = 0, \ldots, k-1, \ \sum_{i=1}^{k} u_i^{(j)}(t) v_i(t) = \delta_{j,k-1} .$$

The argument for Theorem (10) uses that $v_i \in C^{(k-1)}[0,1]$, $i = 1, \ldots, k$, and that, for each $t \in [0,1]$, the set of functions $\{\hat{u}_j \mid j = 1, \ldots, k\}$, given by

$$\hat{u}_{j+1}(s) = \sum_{i=1}^{k} u_i(s) v^{(j)}(t), \quad j = 0, \ldots, k-1 ,$$

is a basis for ker M. This is insured by the assumption that the coefficients of M (cf. (1)) satisfy $a_i \in C^{(i)}[0,1]$, $i = 0, \ldots, k-1$.

In particular, best approximation by the set of generalized spline functions with respect to ker M with m joints (in the sense of Greville [4]) is best approximation by $\{p(\alpha, \tau) \in S^e_{M, m+k} \mid \tau_1 = \ldots = \tau_k = 0\}$, and is covered by Theorem (10) with minor and obvious modifications.

5. Strict monotonicity of the error. In almost all of the examples given in the introduction, the linear span of $\{\gamma(t) \mid t \in T\}$ is dense in X. If X is smooth, this has the perhaps surprising consequence that, for all $f \in X$, $\text{dist}(f, \mathbb{P}_{\gamma,n})$ is strictly decreasing as a function of n. Precisely, one has,

(1) Theorem. If X is smooth, and the linear span of $\{\gamma(t) \mid t \in T\}$ is dense in X, then, for all $f \in X$, $\text{dist}(f, \mathbb{P}_{\gamma,1}) < \|f\|$ unless $f = 0$.

Proof. If $\text{dist}(f, \mathbb{P}_{\gamma,1}) = \|f\|$, then, for all $t \in T$, 0 is a b.a. to f in the linear span of $\gamma(t)$, hence there exists $\lambda_t \in X^*$ such that

$$(2) \qquad \lambda_t f = \|f\|, \quad \|\lambda_t\| = 1$$

and

$$(3) \qquad \lambda_t \gamma(t) = 0 .$$

If $f \neq 0$, then, by the smoothness of X, λ_t is uniquely determined by (2), i.e., λ_t does not depend on t. It then follows from (3), that some nonzero continuous linear functional on X vanishes on the linear span of $\{\gamma(t) \mid t \in T\}$, contradicting the denseness of the linear span of $\{\gamma(t) \mid t \in T\}$ in X. q.e.d.

(4) Corollary. If X is smooth, the linear span of

$\{\gamma(t) \mid t \in T\}$ is dense in X, and $\mathbb{P}^k_{\gamma,n}$ is an existence set, then for all $f \in X$,

$$\text{dist}(f, \mathbb{P}_{\gamma,n+1}) < \text{dist}(f, \mathbb{P}_{\gamma,n}) \quad \text{or} \quad f \in \mathbb{P}^k_{\gamma,n} .$$

Thus, for $1 < p < \infty$, the distance of $f \in L_p[0,1]$ from $S^e_{M,n}$ is strictly decreasing with n unless and until $f \in S^e_{M,n}$ for some n.

6. Characterization. A first step toward the characterization of a b.a. to $f \in X$ in $\mathbb{P}^k_{\gamma,n}$ is the following

(1) Theorem. Assume that $\gamma \in C^{(k)}_X(T)$ and that T is an interval with endpoints a, b. Let

$$p = \sum_{i=1}^{r} \sum_{j=0}^{m_i} a_{ij}\, \gamma^{(j)}(t_i), \quad \text{with } a < t_i < b,\ a_{im_i} \neq 0,\ i = 1, \ldots, r,$$

be an element of $\mathbb{P}^k_{\gamma,n}$, and let S be the linear span of

(2) $\{\gamma^{(j)}(t_i) \mid j = 0, \ldots, m_i+1;\ i = 1, \ldots, r\} \cup \{\gamma(t_{r+i}) \mid i = 1, \ldots, h\}$,

where $h = n - \sum_{i=1}^{r}(m_i + 1)$, and the additional points $t_{r+1}, \ldots, t_{r+h}$ (if any) are arbitrary in T.

If p is a b.a. to $f \in X$ in $\mathbb{P}^k_{\gamma,n}$, then p is a b.a. to f in S.

Proof. Assume by way of contradiction that, for some $q \in S$,

(3) $$\|f - q\| < \|f - p\| .$$

Then q is of the form

$$q = \sum_{i=1}^{r} \sum_{j=0}^{m_i+1} b_{ij}\, \gamma^{(j)}(t_i) + \sum_{i=r+1}^{h} b_i\, \gamma(t_{r+i}) .$$

Assume without loss of generality that $b_{i,m_i+1} \neq 0$, $i = 1, \ldots, m$

while $b_{i,m_i+1} = 0$ for $i > m$. Then we can write q as

$$q = \sum_{i=1}^{m} b_{i,m_i+1} \gamma^{(m_i+1)}(t_i) + \hat{q} ,$$

where $\hat{q}$ has the property that

$$p + \alpha\hat{q} \in \mathbb{P}^k_{\gamma,n}, \qquad \text{for all scalars } \alpha .$$

Since $m_i + 1 \leq k$, all i, and γ is k times continuously differentiable, it follows from (3) and from Lemma 2. (1), that for some $\varepsilon > 0$ and all ε_i with $|\varepsilon_i| \leq \varepsilon$, $i = 1,\ldots,m$,

$$\|f - \hat{q} - \sum_{i=1}^{m} b_{i,m_i+1} (m_i+1)!\, \gamma(t_i,\ldots,t_i,t_i+\varepsilon_i)\| < \|f-p\| .$$

Therefore, for all ε_i with $|\varepsilon_i| \leq \varepsilon$, and all $\theta \in (0,1]$,

$$\|f - u(\theta, \varepsilon_1,\ldots,\varepsilon_m)\| < \|f-p\| ,$$

where

$$u(\theta, \varepsilon_1,\ldots,\varepsilon_m) = (1-\theta)p + \theta\hat{q} + \theta \sum_{i=1}^{m} b_{i,m_i+1}(m_i+1)!\, \gamma(t_i,\ldots,t_i+\varepsilon_i) .$$

In order to reach a contradiction, it is sufficient to show that θ, $\varepsilon_1, \ldots, \varepsilon_m$ can be so chosen that $u(\theta, \varepsilon_1, \ldots, \varepsilon_m) \in \mathbb{P}^k_{\gamma,n}$. One observes that for $\varepsilon_i \neq 0$, the (m_i+1)st divided difference of γ at $t_i,\ldots,t_i$, $t_i + \varepsilon_i$ can be written as

$$\gamma(t_i,\ldots,t_i, t_i+\varepsilon_i) = \varepsilon_i^{-m_i-1} [\gamma(t_i+\varepsilon_i) - \sum_{j=0}^{m_i} \varepsilon_i^j \frac{1}{j!} \gamma^{(j)}(t_i)] .$$

Hence,

$$u(\theta, \varepsilon_1, \ldots, \varepsilon_m) = \sum_{i=1}^{m} c_i \gamma^{(m_i)}(t_i) + v ,$$

where $v \in \mathbb{P}^k_{\gamma,n}$, and

$$c_i = (1-\theta) a_{i,m_i} + \theta [b_{i,m_i} - \varepsilon_i^{-1}(m_i+1) b_{i,m_i+1}], \quad i = 1, \ldots, m$$

Hence, to get $c_i = 0$, all i, one needs that

$$(4) \quad \varepsilon_i = \theta (m_i+1) b_{i,m_i+1} / (\theta b_{i,m_i} + (1-\theta) a_{i,m_i}), \quad i = 1, \ldots, m$$

Since b_{i,m_i+1}, a_{i,m_i} are not zero, $i = 1, \ldots, m$, it is clearly possible to find $\theta \in (0,1]$, and $\varepsilon_1, \ldots, \varepsilon_m$ with $0 < |\varepsilon_i| \leq \varepsilon$, all i, such that (4) is satisfied; q.e.d.

Remark. The linear span of (2) is, at best, of dimension $n+r$ even though the general element of $\mathbb{P}^k_{\gamma,n}$ depends on 2n parameters. This apparent loss of degrees of freedom (when $r < n$) is, in part, due to the fact that the vanishing of a linear parameter eliminates also the corresponding (non-linear) t-parameter. The phenomenon of varisolvence of exponential sums and other γ-polynomials originates this way. But, this loss also occurs when two or more of the τ_i's coalesce without the vanishing of any of the linear parameters. The implications of this fact for the geometry of the set $\mathbb{P}^k_{\gamma,n}$ are quite striking and should make $\mathbb{P}^k_{\gamma,n}$ worthy of the attention of accomplished topologists.

REFERENCES

1. H. B. Curry and I. J. Schoenberg, On Polya frequency functions, IV. The fundamental spline functions and their limits, J. d'Anal. Math. 17(1966), 71-107.

2. N. V. Efimov and S. B. Steckin, Approximative compactness

and Tchebycheff sets, Dokl. Akad. Nauk SSSR 140 (1961), 522-524.

3. R. E. Graves, Riemann integration and Taylor's theorem in general analysis, Trans. Amer. Math. Soc. 29 (1927), 163-177.

4. T.N.E. Greville, Interpolation by generalized spline functions, MRC Techn. Sum. Report No. 476 (1964).

5. C. R. Hobby and J. R. Rice, Approximation from a curve of functions, Arch. Rat. Mech. Anal. 27 (1967), 91-106.

6. T. Muir, A treatise on the theory of determinants, Longmans, Green and Co., London, 1933 (reprinted 1960 by Dover Publications, Inc., New York).

7. J. R. Rice, The approximation of functions, Vol. 2, Chapter 8, Addison Wesley, Reading, Mass., 1969.

8. J. R. Rice, Chebyshev approximation by exponentials, J. Soc. Indust. Math. 10 (1962), 149-161.

9. M. H. Schultz and R. S. Varga, L-splines, Numer. Math. 10 (1967), 345-369.

This work was partially supported by NSF grant GP-07163

Piecewise Bicubic Interpolation and Approximation in Polygons

GARRETT BIRKHOFF

1. Background. This morning's session will have a somewhat different flavor than the others. Whereas the other sessions have been mainly concerned with univariate "spline" functions, it will be concerned with "spline" functions of two or more variables. Here I have put the currently popular word "spline" in quotation marks, because its meaning seems to be generalized by every new researcher in the field.

This morning's session will also emphasize applications, especially to computerising the representation of smooth functions for which no exact formula is known, and to the accurate solution of self-adjoint elliptic boundary value and eigenvalue problems. Accordingly, it will emphasize conclusions rather than sophisticated proofs, and the practical effectiveness of algorithms rather than the rigorous wording of theorems. Hopefully, it will leave you with a realistic broad-brush picture of some recently developed and very powerful techniques, whose precise extent of applicability is still uncertain.

My own talk will emphasize computational problems, moreover. Thus I shall try to correlate theoretical considerations with practical considerations of algorithmic efficiency and computer storage utilization. And I shall try to indicate some of the pitfalls in using digital computers to evaluate numerically the expressions which result from using piecewise bicubic polynomials to interpolate and approximate smooth bivariate functions. And I shall indicate ways of avoiding some of these pitfalls.

Finally, my talk will emphasize the specific. Thus, I shall emphasize piecewise bicubic polynomial functions defined in some polygonal domain $\mathcal{P}$ which has been subdivided into rectangles and boundary triangles by an appropriate partition π. (As I shall show in §8, such a subdivision is not possible in every polygon.) My concern will thus be with subdivided polygons $(\mathcal{P}, \pi)$ of the kind shown in Figure 1.

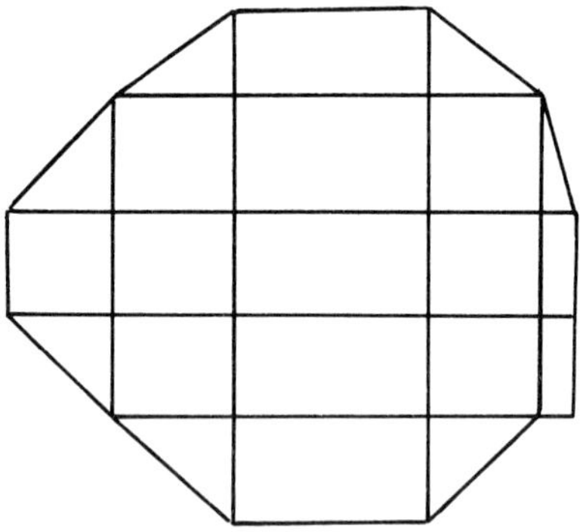

Figure 1

Five Years Ago. The emphasis described above is similar to that of my survey paper written five years ago with Carl de Boor [12, pp. 164-90]. Hence I shall take this paper as my point of departure.

When de Boor and I wrote our paper, we had been working for five years with H. L. Garabedian at the General Motors Research Laboratories on the general geometrical problem of fitting smooth surfaces by bivariate spline functions. We summarized our main conclusions as follows [12, pp. 165-6]:

"To represent smooth curves and surfaces economically, with the help of high-speed digital computers, we recommend the use of piecewise polynomial functions.

Indeed, we will be much more specific. Although piecewise quintic polynomials have proved useful on occasion and, for some applications, approximation by piecewise quadratic or even linear functions is most suitable, in general we recommend piecewise cubic polynomials for fitting smooth curves, and piecewise bicubic polynomials for fitting smooth surfaces, as good bets to be tried first, in the absence of special reasons for trying something else.

. . . .

It can hardly be said that this idea is either very deep or very novel. The use of generalized splines, and other piecewise polynomial functions of higher degree, to approximate smooth functions of one variable, was considered earlier very carefully by I. J. Schoenberg in an important paper where deep results were obtained for the case of a uniform mesh on the infinite line. A thorough study of the literature would probably reveal many other relevant papers.

What is important is that spline interpolation converges rapidly on a wide variety of meshes, that it is insensitive to roundoff, and that it is easy to perform on high-speed computers. The demonstration of these facts seems to be new. What we consider most original about our work is the development of practical schemes of surface-fitting, applicable to wide classes of smooth surfaces."

The general statements about bicubic spline functions just quoted were then elaborated upon in various contexts, including especially the following three:

(a) The accurate interpolation and approximation of smooth functions on rectangles [12, pp. 173-6];

(b) Less accurate interpolation and approximation to functions on L-shaped regions and other rectangular polygons [12, pp. 172-3 and pp. 185-7];

(c) The use of "approximating subspaces" of (piecewise) cubic and bicubic spline functions associated with suitable rectangular meshes on rectangular regions, so as to solve accurately elliptic boundary value and eigenvalue problems [12, pp. 181-3 and pp. 187-9].

Five years ago, though we believed that spline interpolation of degree $2n-1$ had $O(h^{2n-j})$ accuracy for $u^{(j)}(x)$, we were sure of this only in the case of univariate cubic splines, and bounded mesh-length ratio $\theta = \Delta x_{max}/\Delta x_{min}$. In this case, we had proved the following inequalities for $f \in C^4[a,b]$ for the cubic spline interpolation operator P in the uniform norm $\|u\|_\infty = \mathrm{Max}|u(x)|$:

$$(1) \qquad \|Pf - f\|_\infty \leq K_1(\theta) \cdot \|f^{iv}\|_\infty h^4, \quad h = \Delta x_{max},$$

and

$$(2) \qquad \|Pg\| \leq K_2(\theta) \cdot \|g\|_\infty .$$

and even in this case sharper results have since been proved (see §3). In the all-important bivariate case, we had sketched [12, §9] a proof of Theorem 1 below for the special case $j = k = 0$. However, our confidence in bicubic splines was based primarily on empirical evidence: bicubic spline interpolation had successfully fitted a wide variety of "smooth" surfaces.

As regards (b), though we had proposed an algorithm in [12], we had neither tested it experimentally nor studied in detail its theoretical order of accuracy. As regards (c), we had determined that the Rayleigh-Ritz method gave eigenvalues with order of accuracy $O(h^6)$, when applied to two "model problems" using spline functions. However, we had proved nothing theoretically in the general case.

2. Recently Published Research. In describing progres since 1964, I shall begin by summarizing very briefly some relevant research published since then. My purpose is to provide an adequate background for the new results about bivariate splines which I shall describe later, and whose general conclusions may be summarized in the following three statements:

(i) Bicubic spline functions provide very accurate and flexible tools for interpolating and approximating "smooth" functions. †

(ii) In particular, bicubic splines seem to give more accurate numerical solutions of elliptic DE's than five-point difference approximations.

(iii) For many special problems involving analytic functions, other special methods are significantly more accurate.

This last fact became apparent when we compared the numerical eigenvalues for the Mathieu functions obtained with the Rayleigh-Ritz method in [4], using cubic Hermite and spline approximation, with the eigenvalues computed earlier by Bazley and Weinstein using trigonometric approximations. For details, see [7, §5]. ‡

The calculations just referred to concerned functions of one variable, but the general conclusion (iii) to which they led has since been confirmed by other eigenvalue studies [7]. Indeed, a good theory of unvariate splines is a necessary prerequisite for a good theory of bivariate splines. Therefore I shall begin my review of recently published research by describing some basic new theoretical results about univariate splines.

Here I should mention first that Carl de Boor* and C. A. Hall have significantly sharpened (1), by showing that $\sup_\theta K_1(\theta)$ is finite. In the cases of cubic and quintic spline interpolation (the cases $n = 2, 3$ of spline interpolation of degree $2n-1$), for functions $f \in C^{2n}[a,b]$ and for $j = 0, 1$,

---

† Important exceptions are positive functions which, like e^{-x^2} or erf x on $[0,10]$ and e^{-x^2} and $e^{-x^2-y^2}$ on $[0,10]^2$, vary by several orders of magnitude. Such functions should be expressed on a logarithmic scale.

‡ Also, G. Birkhoff and G. Fix, Indian J. Math. 2 (1968).

* C. de Boor, J. Approx. Theory 1 (1968), 219-35 and 452-63; his results were also in his Ph. D. Thesis (Univ. of Michigan, 1966).

Hall [14] has in fact found the best (smallest) known constants $K_{n,j}$ such that

$$(3) \quad |e^{(j)}(x)| = K_{n,j} \|f^{(2n)}\|_{\infty} h^{2n-j}, \quad e(x) = Pf(x) - f(x) .$$

(Arthur Priver and I had previously found the analogous best possible constants $\tilde{K}_{n,j}$ for low-order Hermite interpolation. †)

Second, I have described an algorithm of "matching moments"‡ for approximating any function $u \in C^{2n}[a,b]$ by splines of odd degree $2n-1$ with an error which is $O(h^{2n})$. See Appendix A.

Third, B. Hulme has shown that the variational approximation by splines of odd degree $2n-1$, that is, by the spline approximation which minimizes the mean square error in the n-th derivative (cf. §3), is not only interpolatory but also has an unexpectedly high order of accuracy as regards derivatives at the mesh-points (alias "joints" or "knots"). * Hulme's result has been greatly generalized by Perrin, Price, and Varga.

There has also been major progress over the past few years in the theory of piecewise polynomial interpolation and approximation to smooth bivariate functions. The groundwork for this was laid in [6], in which Varga, Schultz, and I made a thorough analysis of the errors in bivariate Hermite interpolation and approximation of odd degree $2n-1$. In particular, we showed that the order of magnitude of the error in partial derivatives of order j is $O(h^{2n-j})$ for any $u \in C^{(4,0)}(\mathfrak{R}) \cap C^{(2,2)}(\mathfrak{R}) \cap C^{(0,4)}(\mathfrak{R})$, # as might be expected. Our tools were provided by earlier considerations of Peano, Sard, and Stancu

† G. Birkhoff and Arthur Priver, J. Math. and Phys. 46(1967), 440-7.

‡ G. Birkhoff, J. Math. Mech. 16(1967), 984-90. See also C. de Boor, ibid, 17(1968), 729-36.

* B. Hulme, J. Math. Mech. 18(1968), 337-43. The fact that this was the interpolating spline had been previously observed by C. de Boor and R. E. Lynch, J. Math. Mech. 15(1966), 953-

Recently, J. H. Bramble and S. R. Hilbert have shown that it is enough to assume that $u \in C^{(4,0)}(\mathfrak{R}) \cap C^{(0,4)}(\mathfrak{R})$.

The results of [6] have been since shown to be valid also for multivariate spline interpolation and approximation by Martin Schultz: he will tell you about his work, and the references in his bibliography contain much additional information.

I shall talk mostly about the independent, less general, but more computationally oriented work of Hall and Carlson centering about bicubic splines. Their results are most easily described by letting $C^{(r,s)}(\mathfrak{R})$ signify the set of all functions $u(x,y)$ on $\mathfrak{R}$ such that $\partial^{r+s}u/\partial x^r \partial y^s$ exists and is continuous throughout $\mathfrak{R}$, while $C^{(s)}(\mathfrak{R})$ signifies the set of all $u(x,y)$ on $\mathfrak{R}$ such that all partial derivatives $\partial^s u/\partial x^k \partial y^\ell$ $(k+\ell=s)$ exist and are continuous throughout $\mathfrak{R}$.

Hall was the first to prove [4, Theorem 3] the following extension of the result of [12, § 9] already mentioned above.

THEOREM 1 (C. A. Hall). Let $u \in C^{(1,4)} \cap C^{(4,1)}$ in a closed rectangle $\mathfrak{R}$ subdivided by a rectangular partition $\pi \times \pi'$, where π and π' have maximum mesh-length h and h', respectively. Then bicubic spline interpolant to u satisfies

$$\|(s-u)^{(j,k)}\| = O(h^{4-j} + h'^{4-k}), \quad j,k = 0,1,2,3 ,$$

provided that the maximum mesh-length ratio in one direction is bounded.

Sketch of proof (cf. [12, § 9]). By (3), the cubic spline interpolation error $e(x,y_j) = s(x,y_j) - u(x,y_j)$ satisfies

$$e(x,y_j) = K_{2,0}\|u^{(4,0)}\|_\infty h^4, \quad h = |\Delta x_i|_{max} ,$$

on every horizontal mesh-line $y = y_j$ $(j = 0, 1, \ldots, J)$. On the two horizontal edges $y = y_j$ $(j = 0, J)$, the derivative e_y of the cubic spline interpolation error satisfies $e_y(x, y_j) = K_{2,0}\|u^{(4,1)}\|_\infty h^4$, by (3) applied to the univariate spline $u_y(x, y_j)$.

On the other hand, for any fixed $x = x^*$, the bicubic spline interpolant $s_\pi(x^*, y)$ to $u(x^*, y)$ is also the univariate cubic spline interpolant on the mesh π' to the sum $u(x^*, y) + e(x^*, y)$ with end slopes $u_y(x^*, y) + e_y(x^*, y)$, since both functions are cubic spline functions of y over the same mesh. Hence $s_\pi(x^*, y)$ is the sum of the univariate cubic spline interpolant to $u(x^*, y_j)$ and that to $e(x^*, y)$, with the specified end-slopes. Its interpolation error is, correspondingly, bounded by the sum of the maximum possible error in cubic spline interpolation to the $u(x^*, y_j)$ and the maximum magnitude of the cubic spline interpolant to the $e(x^*, y_j)$. By (3) and (2), respectively, this gives a total error bound

$$K_{2,0} \|u^{(4,1)}\|_\infty h'^4 + I(\theta)\,[K_{2,0} \|u^{(4,0)}\| + \|u^{(4,1)}\|_\infty] h^4 .$$

where h' is the maximum mesh and θ' is the maximum mesh-ratio in y. This completes the proof.

I am delighted to report that Carlson and Hall have recently proved the preceding inequality without restriction on θ or θ' (personal communication).

3. Variational Considerations In extending the theory of spline functions to non-rectangular domains, I think one should accept Professor Golomb's view that the spline concept is essentially variational. Thus, as was asserted in [12, p. 175] and proved rigorously in [1, Section 7.6], bicubic spline functions on rectangles minimize the double integral

$$J[u] = \frac{1}{2} \iint_{\Re} u_{xxyy}^2 \, dx\, dy \quad , \tag{4}$$

in the class of all functions which interpolate to the same value at all mesh-points of a rectangular mesh. Moreover, it is state in [1, Section 7.14] that the definitions and proofs have straightforward extensions to rectangular polygons (but see §4 below).

Likewise, Dr. Gordon [13, Theorem 3] has proved that his spline-blended interpolating functions of cubic type minimize the integral (4) in the class of functions $f \in C^{2,2}(\Re)$ with interpolate to the same values on all mesh-lines. This shows, incidentally, that functions which minimize (4) need not be piecewise bicubic polynomial functions (bicubic spline functions), except for special sets of interpolating conditions.

Therefore, it seems reasonable to use the variational condition $J[u] = J_{min}$ (which implies $\delta J = 0$) to define splines on general domains. However, this is not to say that the usual textbook theory of the calculus of variations for functions of two variables applies in a straightforward way. Thus, the usual Euler-Lagrange equation for the variational condition $\delta J = 0$ is $u^{(4,4)} = 0$, and this condition does characterize bicubically spline-blended functions, but it does not characterize bicubic polynomial splines, which are better characterized by the two differential equations $u^{(4,0)} = u^{(0,4)} = 0$.

As a result, to apply variational ideas to the study of spline functions on general domains will require the development of new techniques, extending those worked out in [1, Section 7.6]. I shall next take a first step in this direction by applying the condition $\delta J = 0$ to a general region; it seems previously not to have been applied to domains other than rectangular polygons.

THEOREM 2. In any domain $\mathfrak{R}$,

$$(5) \quad \delta J = \iint u^{(4,4)}(x,y)\, \delta u(x,y)\, dx\, dy + L[\delta u^{(2,2)}, u^{(2,2)}] + L[\delta u^{(1,1)}, u^{(2,2)}] + L[\delta u, u^{(3,3)}]$$

where $L[f,g]$ signifies the boundary integral

$$(5') \qquad L[f,g] = \oint_{\partial\mathfrak{R}} [f(g_y dy - g_x dx) - g(f_y dy - f_x dx)] ,$$

provided $u^{(3,3)}$ is continuous and of bounded variation in $\mathfrak{R}$, and the boundary integrals in (5) converge.

Proof. In the plane, the divergence theorem can be written as

$$(6) \qquad \iint_{\mathfrak{R}} \left[\frac{\partial X}{\partial x} + \frac{\partial Y}{\partial y}\right] dx\, dy = \oint_{\partial\mathfrak{R}} [X dy - Y dx] .$$

Setting $X = fv_y - vf_y$ and $Y = fv_x - vf_x$, we get

(7) $$\operatorname{div}(X,Y) = \frac{\partial X}{\partial x} + \frac{\partial Y}{\partial y} = 2[fv_{xy} - vf_{xy}]$$

and

(7') $$Xdy = fv_y\,dy - vf_y\,dy, \qquad Ydx = fv_x\,dx - vf_x\,dx \ .$$

Putting together the preceding results and using the notation of (5'), we get finally (after dividing by two and transposing):

(8) $$\iint_{\mathfrak{R}} f\ v_{xy}\,dx\,dy = \iint_{\mathfrak{R}} v\,f_{xy}\,dx\,dy + L[f,v] \ .$$

Actually, formula (8) applies in the sense of double Stieltjes integrals to all functions f and v which are both continuous and of bounded variation in $\mathfrak{R}$.

Theorem 2 follows if one applies (8) twice; indeed, one can set

$$u^{(4,4)}(x,y)\,dx\,dy = du^{(3,3)}(x,y)$$

provided $u^{(3,3)}$ is of bounded variation in $\mathfrak{R}$. It has the following corollary.

COROLLARY. If $u^{(3,3)}$ is continuous and of bounded variation in $\mathfrak{R}$, then $J[u] = J_{min}$ implies that $u^{(4,4)} = 0$ at any point not a point of interpolation.

In rectangles, the definition (4) of spline functions has been implemented by effective codes for computing spline functions as piecewise bicubic polynomials. I shall now discuss the corresponding implementation problem for non-rectangular regions, which is much more difficult. Here the most detailed studies are those of Carlson and Hall ([8] and [9]), and the most extensive studies are those of Martin Schultz.† These

† See the bibliography of the paper by Schultz to follow. Schul treats general plane domains, proving no special results about rectangular polygons.

authors define bicubic spline functions on a rectangularly subdivided domain $(\mathcal{D}, \pi)$ as piecewise bicubic polynomial functions of class $C^{(2)}$, thus stressing algebraic (i.e., computational) and non-variational considerations. It would be very interesting to correlate their algebraic definition with the variational definition of §3.

4. Rectangular Polygons. Consider a general subdivided rectangular polygon $(\mathfrak{R}, \pi)$ with horizontal and vertical sides. Such an $\mathfrak{R}$ with $m \geq 0$ reentrant corners always has $4 + m$ convex corners and $4 + 2m$ edges. The case $m = 0$ gives a rectangle, while $m = 1$ gives an L-shaped region, the case which was discussed carefully in [12, Appendix A].

Since Hermite interpolation is local, the results of [6] hold in any subdivided rectangular polygon. However, it is much harder to determine the implications for the error in bivariate spline interpolation of the smoothness constraints which are its basis. One cannot, for example, simply use the concept of tensor product to determine suitable bases for the subspace of all (bicubic or higer odd degree) spline functions, or even to determine the dimensionality of this subspace.

Substantial progress in attacking this difficult problem has been made by Carlson and Hall [8], and I shall restate for you some of their principal results.

First, let $H^n_\pi(\mathfrak{R})$ be the space of piecewise polynomial functions of degree $\leq 2n-1$ in x and y which satisfy the smoothness condition for Hermite interpolation, $u \in C^{(n-1, n-1)}(\mathfrak{R})$. As before, $C^{(r,s)}(\mathfrak{R})$ signifies the set of all functions $u(x, y)$ on $\mathfrak{R}$ such that $\partial^{r+s}u/\partial x^r \partial y^s$ exists and is continuous throughout $\mathfrak{R}$, while $C^{(s)}(\mathfrak{R})$ signifies the set of all $u(x, y)$ on $\mathfrak{R}$ such that all partial derivatives $\partial^s u/\partial x^k \partial y^\ell$ $(k + \ell = s)$ exist and are continuous throughout $\mathfrak{R}$.) Carlson and Hall have shown that, for any $\ell = 0, 1, \ldots, n-1$, the smoothness constraint $u \in C^{n-1+\ell}(\mathfrak{R})$ on functions in $H^n_\pi(\mathfrak{R})$ implies $u \in C^{(n-1+\ell, n-1+\ell)}(\mathfrak{R})$. They first define the subspace of spline functions of order n (and degree at most $2n-1$ in each variable) as the intersection

$$(10) \qquad Sp(\mathfrak{R}, \pi, n) = H^n_\pi(\mathfrak{R}) \cap C^{2n-2}(\mathfrak{R}) \; .$$

Then, they define a set S of four corner points of $\Re$ to be amenable when: (i) some pair but no triple of the points of S lie on the same mesh-line, and (ii) S spans $\Re$, in the sense that successive augmentation of S by mesh-points on mesh-lines in $(\Re, \pi)$ through pairs of mesh-points already included by the augmentation gives, ultimately, all mesh-points of $(\Re, \pi)$. They prove the following result, which is quite satisfactory from a purely algebraic standpoint.

THEOREM 3 (Carlson-Hall). Let there be given: (i) values of $u(x_k, y_j)$ at all mesh-points, (ii) values of normal derivatives of orders $\leq n-1$ at all edge mesh-points not re-entrant corners, and (iii) mixed derivatives of all orders $\leq (n-1, n-1)$ at four "amenable" corner points. Then there exists one and only one piecewise polynomial function which is of degree $2n-1$ or less in both variables, of class $C^{2n-2}(\Re)$, and interpolates to the given values.

For the case of piecewise bicubic polynomial functions, Theorem 3 asserts existence and uniqueness for functions of class $C^{(2,2)}(\Re)$ having given values at all mesh-points, normal derivatives at edge points not on re-entrant corners, and cross-derivatives $\partial^2 f/\partial x \partial y$ at the four "amenable" corner points. In the terminology of [12, p. 169], the interpolation problem is algebraically well-set for the above data.

However, the interpolation problem is not analytically well-set, for the above data. As was essentially pointed out in [12, Appendix A] for $n = 2$, the above "interpolation" process involves spline extrapolation if $m > 0$ (i.e., except for rectangles), and this is unstable.

The proof of Theorem 3 is highly technical; I shall briefly summarize in Appendix B some of the combinatorial considerations which it involves, and indicate how the detailed studies of Carlson and Hall open up a fascinating new area of combinatorial mathematics.

Carlson and Hall have also described a different bicubic spline "interpolation" scheme which is analytically well-set in L-shaped regions. This scheme is closely related to a method of "spline function approximation" by tangent-fitting due to

Loscalzo and Talbot†. Moreover, in the special case of a uniform mesh and sufficiently smooth f, their scheme even has the maximum possible order of accuracy. Specifically, they have proved [8, Theorem 6]:

THEOREM 4. Let $f \in C^s(\mathfrak{R})$, where $(\mathfrak{R}, \pi)$ is the subdivided L-shaped region of Figure 2. Let $s(x, y)$ be a bicubic spline function which interpolates to f at each mesh-point, to the normal derivative at each edge mesh-point not on $\overline{CD}$, to f_{xy} at A, B, F, and along $\overline{CD}$. Then, for $s = 4$ and bounded mesh-ratio, the error $f^{(k,\ell)} - s^{(k,\ell)}$ is $O(h^{2-(k+\ell)})$ if $k + \ell \leq 1$, and for $s = 5$ the error is $O(h^{4-(k+\ell)})$ if the mesh is uniform and $k + \ell \leq 3$.

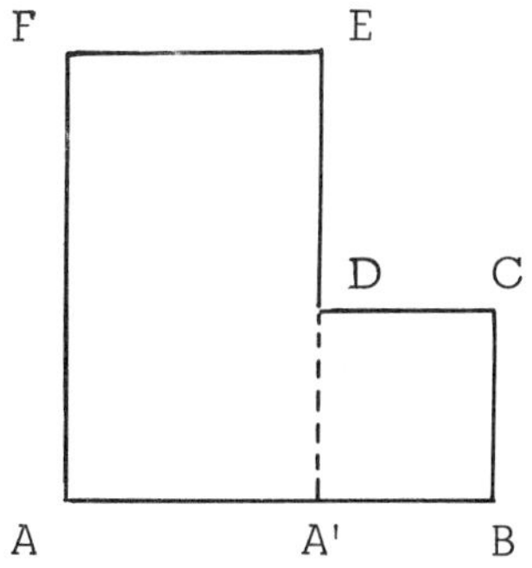

Figure 2

5. Whitney's Extension Theorem. If one is interested only in proving theorems about the order of accuracy of bivariate spline approximation, one can ignore the technical results described above. For proving such theorems, as has been observed by Martin Schultz [17], one can by-pass the need for

† F. Loscalzo and T. Talbot, SIAM J. Num. Anal. 4 (1967), 433-45.

ingenious special interpolation schemes by invoking Whitney's Extension Theorem.† This theorem asserts that any function $u(x, y) \in C^m(A)$, where A is <u>any closed set</u>, can be extended to a function of class C^m in the entire plane.

From Whitney's Extension Theorem, one can derive almost trivially a theoretical solution to the bicubic spline interpolation problem of §4, as follows.‡ Let $u \in C^5(\mathfrak{R})$, where $\mathfrak{R}$ is a given rectangular polygon, and extend u to the circumscribing rectangle $\tilde{\mathfrak{R}}$ of $\mathfrak{R}$ without loss of differentaiability.

If $\mathfrak{R}$ is subdivided by a <u>rectangular partition</u> $\pi \times \pi'$, therefore, and one extends $\pi \times \pi'$ to $\tilde{\mathfrak{R}}$, one can approximate the extension $\tilde{f}(x, y)$ by a bicubic spline approximation $\tilde{u}(x, y)$ with error $O(h^{4-j} + h'^{4-k})$ in $\tilde{f}^{(j,k)}$ for $j, k = 0, 1, 2$, all by Theorem 1. The restriction $u(x, y)$ of $\tilde{u}(x, y)$ to $\mathfrak{R}$ therefore approximates $f(x, y)$ with at most the same error. In outline, this proves

THEOREM 5. Let $f(x, y) \in C^{4,1}(\mathfrak{R}) \cap C^{1,4}(\mathfrak{R})$ in a rectangular polygon $\mathfrak{R}$ subdivided by a rectangular mesh $\pi \times \pi'$ with bounded mesh-ratios. Then there exists a bicubic spline function $u(x, y)$ in $(\mathfrak{R}, \pi \times \pi')$ such that $u^{(j,k)}$ approximates $f^{(j,k)}$ in $\mathfrak{R}$ with an error $O(h^{4-j} + h'^{4-k})$ for $j, k = 0, 1, 2$.

Similar results hold in n dimensions for spline approximations of any odd degree; see [17].

<u>Boundary conditions</u>. The preceding existence theorem, though very elegant, does not generally provide effective computational techniques for satisfying <u>boundary conditions</u> on f. Thus, it does not provide an algorithm for selecting

---

† H. Whitney, Trans. Am. Math. Soc. 36 (1954), 63-89. See especially his Theorem 1 on p. 65.

‡ A few points are slurred over here: there exist partitions of rectangular polygons into subrectangles which are not of the form $\pi \times \pi'$, and one can weaken the hypothesis $u \in C^5(\mathfrak{R})$. However, I do not wish to distract attention from the main point by elaborating on such technical details.

functions which satisfy $f \equiv 0$ on $\partial\mathfrak{R}$, the boundary of a rectangular polygon. Again, it does not assert that the stated order of accuracy can be achieved by interpolating to mixed boundary conditions — as it can if $\mathfrak{R} = \tilde{\mathfrak{R}}$ is a rectangle. (However, Dirichlet-type boundary conditions on $\partial\mathfrak{R}$ are approximately satisfied by the bicubic spline interpolant to $\tilde{f}(x, y)$ on $(\mathfrak{R}, \pi \times \pi')$.)

6. Remarks on Applicability. Indeed, from the standpoint of applications, the elegant construction leading to Theorem 5 is of doubtful utility. Let me now make a few remarks on the realities of bicubic spline interpolation and approximation, as seen by an applied mathematician. From this standpoint, the scheme suggested by Carlson and Hall is only one of many possible theoretically interesting schemes which one can devise for generalized "spline interpolation" to approximate smooth functions in rectangular polygons. Moreover, the relative utility of such schemes depends considerably on the application intended.

In geometrical applications to surface fitting (whether for surface design or surface representation), one typically knows offsets of surfaces along plane sections very accurately, whereas one has relatively crude data regarding slopes.† Therefore, the interpolation scheme which de Boor and I suggested in [12, pp. 185-7] for surface fitting may be as good or better for this purpose than the analytically well-set interpolation scheme for the spline subspace $(\mathfrak{R}, \pi)$ of an L-shaped region $\mathfrak{R}$ invented by Carlson and Hall.

For physical applications to elliptic boundary-value and eigenvalue problems, including higher-order "finite element" methods, the situation is again very different. Here, one is primarily interested in approximation in Sobolev norms such as $\int_R\int [u^2 + \nabla u \cdot \nabla u]\, dx\, dy$, and the ultimate objective is to get

† In principle, slopes could also be measured very accurately by optical instruments, but few instruments for making such measurements seem to be available.

more accurate results than are obtainable in practice with the finite difference methods which have been mainly used on computers hitherto. The basic idea of the approach, which was sketched in [12, pp. 180-84], is to apply Rayleigh-Ritz or Galerkin methods to approximating subspaces which contain all (bicubic, say) Hermite or spline functions defined over a fixed rectangular mesh.

For such physical applications, which typically involve first or second derivatives in "energy" expressions, I would expect predicted orders of accuracy of approximation by interpolation schemes based on the idea of Loscalzo and Talbot to be better than those based on the interpolation scheme which de Boor and I proposed.

However, for relatively simple rectangular polygons, I think that it would be preferable to use an augmented subspace of piecewise bicubic functions, in which the constraint of continuity of second derivatives is relaxed along part or all of one or more mesh-lines. For example, in the L-shaped polygon of Figure 2, one might relax this requirement on the dashed continuations of one or both of the reentrant edges.

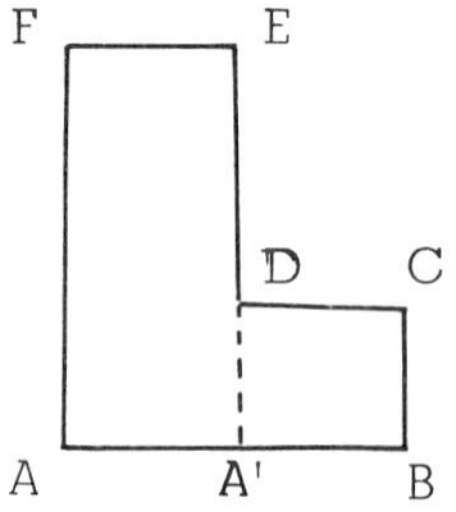

Figure 2

Pros and cons of splines. Indeed, it is tempting to go the whole way, and to simply use the space $H_{\pi}^{n}(\Re)$ defined in §4, which is so much easier to treat theoretically than the spline subspace $Sp(\Re, \pi)$. Unfortunately, there are two major practical considerations which make it undesirable to use $H_{\pi}^{n}(\Re)$ when really accurate approximations are wanted.

First, as was already observed above, in geometrical problems of surface fitting, the offsets required for spline interpolation are usually accurately known, whereas the slopes and cross-derivatives also required for Hermite interpolation are not commonly measured at all.

In applications to the approximate solution of boundary value and eigenvalue problems, the preceding difficulty does not arise. Here the main advantage of using an approximating subspace of bicubic splines over using $H_\pi^2(\mathfrak{R})$ (bicubic Hermite approximation) consists in the fact that the spline subspace has about one-fourth as many dimensions, for about the same accuracy. Thus, for a rectangle subdivided into $M \times N$ subrectangles, the dimension-ratio is $(M+3)(N+3)/4(M+1)(N+1)$.

There are, however, important and partially compensating practical disadvantages to using spline functions. These concern numerical linear algebra, and depend on the choice of basis in the spline subspace $Sp(\mathfrak{R}, \pi)$. If bases of functions which are simplest for many theoretical purposes are used, like $[(x-x_i)_+^3(y-y_i)_+^3]$, the resulting matrices are both dense and extremely ill-conditioned. Because the resulting matrices are ill-conditioned, the roundoff errors amplify to such an extent that very inaccurate if not worthless "approximations" are computed in large-scale problems. And because the matrices are dense, core storage may become impossible.

Patch bases. Applied mathematicians having to make substantial computations resolve both of these difficulties by using "patch bases", whose "support" (the locus in which non-zero values of the function arise) is small. For bicubic splines, the support typically consists of a 4×4 array of mesh-rectangles, for reasons to be explained in Appendix A.

7. Corner Singularities. Perhaps the greatest limitation on the accuracy of piecewise polynomial approximations to solutions of specific elliptic boundary value and eigenvalue problems in polygons has a quite different origin, namely, boundary singularities. This limitation applies to Hermite as well as spline approximation, and to nonrectangular as well as to rectangular polygons.

Thus, even a uniformly loaded rectangular membrane has singularities at the corners of the order of $r^2 \ell n\, r$. (In mathematical terms, the displacement of such a membrane is the solution of the Poisson DE $-\nabla^2 u = 1$ in $\mathfrak{R}$, with boundary condition $u \equiv 0$ on $\partial\mathfrak{R}$.)

Although the eigenfunctions of a rectangular membrane (i.e., of the Helmholtz equation in a rectangle) are analytic, the eigenfunctions of an L-shaped membrane are not. As a result, one does not achieve high orders of accuracy by applying the Rayleigh-Ritz method to approximating subspaces having bases which consist of piecewise polynomial functions alone.

However, one can achieve high orders of accuracy by supplementing spline functions with singular functions whose multiples include the leading singular terms of the expansion in power series and logarithmic terms of the exact solution. These singular functions can often be determined by elementary computation; the general case is covered by a theorem of S. Lehman.†

In his Harvard Doctoral Thesis [11], George Fix obtained quite accurate eigenvalues of an L-shaped membrane using this "method of supplementary singular functions." Considerably improved versions of his computer program have since been written, and are described in Appendix C.

8. Spline Approximation in General Polygons. It is a familiar fact that plane domains with smooth boundaries can be approximated by polygons in both length and area, up to an error which is $O(h^2)$ where h is the side-length. Moreover smooth functions on these domains can be approximated by piecewise linear and bilinear functions, if these approximating polygons are subdivided into rectangles and triangles. Furthermore, many engineering structures have polygonal cross-sections to begin with, so that polygonal domains are interesting in their own right.

Partly for these reasons, various finite element methods have been developed in the past decade to handle problems in solid mechanics, in which the real physical displacement is approximated by such functions, the "best" approximation being taken to be that of least energy. There is now a large technical literature dealing with such methods.‡

† See [7, §2] for a discussion of Lehman's Theorem in the present context; Appendix B of my article in [2] gives an elementary treatment of the case of a uniformly loaded rectangular membrane.

‡ See the review articles by Clough and Felippa and by Pian i[illegible] as well as O.C. Zienkewicz and Y.C. Chaung, "The Finite Ele[illegible] Method in Structural and Continuum Mechanics", McGraw-Hill

PIECEWISE BICUBIC INTERPOLATION

In an effort to obtain better than $O(h^2)$ accuracy, piecewise bicubic Hermite approximation has been used by Bogner, Fox, Schmit and others for some problems. And for the past 2-3 years, C. A. Hall and R. E. Carlson at the Bettis Atomic Laboratory and B. L. Hulme at Harvard have been studying such "higher-order" finite element methods systematically. I want to tell you about some of their most significant findings; I feel qualified to do this because I have been advising them in this work. For further details, I refer you to their reports [15], [9], and [16]. They consider only subdivided polygons $(\mathcal{P}, \pi)$ whose subdivisions are rectangles and boundary triangles.

Their methods are not easy to analyze rigorously. For example, "almost no" polygons can be exactly subdivided by a rectangular partition having one mesh-line through each vertex; Figure 3 indicates why such an exact partition is usually impossible: for almost every set of edge-slopes, the sequence of alternating horizontal and vertical line segments bounded by edges is an infinite non-terminating sequence. Hence, to get $O(h^2)$ accuracy, one must approximate the polygonal domain as well as the functions.

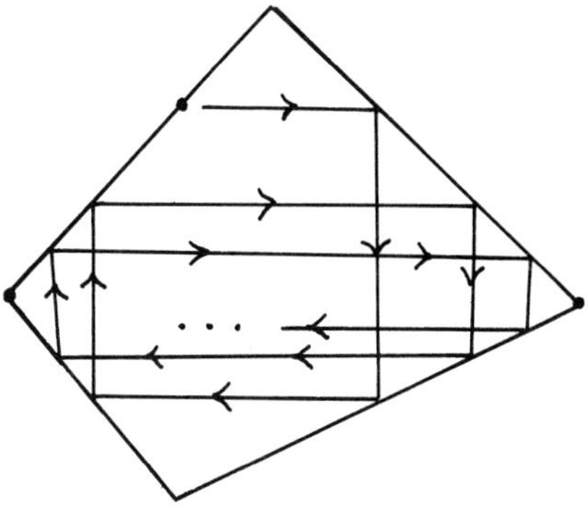

Figure 3

In [15], Hall describes three schemes for supplementing the Hermite approximation scheme of [6] by bicubic polynomials in the boundary triangles (see §1). Hall concluded that his Scheme B was the best adapted to Dirichlet boundary conditions of the form $u = f$. In [16], Hulme modified it so as to satisfy exactly "Neumann-type" boundary conditions of the form $\partial u/\partial n = 0$. As i had previously noted (Hall's Scheme C), one can alternatively use cubic (instead of bicubic) polynomials to fit boundary values and slopes at the vertices.

In [9], Carlson and Hall have extended the considerations of [8] to polygons $\mathcal{R}$ which are partitioned by a rectangular network $\pi \times \pi'$ into a union of rectangles and boundary right triangles. On the rectangular polygon $\mathcal{R}_1$ obtained from $\mathcal{R}$ by deleting the boundary triangles, they use the interpolating subspace (for any choice of u_{xy} at four "amenable" corner points) of "bicubic spline" functions defined in [8], already discussed in §4.

They then extend these functions to the rest of $\mathcal{R}$. For the given order of fit on the boundary mesh-points of $\mathcal{R}_1$, this has order of accuracy $O(h^4 + h'^4)$, always assuming that the function fitted is sufficiently smooth, and that the mesh ratios are bounded.

Note that the algorithm for making this extension is explicit and that the extension is unique — very unlike that which simply appeals to Whitney's Extension Theorem. Further more, it is possible to impose arbitrary boundary conditions as <u>exact constraints</u> at mesh-points or (in many cases; see Appendices B - C) along mesh-lines.

However, I do not want to give you the impression that such bicubic Hermite or spline approximations are always the best ones to use, even in polygons. For example. they are probably not best on a <u>rhombus</u>, which is one of the simplest polygonal regions. Indeed, Fox, Henrici, and Moler used other methods when they wished to compute accurately the eige values of a homogeneous rhombic membrane.† So did George Fix and I [7, §10]. Our methods included the augmentation by "singular functions" of a polynomial basis modulated by a <u>mult plier function</u> vanishing on the boundary — a technique which will be discussed further by Professor Schultz. We did <u>not</u> use <u>spline functions</u>, nor did we have any reason to believe that their use would have been advantageous in this special case.

9. <u>Babuška Paradox</u>. As I implied earlier, piecewise bi cubic polynomial approximations have been successfully appli to second-order and fourth-order (e.g., coupled second-order elliptic equations in many polygonal domains. Furthermore, such polygonal domains have been successfully used to appro mate other plane domains with curved boundaries. This raises

† L. Fox, P. Henrici, and C. Moler, SIAM J. Num. Analysis 4 (1967), 89-102.

the question of the universality of the preceding method.

In this connection, Fix and I have sketched theoretical arguments in [7], based on variational and similarity considerations, which plausibly imply the following result. Consider the approximations to a domain D with piecewise smooth boundary ∂D, by subdivided polygons $(\mathcal{P}, \pi)$ with maximum mesh-length h. Then one can so choose $\mathcal{P}$ and π that the error made in replacing second-order elliptic problems in D by the corresponding problems in $\mathcal{P}$ is $O(h^2)$.

However, for fourth-order elliptic problems, this is not true in any simple general sense. This fact is illustrated dramatically by the following important paradox due to Babuška [3], which shows how dangerous it can be to accept uncritically conclusions concerning approximations.

Physically, consider the problem of a simply supported, uniformly loaded plate. Or, equivalently, consider mathematically the boundary value problem defined by the biharmonic DE $\nabla^4 u = 1$, and the homogeneous linear boundary conditions $u = \partial^2 u/\partial n^2 = 0$. Let Δ be the unit disc, and let Δ_n be the regular n-sided (convex) polygon inscribed in this disc. Intuitively, one would expect the solutions $f_n(x, y)$ of the preceding boundary value problem in the polygons Δ_n to approach its solutions

$$(11) \qquad f(x, y) = \frac{1}{64}[(1-r^2)(5-r^2)], \quad r^2 = x^2 + y^2 ,$$

in the disc Δ, as $n \to \infty$. The Babuška paradox consists in the fact that they do not.

I have devoted considerable thought to the explanation of this Babuška paradox, which I have discussed with Babuška himself, as well as with D. S. Griffin at Bettis and with Hulme. I think that we now understand its origin fairly well (see also [16]).

A partial explanation of the Babuška paradox is provided by the notion of natural boundary conditions for a variational problem. Physically, the exact solution of the equations of elasticity for a "simply supported plate" with load density $p(x, y)$ and Poisson ratio ν minimizes the sum of the gravitational energy and the strain energy

$$(12)\qquad J[u] = \iint_R \tfrac{1}{2}[(\nabla^2 u)^2 - 2(1-\nu)(u_{xx}u_{yy} - u_{xy}^2)] - pu \; dx\,dy ,$$

subject to the boundary condition $u = 0$ on $\partial\mathfrak{R}$. The corresponding <u>natural</u> boundary conditions are,

$$(13a)\qquad \partial^2 u/\partial r^2 + \frac{\nu}{r}\frac{\partial u}{\partial r} = 0 \quad \text{on} \quad \partial\mathfrak{R} ,$$

in the case of a circular disc, whereas they are

$$(13b)\qquad \partial^2 u/\partial n^2 = 0 \quad \text{on} \quad \partial\mathfrak{R}$$

along the edges of a (regular) polygon. Hence the boundary condition $\partial^2 u/\partial n^2 = 0$ is <u>unnatural</u> for a simply supported disc unless $\nu = 0$, which is physically unstable.

A more careful examination of (12) brings out another interesting fact. The Euler-Lagrange equation equivalent to the variational condition $\delta J = 0$ in (12) is $\nabla^4 u = 0$, regardless of what the Poisson ratio ν is. Stated in another way, the term $2(1-\nu)(u_{xx}u_{yy} - u_{xy}^2)$, which is the linearized Gaussian curvature, does not affect the Euler-Lagrange DE for a loaded plate. This is because, as has been observed by Langhaar,† the Gauss-Bonnet Theorem of differential geometry implies

$$(14)\qquad \iint_R (u_{xx}u_{yy} - u_{xy}^2)\,dx\,dy = 2\pi - \sum \Delta\theta_i - \int_{\partial\mathfrak{R}} K\,ds ,$$

where the $\Delta\theta_i$ are the exterior vertex angles, and K is the (linearized) geodesic curvature of the surface $z = u(x,y)$. From this it follows that the deflection of a simply supported,

† H. L. Langhaar, J. Appl. Mech. 9 (1952), 228-
See also S. G. Mikhlin, "Variational Methods in Mathematical Physics", Pergamon, 1964, Sec. 17, and S. M. Key, Ph.D. Thesis, University of Washington, 1966.

uniformly loaded polygonal plate is independent of ν, whereas that for a disc depends on ν, as (13a) suggests.

ACKNOWLEDGEMENT

The research reported here was done under AEC Contract AT(30-1)-3971. The exposition owes much to friendly suggestions by R. Barnhill, Ralph Carlson, George Fix, W. J. Gordon, C. A. Hall, and Martin Schultz.

REFERENCES

[1] J.H. Ahlberg, E.N. Nilson, and J.L. Walsh, The Theory of Splines and Their Applications, Academic Press, 1967.

[2] A.K. Aziz (editor), Numerical Solution of Differential Equations, van Nostrand, 1969.

[3] I. Babuška (editor), Differential Equations and their Applications, Academic Press, New York, 1963. (Cf. Math. Revs. 23A (1962), p. 499, Review A2629.)

[4] G. Birkhoff, C. de Boor, B. Swartz, and B. Wendroff, "Rayleigh-Ritz Approximation by Piecewise Cubic Polynomials", J. SIAM Numer. Anal. 3 (1966), 188-203.

[5] G. Birkhoff and H.L. Garabedian, "Smooth Surface Interpolation", J. Math. and Phys. 39 (1960), 353-68.

[6] G. Birkhoff, M.H. Schultz, and R.S. Varga, "Piecewise Hermite Interpolation ... with Applications to Partial Differential Equations", Numerische Math. 11 (1968), 232-56.

[7] G. Birkhoff and R.S. Varga (editors), Numerical Solution of Field Problems in Continuum Physics, Proc. 2nd SIAM-AMS Symposia Appl. Math. Am. Math. Soc., 1969.

[8] R. E. Carlson and C. A. Hall, "Piecewise Polynomial Approximation of Smooth Functions in Rectangular Polygons", Report WAPD-T-2160, March, 1969.

[9] R. E. Carlson and C. A. Hall, "Bicubic Spline Approximation of Smooth Functions in Polygonal Regions", Report WAPD-T-2177, July, 1968.

[10] P. G. Ciarlet, M. H. Schultz, and R. S. Varga, "Numerical Methods ..., I. One-dimensional Problems", Numerische Math. 9 (1967), 394-430.

[11] George J. Fix, "Higher-order Rayleigh-Ritz Approximations", J. Math. Mech. 18 (1969), 645-58. For fuller details, see Fix's Ph. D. Thesis, Harvard Univ., 1968.

[12] H. L. Garabedian (editor), Approximation of Functions, Elsevier, 1965. (Section references are to Carl de Boor's and my article on pp. 164-90.)

[13] W. J. Gordon, "Spline-blended interpolation through curve networks", J. Math. Mech., 18(1969), 931-52.

[14] C. A. Hall, "Error Bounds for Spline Interpolation", J. Approximation Theory 1 (1968), 209-18.

[15] C. A. Hall, "Bicubic Interpolation over Triangles", J. Math. Mech. (to appear)

[16] B. L. Hulme, "Piecewise Bicubic Methods for Plate Bending Problems", Ph. D. Thesis, Harvard University, 1969.

[17] M. H. Schultz, "L^{∞}-multivariate Approximation Theory", to appear in SIAM J. Num. Analysis.

[18] J. H. Wilkinson, The Algebraic Eigenvalue Proglem, Oxford University Press, 1964.

APPENDIX A. SPLINE APPROXIMATION BY MOMENTS

It is instructive to consider bicubic spline functions vanishing on the boundary as solutions of the inhomogeneous generalized Mangeron (draftsman's) equation†

(A1) $$u^{(4,4)}(x,y) = \sum m_{k\ell}\ \delta(x_k, y_\ell)$$

in a rectangle $\mathfrak{R}$, and to consider smooth functions with fourth cross-derivative

(A2) $$u^{(4,4)}(x,y) = f(x,y) \quad ,$$

as given in terms of their Green's function (Peano-Sard kernel) by

(A3) $$u(x,y) = \iint_{\mathfrak{R}} G_{4,4}(x,y;\xi,\eta)\ f(\xi,\eta)\,d\xi\ d\eta \quad .$$

One can then interpret any scheme for <u>approximating</u> smooth functions in a rectangle by bicubic splines as a scheme for approximating the continuously distributed "loads" $f(x,y)$ associated with (A2) by point-concentrated loads of the type described in (A1).

In any rectangular domain, one can do this systematically by a "method of moments" whose univariate analog has been described elsewhere.‡ Namely, one can replace the continuously distributed load in any mesh-rectangle R (the shaded rectangle in Figure A-1) by loads concentrated at the 16 mesh-points of any 3×3 array S of subrectangles which contains R, in such a way as to <u>match moments</u>. That is, we can so assign point loads m_k to the 16 mesh-points (x_k, y) as to match the following 16 moments:

† See G. Birkhoff and W. J. Gordon, "The Draftsman's Equation", J. Approx. Theory 1 (1968), 199-208.

‡ G. Birkhoff, J. Math. Mech. 16 (1967), 987-90; C. de Boor, ibid (1968).

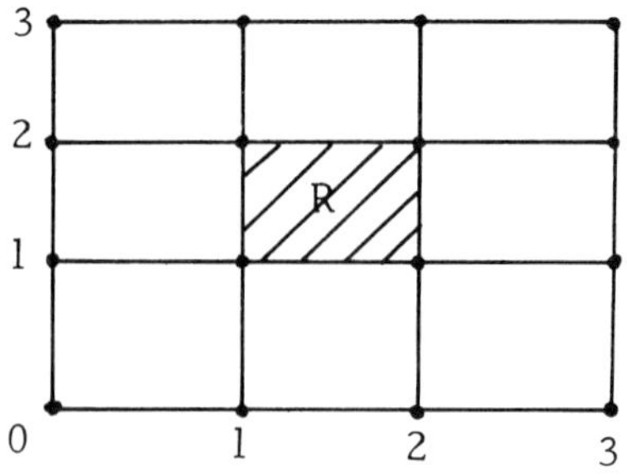

Figure A-1

$$\text{(A4)} \quad \iint_R x^i y^j f(x,y)\,dx\,dy = \sum_{k,\ell} x_k^i y_\ell^j m_{k\ell} \qquad (i,j = 0,1,2,3) \;.$$

Furthermore, this assignment is unique: given R, f, and the (x_k, y_ℓ), the 16 equations (A4) have unique solutions $m_{k\ell}$.

Moreover, the proof of Lemma 1 of my paper can be extended, to prove the following result. If one uses this "method of moments", to approximate any smooth function with $f(x,y) \equiv 0$ on ∂R for which $f(x,y)$ has support in R, then the approximating function has support in S. In this sense, the approximating function has a minimum support, just like the patch basis to be described in Appendix C. This method of moments specifically approximates the Green's function for the partial DE (A2) in the unit square, which is

$$\text{(A5)} \qquad G_{4,4}(x,y;\xi,\eta) = G_4(x,\xi)\, G_4(y,\eta) \;,$$

by the sum of a bicubic polynomial and a function of this patch basis.

Note that the preceding scheme of bicubic spline approximation by moments is also a method of projection: it is exact for any bicubic spline function — i.e., if $v = P_\pi[u]$ is any function in the approximating subspace, then $P_\pi[v] = v$.

APPENDIX B

COMBINATORIAL ASPECTS OF INTERPOLATION ALGEBRA

In this appendix, I shall describe a new approach to the interpolation algebra of piecewise polynomial functions with prescribed smoothness (e.g., bicubic spline functions). This approach is essentially combinatorial. Formulated abstractly, it concerns a new and interesting family of geometric lattices whose study will, I hope, become a new chapter of combinatorial algebra. However, my emphasis here will be on a few concrete examples, whose careful analysis by combinatorial methods will, I believe, have important applications to the construction of general digital computer programs for bivariate and multivariate spline functions in rectangular polygons and polyhedra.

Example B1. My first example concerns what has come to be called "G. D. Birkhoff's Interpolation Problem"†. In the vector space P_n of all univariate polynomial functions of degree at most n, for a given choice of $\pi: x_0 < x_1 < \ldots < x_n$, consider the linear functionals

$$\text{(B1)} \qquad \alpha_{k,i}: f \mapsto f^{(k)}(x_i) = \alpha_{k,i}[f] \qquad (k, i = 0, 1, \ldots, n) .$$

The question is: which subsets of these $(n+1)^2$ linear functionals are linearly independent in P_n', the dual of P_n. In which cases is this true for all π?

Considered with respect to linear dependence, these linear functionals define a geometric lattice of length $n+1$ with $(n+1)^2$ "atoms".‡ The basic question is to determine which subsets of these atoms are independent.

† G. D. Birkhoff, Trans. Am. Math. Soc. 7 (1906), 107-36; I. J. Schoenberg, J. Math. Anal. Appl. 16 (1966), 538-43; D. Ferguson, J. Approx. Theory 2 (1969), 1-28.

‡ For a very general abstract discussion of this fact, see G. Birkhoff, "Lattice Theory", 3d edition, p. 84.

Example B2. My next example concerns piecewise polynomial functions of degree $2p-1$ or less in $[x_0, x_n]$ with joints at the x_i of some partition π: $x_0 < x_1 < \ldots < x_n$. This time for $k = 0, 1, \ldots, 2p-1$ we let

$$\text{(B2)} \qquad \alpha_{k,i}: f \mapsto f^{(k)}(x_i^-) \quad \text{and} \quad \beta_{k,i}: f \mapsto f^{(k)}(x_i^+) \;.$$

These linear functionals can be taken as the $4pn$ "atoms" of a geometric lattice of length $2pn$. The periodic case with $x_0 = x_n$ is especially interesting, and applications to interpolation theory again rely on relations of linear independence which are independent of the choice of π.

Thus Hermite interpolation of order p is possible and unique because the set S of the $2pn$ $\alpha_{k,i}$ and $\beta_{k,i}$ with $k = p-1$ is linearly independent. Also, we are interested in the $2pn$ linear functionals $\gamma_{k,i} = \beta_{k,i} - \alpha_{k,i}$. The Hermite subspace is defined by the conditions $\gamma_{k,i} = 0$ for $k = 0, \ldots, p-1$ and all i: the spline subspace by the conditions $\gamma_{k,i} = 0$ for $k = 0, \ldots, 2p-1$ and all i: the geometric lattice of all intermediate subspaces defined by subsets of the $\alpha_{k,i}$, $\beta_{k,i}$, $\gamma_{k,i}$ is very interesting; some of these are discussed in [10]. Again, the essential observation underlying the basic computational formula of [12, p. 167, (3.1)] is that for any j the 9 atoms $\alpha_{k,i}$ with $k = 0, 1, 2$ and $i = j-1, j, j+1$ have a span of height 8 in the cubic spline subspace.

Rectangular Polygons. Though the preceding examples are very interesting as examples of geometric lattices, sophisticated combinatorial considerations are not needed to handle the relevant linear independence relations. But I suspect that it will be helpful for programming problems involving spline functions in rectangular polygons, to which I shall now turn my attention.

First, I observe that a given rectangular polygon is specified by any cyclic sequence of its corners: $C_1, C_2, \ldots, C_{2m+4}$. These are connected by $m+2$ horizontal edges H_i and $m+2$ vertical edges V_i, with $C_{2k-1} = H_k \cap V_k$ and

$C_{2k} = V_k \cap H_{k+1}$. The number of reentrant corners is m.

Now suppose $\mathfrak{R}$ subdivided by a rectangular partition π; let M denote the set of all mesh-points, and H and V the sets of all those mesh-points which lie on horizontal and vertical edges, respectively. Then $\mathbb{C} = H \cap V = \bigcup_{i=1}^{2m+4} C_i \subset M$ is the set of all corner mesh-points, while $H = \bigcup_{k=1}^{m+2} (H_k \cap M)$ and $V = \bigcup_{k=1}^{M+2} (V_k \cap M)$, where the symbol $\bigcup$ designates a sum of disjoint subsets.

Let finally G be the rectangular graph defined by $(\mathfrak{R}, \pi)$. I shall not try to define this class of planar graphs abstractly in its full generality; on the contrary, I shall suppose that its mesh-points can be represented as a subset $M \subset \underline{Z} \times \underline{Z}$, so that adjacent mesh-points (i,j) and (i',j') have "adjacent" coordinates with $|i-i'| + |j-j'| = 1$.

Basic index-sets. In his pioneer paper (presented at the age of nineteen!), my father considered sets S of index-pairs (k,i) such that the values of $f^{(k)}(x_i) = \alpha_{k,i}[f]$ for $(k,i) \in S$ uniquely determined a polynomial of degree n in x. Such sets are "basic index-sets" in the following sense.

Definition. A subset S of indices σ of a set of indexed linear functionals λ_σ on a function space V is a basic-index-set when the conditions $\lambda_\sigma(f) = c_\sigma$ uniquely determine $f \in V$.

Thus, when the λ_σ are interpolation functionals, like those of (B1) or (B2), an index-set is basic if and only if it defines an algebraically well-set interpolation problem. I think that the device of specifying functions through basic index-sets of linear functionals will be a very useful device for programming systematically piecewise polynomial interpolation.

Example B3. Varga's construction of a cubic spline "patch function" $u_p(x)$ on a subdivided interval $[x_0, x_I]$, vanishing identically outside $[x_{p-2}; x_{p+2}]$ is easily defined in the above notation. A basic index set of $\alpha_{k,i}$ consists of the $(0,i)$ for all $i \notin \{p-1, p+1\}$ and the $(1,i)$ with $i \in \{0, p-2, p+2, I\}$. We omit the details.

In discussing bivariate functions, it is natural to consider index-quadruples $(k,\ell;i,j)$ associated with linear functionals $\alpha_{k,\ell;i,j}[f] = f^{(k,\ell)}(x_i, y_j)$. Carl de Boor's scheme for interpolating bicubic spline functions in a rectangle $[x_0, x_I] \times [y_0, y_J]$ essentially asserts that the union C of the following four sets of index-quadruples $(k,\ell; i,j)$ is a basic index set for bicubic splines: (i) all $(0,0;i,j)$, (ii) all $(1,0;0,j)$ and all $(1,0;I,j)$, (ii') all $(0,1;i,0)$ and all $(0,1;i,j)$ for all $(i,j) \in \{(0,0), (I,0), (0,J), (I,J)\}$.

In the case of <u>subdivided rectangles</u> (that is, if $m = 0$ in the classification of rectangular polygons above), tensor product techniques enable one to deduce an enormous number of theorems about algebraically well-set bivariate interpolation schemes in $\Re$ from corresponding theorems about well-set univariate interpolation schemes on lines. For example, we can extend de Boor's interpolation scheme as follows.

THEOREM B1. For $(\Re, \pi)$ a subdivided rectangle, let there be given: (i) $u(x_i, y_j)$ at all mesh-points, (ii) $u_x(x_i, y_j)$ on two vertical mesh-lines, and (iii) $u_y(x_i, y_j)$ on two horizontal mesh-lines. The data (i)-(iii) and the values of $u_{xy}(x_i, y_j)$ at four mesh-points P, Q, R, S uniquely determine $u \in Sp(\Re, \pi, 2)$ in each of the following cases: (a) $\overline{PQ}$ and $\overline{RS}$ any two horizontal mesh-lines or any two vertical mesh-lines, (b) $\overline{PQ}$ and $\overline{PR}$ any horizontal and vertical mesh-lines, and S on neither.

The conditions of Theorem B1 always define <u>algebraically</u> well-set problems. Unfortunately, the only case in which they define <u>analytically</u> well-set problems is the case that P, Q, R, S are the four corners of $\Re$. This is the case covered by de Boor's Theorem (and Theorem 1).

<u>The case m = 1</u>. For a general subdivided rectangular polygon $(\Re, \pi)$, we define the <u>spline subspace</u> as in (4):

$$\text{(B3)} \qquad Sp(\Re, \pi, q) = H_\pi^q(\Re) \cap C^{(2q-2)}(\Re) \quad ,$$

where $H_\pi^q(\Re)$ is the Hermite subspace of all piecewise polyno function of degree $(2q-1, 2q-1)$ or less and class $C^{(q-1)}(\mathfrak{R})$ on $(\mathfrak{R}, \pi)$. An important theorem proved in [8] asserts that

$$\text{(B4)} \qquad Sp(\mathfrak{R}, \pi, q) = H_\pi^q(\mathfrak{R}) \cap C^{(2q-2, 2q-2)}(\mathfrak{R}) .$$

Further, Theorem 2 of Carlson and Hall implies the following dimensional count:

$$\text{(B5)} \qquad \dim Sp(\mathfrak{R}, \pi, q) = |M| + (q-1)(|H| + |V| - 2m) + 4(q-1)^2 .$$

(In [8], there is given also a dimensional count for the "intermediate" spline subspaces $H_\pi^q(\mathfrak{R}) \cap C^{(s)}(\mathfrak{R})$ with $q \leq s \leq 2q-3$, $q > 2$.)

For example, if $\mathfrak{R} = \mathfrak{L}$ is an L-shaped region with mesh-points (x_i, y_j), $i = 0, 1, \ldots, I$, $j = 0, 1, \ldots, J$, whose reentrant corner is at (x_k, y_ℓ), then $|M| = (I+1)(J+1) - (I-k)(J-\ell)$, $|H| + |V| = 2I + 2J + 2$, $m = 1$, and hence the dimension of the subspace of bicubic splines on $(\mathfrak{L}, \pi)$ is

$$\text{(B6)} \qquad \dim Sp(\mathfrak{L}, \pi, 2) = I\ell + Jk - k\ell + 3I + 3J + 9 .$$

The dimension of the subspace of bicubic spline functions which satisfy $u \equiv 0$ on the boundary $\partial\mathfrak{L}$ is, similarly,

$$\text{(B7)} \qquad I\ell + Jk - k\ell + I + J - 3 ,$$

since there are $2I + 2J$ edge points where u must vanish and 6 corner points where the two first partial derivatives must vanish to make $u \equiv 0$ on $\partial\mathfrak{L}$. Note that for such problems, the boundary conditions at any reentrant corner are compatible (for bicubic splines) with the single boundary condition $u = 0$ at the end of the mesh-line going along each edge adjacent to the reentrant corner.

Formula (B7) also applies to any L-shaped region which constitutes one quadrant of a hollow square, with the boundary conditions indicated in Figure B1 below (where $I = J = 6$ and Gary Birkoff has noted that the dimension of each of the above spaces (of bicubic splines with homogeneous boundary conditions) is exactly 4 less than the number of mesh-points, or $|M| - 4$.

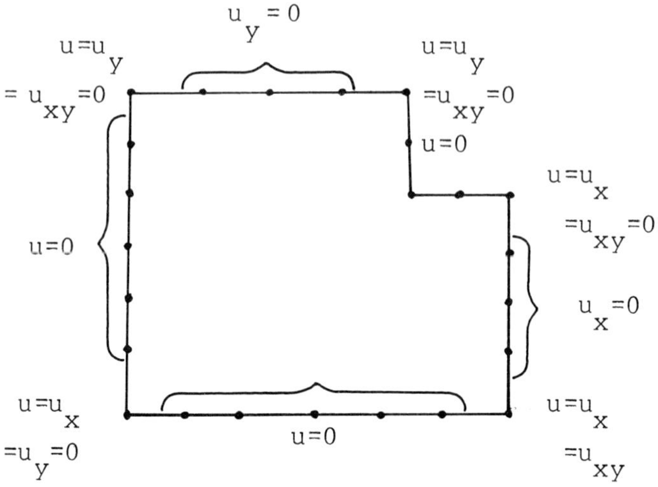

Figure B1

Carlson, Fix, and Hall have proved

(B8) $$\dim \mathcal{S}_0(\mathcal{R}, \pi) = |M| - 4m \quad .$$

However, even more challenging questions concern the degrees of approximation and convergence by <u>bivariate</u> spline functions and I shall now turn your attention to some observations of George Fix and Gary Wakoff about these.

APPENDIX C

EIGENVALUES OF RECTANGULAR POLYGONAL MEMBRANES

by G. J. Fix and G. I. Wakoff

In [11] (see also [7] for a brief summary), one of us described a computer program for calculating eigenvalues of L-shaped membranes. This program applies the Rayleigh-Ritz method to an approximating "Hermite" subspace spanned by a basis of piecewise bicubic polynomial "patch functions," supplemented by singular functions to match the leading singular terms in the analytical expansion of the exact solution.

We have since extended this program so as to cover the following rectangular polygons: a hollow square, a T-shaped domain and an H-shaped domain. Using vertical symmetry and (where it exists) horizontal symmetry, and confining attention to eigenvalues of even (even-even) eigenfunctions, the domains of the above problems can be reduced to those indicated in Figures C1 and C2.

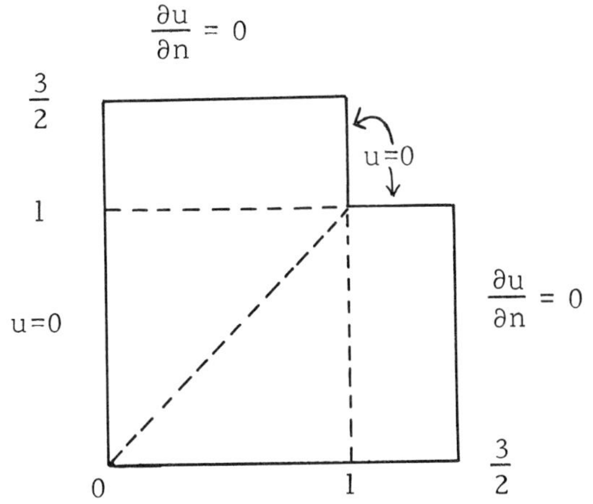

Figure C1

(Hollow Square)

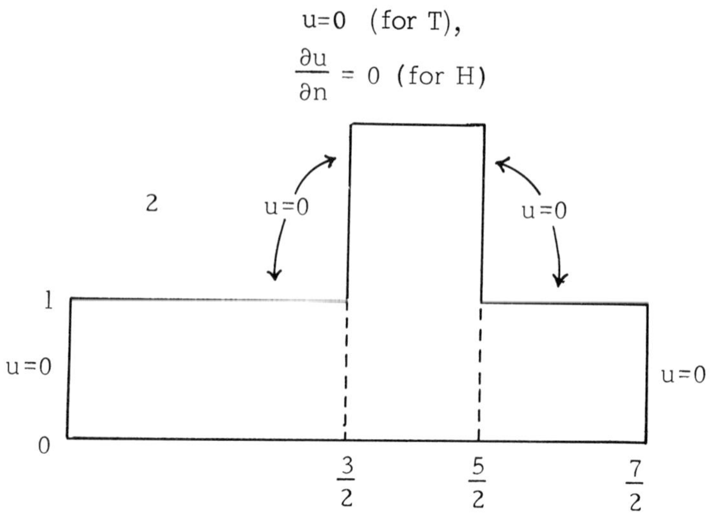

Figure C2

(T-Shaped and H-Shaped Regions)

The program assumes given a basis of functions φ_i (e.g. a basis of piecewise polynomial patch functions) which span an approximating Hermite subspace of the space of all smooth functions which satisfy the specified boundary and symmetry conditions. The program then evaluates the inner product matrices $A = \|a_{ij}\|$ and $B = \|b_{ij}\|$, where $a_{ij} = (\nabla\varphi_i, \nabla\varphi_j)$ and $b_{ij} = (\varphi_i, \varphi_j)$. Finally, it uses the Rayleigh-quotient iteration scheme to compute the lowest eigenvalue of the resulting matrix eigenvalue problem $A\underline{x} = \lambda B\underline{x}$.†

† Forsythe and Wasow, "Finite Difference Methods for Partial Differential Equations," Wiley, 1965, pp. 353-7 and 375-6.

The program can also be modified to handle any rectangular polygon. One need only specify certain input information, namely, reentrant corners, connectors, and annihilators. These specify the geometry and boundary conditions of the region in question.

The reentrant corners specify where and in what order singular functions must be added to the basis.

Connectors relate the support of the basis functions to the rectangular polygon in question. Each bicubic Hermite "patch basis" function is non-zero on a subdomain consisting of four subrectangles with a common vertex. Our "connectors" associate a 1 with each subsquare of this square which is in the rectangular polygon and a 0 with each subsquare that is not. In this way, the connectors also designate the geometry of the region.

Annihilators cover the boundary conditions. In bicubic Hermite interpolation, four quantities u, u_x, u_y, and u_{xy} at every mesh-point are used to specify each approximating function. However, at boundary mesh-points, one or more of these is specified by the boundary condition ($u = 0$ or $\partial u/\partial n = 0$). Annihilators indicate by a 0 which of u, u_x, u_y, and u_{xy} is (are) specified by the boundary condition.

Using this program the numerical results obtained by the above programs for the lowest eigenvalue were as follows (on an IBM 7094):

TABLE 1

Hollow Square	h	λ	Dims of Subspace	Time (Min)
	1/2	8.71301	33	1.1
	1/4	8.61132	131	2.8
	1/8	8.60909	515	7.7

T-Shape	h	λ	Dims of Subspace	Time (Min)
	1/2	8.17869	78	1.6
	1/4	8.11668	294	5.5

H-Shape	h		Dims of Subspace	Time (Min)
	1/2	7.98949	78	1.6
	1/4	7.92727	294	5.6

In the case of the hollow square, the results indicate $O(h^6)$ convergence. As in [7], this was obtained by defining the singular functions over one mesh length around the re-entrant corner (see Figure C3) and adding more singular functions as h was reduced (three singular functions were used for h = 1/2, five for h = 1/4, 1/8).

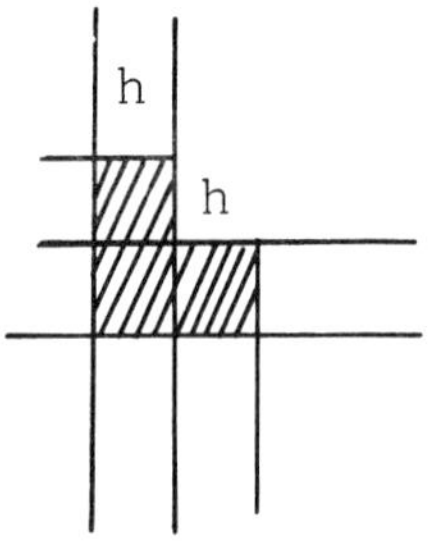

Figure C3

Domain of Singular Functions

The results for the T-and H-shaped regions were obtained using five singular functions per reentrant corner. By analogy with the L-shaped region and the hollow square, we believe that the values for h = 1/4 are accurate to three to four significant figures. The singular functions used are described explicitly in Appendix A of the article by Birkhoff and Fix in [7].

As can be seen from Table I, the dimensionality of the subspaces involved increases drastically as h decreases. This limits the accuracy that can be obtained due to cumulative rounding error and limited computer storage. Since bicubic spline approximation gives the same order of accuracy with only about 1/4 as many basis functions, we plan to improve the efficiency of our program by using the subspace of (piecewise) bicubic spline functions.

Our next problem will be to construct a suitable "patch basis" for the subspace S_0 of spline functions which exactly satisfy the typical boundary conditions $u = 0$ and $\partial u/\partial n = 0$, in an L-shaped region (say). Work is in progress on developing a program to do this, and to evaluate the associated inner products with singular functions. We hope thereby to approximate every eigenfunction which satisfies the same boundary conditions with an error which is $O(h^3)$, and the eigenvalues to $O(h^6)$.

Editorial comments by I. J. Schoenberg: Professor Birkhoff's discussion of "Pros and cons of splines" on pages 200-201 seems fairly evenhanded. However, examples of what he calls the "Patch bases" of the applied mathematicians were already used by me in the reference [63] on page 346 of these Proceedings, for the case of a single variable and equidistant knots. In 1947 these were generalized by H. B. Curry and myself, in the reference [1] on page 182, to non-equidistant knots and called "fundamental spline functions", a name that was later changed to "B-splines". These were used in [5], on page 117, to solve the spline interpolation problem by means of a well-conditioned matrix. The B-splines have remarkable variation diminishing properties and led naturally to the variation diminishing spline approximation methods. In the second part of reference [1], on page 182, it is shown how the B-splines lead to the Polya frequency functions. True, all these remarks refer to functions of a single real variable. Just the same I think that they should be mentioned in this connection.

Distributive Lattices and the Approximation of Multivariate Functions

WILLIAM J. GORDON

Abstract: The purpose of this paper is to develop the basic notions of an algebraic approach to the approximation of multivariate functions. We show that any commutative collection of linear projection operators $\mathcal{P}_1, \ldots, \mathcal{P}_M$ (approximation operators) generates a distributive lattice Ψ of projectors and that any operator $\mathcal{A}$ in the lattice provides an approximation $\mathcal{A}[F]$ to the function $F(x_1, x_2, \ldots, x_N)$. In particular, the lattice has a unique "maximal" projector $\mathfrak{M}$ and a unique "minimal" projector $\mathfrak{L}$. The maximal projector has all the algebraic attributes of the entire collection $\mathcal{P}_1, \ldots, \mathcal{P}_M$. The minimal projector, on the other hand, has only those attributes common to all the $\mathcal{P}_m$ $(m = 1, 2, \ldots, M)$. Consistent with the definition of the ordering relation "$\leq$" of the lattice, we term the function $\pi = \mathfrak{M}[F]$ the algebraically best approximation to F.

For any commutative collection of projectors $\mathcal{P}_m$ $(m = 1, 2, \ldots, M)$, we consider in Section II the associated set of remainder operators $\mathcal{R}_m = I - \mathcal{P}_m$ and show that the remainders also generate a distributive lattice Ψ'. Hence, for any projector $\mathcal{A} \in \Psi$, there is a uniquely associated remainder $\mathcal{A}' \in \Psi'$ such that $I = \mathcal{A} + \mathcal{A}'$. Specifically, we show that the remainder operator $\mathfrak{M}'$ associated with the maximal projector $\mathfrak{M}$ is simply the product of the $\mathcal{R}_m$; i.e., $\mathfrak{M}' = \mathcal{R}_1 \mathcal{R}_2 \ldots \mathcal{R}_M$. Our principal results are summarized in Theorem 4.

The algebraic formalism of Sections I and II provides an operational calculus for the analysis of multivariate functions. In the third section, we apply that operational calculus to some specific collections of projectors. Example 1 analyzes what is perhaps the simplest, nontrivial class of multivariate projectors which satisfies the essential commutativity hypotheses. These projectors are obtained by "parametric extension" of univariate projection operators such as those of Taylor, Lagrange, and Fourier. We show that the algebraically minimal projectors $\mathfrak{L} = \mathcal{P}_1\mathcal{P}_2 \cdots \mathcal{P}_M$ correspond to the well-known finite dimensional tensor product (alias cross product, Cartesian product, Kronecker product) methods of multivariate approximation. The second example involves expansion of a trivariate function in terms of its values on the bounding surface of a right circular cylinder of unit height. There, we use the fact that the Poisson integral operator (i.e., the inverse of the Laplacian ∇^2 for the Dirichlet problem on a circle) is a projector. Finally, in Example 3, we consider the ad hoc construction of a formula for interpolation to a bivariate function $F(x, y)$ on the perimeter of a triangular domain.

The algebraic theory developed in the text provides approximations to an N-variate function in terms of functions of fewer variables. In an appendix, we consider the practical problem of numerical approximation; i.e., approximation of $F(x_1, \ldots, x_N)$ in terms of a finite number of scalar samples of the function. Two extreme alternative methods for numerical approximation are compared, one of which ("successive decomposition") uses the maximal projector $\mathfrak{M}$ and the other of which (tensor product) uses the minimal element $\mathfrak{L}$. With cubic spline projectors as an example, we show that a specified accuracy of approximation can be far more economically achieved, in terms of the total number of data points required, by using the notion of successive decomposition than by the analogous tensor product method. Tables I-VII summarize the comparisons of the two methods.

On the basis of the results of Appendix A and of various related unpublished investigations, one sees that the algebraic formalism of the text, together with the notion of successive decomposition outlined at the beginning of Appendix A, provide a promising new approach to problems of multivariate numerical analysis.

DISTRIBUTIVE LATTICES AND MULTIVARIATE FUNCTIONS

I. Lattices of Approximation Operators

In this section we develop the rudiments of an algebraic approach to multivariate approximation theory. The linear space $\mathfrak{F}$ is any space of functions of the N independent variables $x_1, x_2, \ldots, x_N$ over which the M projectors $\mathcal{P}_m$ $(m = 1, 2, \ldots, M)$ are defined. In general, the number of projectors is independent of the number of variables.

Definition 1: A projector $\mathcal{P}$ is an idempotent linear transformation from a linear space $\mathfrak{F}$ onto a subspace Φ.

A function $\pi = \mathcal{P}[F]$ which is the image of F under $\mathcal{P}$ is called an approximation to F; and the function $(F-\pi)$ is the remainder (error). The remainder is obviously in the subspace of $\mathfrak{F}$ complementary to Φ.

Consider now a collection of M projectors $\mathcal{P}_1, \mathcal{P}_2, \ldots, \mathcal{P}_M$ from the linear space $\mathfrak{F}$ onto the subspaces $\Phi_1, \Phi_2, \ldots, \Phi_M$, respectively. The intersection of any subset of the ranges $\Phi_m (m = 1, 2, \ldots, M)$ will, in general, be nonempty, i. e., the ranges of the projectors are not disjoint.

The notions of addition and multiplication of projectors are defined in the usual way for linear operators, so that the following associative and distributive rules are implicitly satisfied by any collection of projectors $\mathcal{P}_m$ $(m = 1, 2, \ldots, M)$ defined over the same domain $\mathfrak{F}$:

(A1) $$\mathcal{P}_k(\mathcal{P}_\ell \mathcal{P}_m) = (\mathcal{P}_k \mathcal{P}_\ell)\mathcal{P}_m$$

(A2) $$\mathcal{P}_k(\mathcal{P}_\ell + \mathcal{P}_m) = \mathcal{P}_k\mathcal{P}_\ell + \mathcal{P}_k\mathcal{P}_m$$

(A2') $$(\mathcal{P}_k + \mathcal{P}_\ell)\mathcal{P}_m = \mathcal{P}_k\mathcal{P}_m + \mathcal{P}_\ell\mathcal{P}_m$$

The crucial property which we need to assume in order to construct a lattice from the collection $\{\mathcal{P}_m\}$ is that of commutativity:

(A3) $$\mathcal{P}_k\mathcal{P}_m = \mathcal{P}_m\mathcal{P}_k \quad \text{for all } k, m = 1, 2, \ldots, M .$$

The product of any two commutative projectors, $\mathcal{A} = \mathcal{P}_k\mathcal{P}_\ell$, is again a projector since $\mathcal{A}\mathcal{A} = \mathcal{A}$. The binary operation of addition has, however, the unpleasant aspect that the sum of two commutative projectors $\mathcal{P}_k$ and $\mathcal{P}_\ell$ is not a projector since $(\mathcal{P}_k + \mathcal{P}_\ell)(\mathcal{P}_k + \mathcal{P}_\ell) \neq \mathcal{P}_k + \mathcal{P}_\ell$, i.e., the linear operator $\mathcal{A} = \mathcal{P}_k + \mathcal{P}_\ell$ is not idempotent. For this reason, we shall use, instead of ordinary operator addition, the "$\oplus$ addition" defined below.

Definition 2 (Boolean Addition): The Boolean sum of two projectors is:

$$\mathcal{A} \oplus \mathcal{B} = \mathcal{A} + \mathcal{B} - \mathcal{A}\mathcal{B} \quad . \tag{1}$$

Note that $\oplus$ addition coincides with ordinary operator addition if and only if $\mathcal{A}\mathcal{B} = 0$ (i.e., $\mathcal{A}$ and $\mathcal{B}$ are mutually orthogonal).†

Lemma 1: The $\oplus$ sum, $\mathcal{P}_k \oplus \mathcal{P}_\ell$, and the product $\mathcal{P}_k\mathcal{P}_\ell$ of any two projectors from the collection $\{\mathcal{P}_m\}_{m=1}^{M}$ are also projectors.

Some approximations to F are better than others. To make precise the notion of "better," we introduce an ordering relation $\leq$. Then, if $\mathcal{A} \leq \mathcal{B}$, we can think of the projector $\mathcal{B}$ as being, in an algebraic sense, larger than $\mathcal{A}$.

Definition 3: $\mathcal{A} \leq \mathcal{B}$ means $\mathcal{A}\mathcal{B} = \mathcal{A}$.

Let Ψ denote the set of all projectors $\mathcal{A}, \mathcal{B}, \ldots$ which can be built up as combinations of the projectors $\mathcal{P}_m$ $(m = 1, 2, \ldots, M)$ using the binary operations of $\oplus$ addition and operator multiplication. That is, Ψ is the set of all Boolean polynomials in the M "variables" $\mathcal{P}_m$ $(m = 1, 2, \ldots, M)$.

† The notion of two projectors being mutually orthogonal should not be confused with the concept of an "orthogonal projection" from an inner product space onto an orthogonal subspace (cf. Corollary 4, Section III).

Theorem 1: The set Ψ is a distributive lattice under the partial ordering $\leq$.

Proof: The proof consists of a straightforward verification of the postulates which define a distributive lattice ([4, pp. 351-354][12, pp. 482-502]). In terms of the projectors $\mathfrak{A}, \mathfrak{B}, \mathfrak{C}, \ldots$ of Ψ and the operations of $\oplus$ addition and multiplication, this means:

i) Reflexive $\mathfrak{A} \leq \mathfrak{A}$ for all $\mathfrak{A} \in \Psi$

ii) Anti-symmetric $\mathfrak{A} \leq \mathfrak{B}$ and $\mathfrak{B} \leq \mathfrak{A}$ imply $\mathfrak{A} = \mathfrak{B}$

iii) Transitive $\mathfrak{A} \leq \mathfrak{B}$ and $\mathfrak{B} \leq \mathfrak{C}$ imply $\mathfrak{A} \leq \mathfrak{C}$

iv) Idempotent $\mathfrak{A} \oplus \mathfrak{A} = \mathfrak{A}$ and $\mathfrak{A}\mathfrak{A} = \mathfrak{A}$

v) Commutative $\mathfrak{A} \oplus \mathfrak{B} = \mathfrak{B} \oplus \mathfrak{A}$ and $\mathfrak{A}\mathfrak{B} = \mathfrak{B}\mathfrak{A}$

vi) Associative $\mathfrak{A}(\mathfrak{B}\mathfrak{C}) = (\mathfrak{A}\mathfrak{B})\mathfrak{C}$

$\mathfrak{A} \oplus (\mathfrak{B} \oplus \mathfrak{C}) = (\mathfrak{A} \oplus \mathfrak{B}) \oplus \mathfrak{C}$

vii) Distributive $\mathfrak{A}(\mathfrak{B} \oplus \mathfrak{C}) = \mathfrak{A}\mathfrak{B} \oplus \mathfrak{A}\mathfrak{C}$

$\mathfrak{A} \oplus (\mathfrak{B}\mathfrak{C}) = (\mathfrak{A} \oplus \mathfrak{B})(\mathfrak{A} \oplus \mathfrak{C})$

(viii) Consistency The three conditions $\mathfrak{A} \leq \mathfrak{B}$, $\mathfrak{A} \oplus \mathfrak{B} = \mathfrak{B}$ and $\mathfrak{A}\mathfrak{B} = \mathfrak{A}$ are equivalent.

All of these relations are quite easy to verify and, therefore, we omit the details.

Since a lattice is only partially ordered, not every two projectors $\mathfrak{A}$ and $\mathfrak{B}$ in Ψ are comparable; i.e., it is not necessarily true that either $\mathfrak{A} \leq \mathfrak{B}$ or $\mathfrak{B} \leq \mathfrak{A}$. But, by the definition of a lattice [4, p. 352], any two elements of Ψ do have a "least upper bound" and a "greatest lower bound".

Definition 4: The least upper bound (l.u.b.) of the projectors $\mathfrak{A}$ and $\mathfrak{B}$ in the lattice Ψ is the smallest (with respect to the order relation $\leq$) projector $\mathcal{C}$ such that $\mathcal{A} \leq \mathcal{C}$ and $\mathcal{B} \leq \mathcal{C}$. The greatest lower bound of $\mathcal{A}$ and $\mathcal{B}$ is the largest $\mathcal{C}$ such that $\mathcal{C} \leq \mathcal{A}$ and $\mathcal{C} \leq \mathcal{B}$.

Lemma 2: If $\mathcal{A}$ and $\mathcal{B}$ are projectors in Ψ, then

$$\text{l.u.b. } \{\mathcal{A}, \mathcal{B}\} = \mathcal{A} \oplus \mathcal{B} \tag{2}$$

$$\text{g.l.b. } \{\mathcal{A}, \mathcal{B}\} = \mathcal{A}\mathcal{B} \ . \tag{3}$$

Let the projector $\mathcal{A}$ be any element of the lattice Ψ generated by the $\{\mathcal{P}_m\}_{m=1}^{M}$. Then, $\pi = \mathcal{A}[F]$ is an approximation to $F \in \mathfrak{F}$ and the error is $(F - \pi)$. Hence, every $\mathcal{A} \in \Psi$ provides a method for approximating $F \in \mathfrak{F}$. Among all of these, we seek that projector $\mathfrak{M}$ which is maximal and provides the algebraically best approximation. By maximal we mean that $\mathcal{A} \leq \mathfrak{M}$ for all $\mathcal{A} \in \Psi$. Similarly, the minimal projector $\mathfrak{L}$ is that lattice element for which $\mathfrak{L} \leq \mathcal{A}$ for all $\mathcal{A} \in \Psi$. The proof of the next result involves repeated application of Lemma 2.

Lemma 3:

$$\begin{aligned} \mathfrak{M} &= \text{l.u.b. } \{\mathcal{P}_m\}_{m=1}^{M} \\ &= \mathcal{P}_1 \oplus \mathcal{P}_2 \oplus \dots \oplus \mathcal{P}_M \end{aligned} \tag{4}$$

and

$$\begin{aligned} \mathfrak{L} &= \text{g.l.b. } \{\mathcal{P}_m\}_{m=1}^{M} \\ &= \mathcal{P}_1 \mathcal{P}_2 \dots \mathcal{P}_M \ . \end{aligned} \tag{5}$$

The maximal projector $\mathfrak{M}$ combines the approximation properties of the complete collection $\{\mathcal{P}_m\}$. Heuristically, if $\mathcal{S}_m$ represents the set of algebraic attributes of the projector $\mathcal{P}_m$, then $\mathfrak{M}$ has the set of attributes $\mathcal{S}_1 \cup \mathcal{S}_2 \cup \dots \cup \mathcal{S}_M$

At the other extreme, the minimal projector $\mathcal{L}$ has only those properties which are shared by all the projectors $\mathcal{P}_m$ $(m = 1, 2, \ldots, M)$. Namely, $\mathcal{L}$ has the set of attributes $\mathcal{S}_1 \cap \mathcal{S}_2 \cap \ldots \cap \mathcal{S}_M$. These remarks are one interpretation of the following lemma.

<u>Lemma 4</u>: The maximal projector $\mathfrak{M}$ has the properties that

(6) $$\mathcal{P}_m \mathfrak{M} = \mathcal{P}_m \quad \text{for all } m = 1, 2, \ldots, M .$$

The minimal projector $\mathcal{L}$ satisfies

(7) $$\mathcal{P}_1 \mathcal{P}_2 \ldots \mathcal{P}_M \mathcal{L} = \mathcal{P}_1 \mathcal{P}_2 \ldots \mathcal{P}_M .$$

The proofs of these results are trivial consequences of Lemma 3 and the definitions of g.l.b. and l.u.b. in terms of the ordering relation $\leq$. Nevertheless, the conclusions (6) and (7) have important implications for multivariate approximation theory.

<u>The expression for the maximal projector $\mathfrak{M}$ can also be given in recursive form</u>. Instead of $\oplus$ addition, the formulas below involve ordinary operator addition:

(8) $$\mathfrak{M} = \mathcal{P}_1^* + \mathcal{P}_2^* + \ldots + \mathcal{P}_M^*$$

where

(9) $$\mathcal{P}_1^* = \mathcal{P}_1$$
$$\mathcal{P}_m^* = \mathcal{P}_m - \mathcal{P}_m \sum_{i=1}^{m-1} \mathcal{P}_i^* \qquad (m = 2, 3, \ldots, M) .$$

Note the resemblance of these recursion relations to those of Gram-Schmidt orthogonalization. Note too that $\mathcal{P}_k^* \mathcal{P}_m^* = 0$ for $k \neq m$; and, of course, $\mathcal{P}_m^* \mathcal{P}_m^* = \mathcal{P}_m^*$ since the "orthogonalized" operators $\mathcal{P}_m^*$ are projectors.

The algebraic structure discussed above is isomorphic to the familiar "algebra of sets" under the binary operations of union and intersection. In effect, multiplication and $\oplus$

addition of the projectors $\{\mathcal{P}_m\}$ induces a set algebra on their ranges $\{\Phi_m\}$.

Theorem 2: Under the correspondences $\mathcal{P}_m \to \Phi_m$, $\cdot$ (multiplication) $\to \cap$, $\oplus \to \cup$ and $\leq \to \subseteq$ (set inclusion), the lattice Ψ is isomorphic to the distributive lattice generated by the ranges Φ_m $(m = 1, 2, \ldots, M)$.

Hence, given a projector $\mathcal{Q} \in \Psi$, there is never any doubt as to its range: In the Boolean expression for $\mathcal{Q}$ we simply make the replacements indicated in Theorem 2 and the resulting Boolean expression in terms of $\cup$, $\cap$ and $\{\Phi_m\}$ defines the range of $\mathcal{Q}$. In particular, the image function $\pi = \mathfrak{M}[F]$ is in the largest space formed as combinations of the subspaces $\Phi_m (m = 1, 2, \ldots, M)$ — namely, the space $\Phi_1 \cup \Phi_2 \cup \ldots \cup \Phi_M$. And, the image of F under $\mathfrak{L}$ is in the smallest space — $\Phi_1 \cap \Phi_2 \cap \ldots \cap \Phi_M$.

II. Complementarity, Duality and Remainder Theory

In this section we will study the lattice of remainders Ψ' associated with the lattice Ψ generated by $\{\mathcal{P}_m\}$. We first note the properties of the identity operator I and the null operator O on the space $\mathfrak{F}$. Clearly both these operators are projectors and, for any projector $\mathcal{Q}$, they satisfy:

ix) $$O \leq \mathcal{Q} \leq I$$

x) $$\mathcal{Q}O = O\mathcal{Q} = O \quad \text{and} \quad \mathcal{Q}I = I\mathcal{Q} = \mathcal{Q}$$

xi) $$O \oplus \mathcal{Q} = \mathcal{Q} \quad \text{and} \quad I \oplus \mathcal{Q} = I .$$

Definition 5: If $\mathcal{Q}$ is a projector, the remainder operator $\mathcal{Q}'$ associated with $\mathcal{Q}$ is defined to be the complement of $\mathcal{Q}$:

$$\mathcal{Q}' \equiv I - \mathcal{Q} \tag{10}$$

where I is the identity operator on the space $\mathfrak{F}$.

Lemma 5: If $\mathfrak{A}$ is a projector from $\mathfrak{F}$ onto a subspace Φ, then $\mathfrak{A}'$ is a projector from $\mathfrak{F}$ onto the subspace Φ' — the complement of Φ relative to $\mathfrak{F}$.

In view of the foregoing definitions and lemmas, it is easy to see that projectors and their complements obey the following rules of composition:

xii) Complementation $\mathfrak{A}\mathfrak{A}' = \mathfrak{A}'\mathfrak{A} = 0$ and $\mathfrak{A} \oplus \mathfrak{A}' = I$

xiii) Dualization $(\mathfrak{A} \oplus \mathfrak{B})' = \mathfrak{A}'\mathfrak{B}'$

$(\mathfrak{A}\mathfrak{B})' = \mathfrak{A}' \oplus \mathfrak{B}'$

xiv) Involution $(\mathfrak{A}')' = \mathfrak{A}$.

Note that the lattice Ψ generated by $\{\mathcal{P}_m\}$ does not contain the complements of projectors and therefore does not, in general, contain the projectors O and I . But, if in addition to the binary operations of multiplication and $\oplus$ addition, we also allow the unary operation of complementation, then the resulting lattice Ω is a Boolean algebra. We recall the definition ([4, p. 354], [12, p. 500]):

Definition 6: A Boolean algebra is a complemented distributive lattice with an identity I.
(Note: "Complemented" means that if $\mathfrak{A} \in \Omega$, then $\mathfrak{A}' \in \Omega$.)

We now see that Ψ contains only elements which are compositions of the projectors $\mathcal{P}_m$, whereas the Boolean algebra contains all possible Boolean compositions of the projectors $\mathcal{P}_m$ and the corresponding remainders $\mathcal{P}'_m$.

In order to emphasize the significance of subsequent algebraic formulae to multivariate approximation theory, we shall henceforth designate the M remainder operators $\mathcal{P}'_m \equiv I - \mathcal{P}_m$ as $\mathfrak{R}_m$ $(m = 1, 2, \ldots, M)$.

In addition to the sublattice $\Psi \in \Omega$, there is also another important sublattice Ψ' which is the dual of Ψ .

Lemma 6: The set Ψ' of all Boolean polynomials in the "variables" $\mathfrak{R}_m$ $(m = 1, 2, \ldots, M)$ which are formed using the binary operations of multiplication and $\oplus$ addition is a distributive lattice. The lattices Ψ and Ψ' are isomorphic under the correspondence $\mathfrak{A} \in \Psi \rightarrow \mathfrak{A}' \in \Psi'$.

For any projector $\mathfrak{A} \in \Psi$ there is a uniquely associated <u>dual projector</u> $\hat{\mathfrak{A}} \in \Psi'$ which is obtained from $\mathfrak{A}$ by replacing all of the $\mathcal{P}_m$ in the Boolean expression for $\mathfrak{A}$ by the corresponding $\mathfrak{R}_m (\equiv \mathcal{P}'_m)$. <u>Note that the dual $\hat{\mathfrak{A}}$ is not, in general, the same as the complement $\mathfrak{A}'$</u>. For instance, if $\mathfrak{A} = \mathcal{P}_1 \oplus \mathcal{P}_2$, its dual is $\hat{\mathfrak{A}} = \mathfrak{R}_1 \oplus \mathfrak{R}_2$ and its complement is $\mathfrak{A}' = \mathfrak{R}_1 \mathfrak{R}_2$.

For every result derived in Section I, for the lattice Ψ, there is a dual result for Ψ'. For example, the dual of Lemma 4 reads:

$$\hat{\mathfrak{M}} = \text{l.u.b.}\{\mathfrak{R}_m\}_{m=1}^{M} = \mathfrak{R}_1 \oplus \mathfrak{R}_2 \oplus \ldots \oplus \mathfrak{R}_M \tag{4'}$$

and

$$\hat{\mathfrak{L}} = \text{g.l.b.}\{\mathfrak{R}_m\}_{m=1}^{M} = \mathfrak{R}_1 \mathfrak{R}_2 \ldots \mathfrak{R}_M \ . \tag{5'}$$

Next, we examine the relationships between the maximal and minimal elements of the two sublattices Ψ and Ψ'. From (4'), (5') and the following lemma it follows by property (xiv) that $\mathfrak{M}' = \hat{\mathfrak{L}}$ and $\mathfrak{L}' = \hat{\mathfrak{M}}$.

<u>Lemma 7</u>:

$$\mathfrak{M} = \text{l.u.b.}\{\mathcal{P}_m\} = (\text{g.l.b.}\{\mathfrak{R}_m\})' \tag{11}$$

$$\mathfrak{L} = \text{g.l.b.}\{\mathcal{P}_m\} = (\text{l.u.b.}\{\mathfrak{R}_m\})' \ . \tag{12}$$

Proof: We need only prove one of these relations, since the other will then follow by duality. Consider the first relation.

$$(\text{g.l.b.}\{\mathfrak{R}_m\})' = \left(\prod_{m=1}^{M} \mathfrak{R}_m\right)'$$

$$= \mathfrak{R}'_1 \oplus \left(\prod_{m=2}^{M} \mathfrak{R}_m\right)'$$

$$(\text{g.l.b.}\{\mathfrak{R}_m\})' = \ldots$$

$$= \mathfrak{R}_1' \oplus \mathfrak{R}_2' \oplus \ldots \oplus \mathfrak{R}_M'$$

$$= \mathcal{P}_1 \oplus \mathcal{P}_2 \oplus \ldots \oplus \mathcal{P}_M$$

$$= \text{l.u.b.}\{\mathcal{P}_m\}$$

where we have used the dualization relation $(\mathfrak{A}\mathfrak{B})' = \mathfrak{A}' \oplus \mathfrak{B}'$ and the definition of $\mathfrak{R}_m \equiv \mathcal{P}_m'$.

The identity element I of the Boolean algebra Ω can be decomposed as $I = \mathfrak{A} \oplus \mathfrak{A}'$ where $\mathfrak{A}$ is any projector in the sublattice Ψ and $\mathfrak{A}'$ is its complement in Ψ' . Obviously, the remainder $\mathfrak{A}'$ will be minimal whenever the projector $\mathfrak{A}$ is maximal, and conversely. Therefore, by Lemma 7 and property (xiv), we have the following theorem.

<u>Theorem 3</u>: With respect to the commutative set of projectors $\{\mathcal{P}_m\}_{m=1}^{M}$, the identity operator has the <u>maximal decomposition</u>

$$I = \mathfrak{M} \oplus \mathfrak{M}'$$

$$(13) \qquad = \mathfrak{M} + \mathfrak{M}'$$

$$= \mathcal{P}_1 \oplus \mathcal{P}_2 \oplus \ldots \oplus \mathcal{P}_M + (\mathfrak{R}_1\mathfrak{R}_2 \ldots \mathfrak{R}_M) \ ;$$

and has the <u>minimal decomposition</u>

$$I = \mathfrak{L} \oplus \mathfrak{L}'$$

$$(14) \qquad = \mathfrak{L} + \mathfrak{L}'$$

$$= \mathcal{P}_1\mathcal{P}_2 \ldots \mathcal{P}_M + (\mathfrak{R}_1 \oplus \mathfrak{R}_2 \oplus \ldots \oplus \mathfrak{R}_M) \ .$$

(Note: Since $\mathfrak{M}\mathfrak{M}' = 0$ and $\mathfrak{L}\mathfrak{L}' = 0$, $\mathfrak{M} \oplus \mathfrak{M}' = \mathfrak{M} + \mathfrak{M}'$ and $\mathfrak{L} \oplus \mathfrak{L}' = \mathfrak{L} + \mathfrak{L}'$.)

To conclude the algebraic development, we translate and summarize the above results in the terminology of approximation theory.

Theorem 4: Let $\{\mathcal{P}_m\}_{m=1}^{M}$ be a set of commutative projectors (approximation operators) defined on the linear space $\mathfrak{F}$ of functions of the N independent variables $x_1, x_2, \ldots, x_N$. The range of $\mathcal{P}_m$ is the subspace Φ_m. If the collection $\{\mathcal{P}_m\}$ satisfies (A1) - (A3), then $F \in \mathfrak{F}$ can be expanded as

$$F = \mathcal{A}[F] + \mathcal{A}'[F] \tag{15}$$

where $\mathcal{A}$ is any approximation operator in the lattice Ψ and $\mathcal{A}'$ is its complement in Ψ'. Among all approximations $\mathcal{A}[F]$ to $F \in \mathfrak{F}$, the algebraically best approximation is the function $\pi = \mathfrak{M}[F]$ where $\mathfrak{M} = \mathcal{P}_1 \oplus \mathcal{P}_2 \oplus \ldots \oplus \mathcal{P}_M$. The maximal approximation π is in the linear space $\Phi_1 \cup \Phi_2 \cup \ldots \cup \Phi_M$ and satisfies

$$\mathcal{P}_m[\pi] = \mathcal{P}_m[F] \quad \text{for all} \quad m = 1, 2, \ldots, M \; . \tag{16}$$

The algebraically minimal approximation to F is the function $\pi' = \mathfrak{L}[F]$ where $\mathfrak{L} = \mathcal{P}_1 \mathcal{P}_2 \ldots \mathcal{P}_M$. The minimal approximation π' is in $\Phi_1 \cap \Phi_2 \cap \ldots \cap \Phi_M$ and satisfies

$$\mathcal{P}_1 \mathcal{P}_2 \cdots \mathcal{P}_M[\pi'] = \mathcal{P}_1 \mathcal{P}_2 \cdots \mathcal{P}_M[F] \; . \tag{17}$$

The errors (remainders) associated with the maximal and minimal approximations are, respectively:

$$F - \pi = \mathcal{R}_1 \mathcal{R}_2 \cdots \mathcal{R}_M[F] \tag{18}$$

and

$$F - \pi' = (\mathcal{R}_1 \oplus \mathcal{R}_2 \oplus \ldots \oplus \mathcal{R}_M)[F] \; . \tag{19}$$

III. Examples of Multivariate Approximations

In this section, we consider three essentially different classes of examples which illustrate the algebraic theory developed in the previous sections. Our principal purpose is to indicate the sense in which those algebraic results provide an operational calculus for the derivation of multivariate approxima formulas and, for specific projectors, to examine in greater detail the meaning of the algebraic formalism.

Example 1: Extensions of Univariate Projectors

The simplest and probably the most important class of projectors which satisfy the hypotheses of Theorem 4 are those obtained by extension of a univariate projector $\tilde{\mathcal{P}}_m$ whose domain is some space $\tilde{\mathfrak{F}}_m$ of univariate functions of the independent variable x_m and whose range is a finite dimensional linear space $\tilde{\Phi}_m$, i.e. $\tilde{\mathcal{P}}_m : f(x_m) \mapsto \tilde{\pi}_m(x_m)$ where $f \in \tilde{\mathfrak{F}}_m$ and $\tilde{\pi}_m \in \tilde{\Phi}_m$. Indeed, with the exception of the important recent paper by Sard [19], all of the papers on multivariate approximation cited in the bibliography deal with this class of projectors. Typically, the finite dimensional linear space $\tilde{\Phi}_m$ is taken to be the space of polynomials of degree k_m and the projector $\tilde{\mathcal{P}}_m$ is of the generic form

$$(20) \qquad \tilde{\mathcal{P}}_m = \sum_{\mu=0}^{k_m} \varphi_\mu^m(x_m)\lambda_\mu^m$$

where $\{\lambda_\mu^m\}_{\mu=0}^{k_m}$ is some set of linearly independent linear functionals, and the set $\{\varphi_\mu^m(x_m)\}_{\mu=0}^{k_m}$ is the unique basis for $\tilde{\Phi}_m$ with the properties that

$$(21) \qquad \lambda_\mu^m(\varphi_\nu^m) = \delta_{\mu\nu} \quad \text{(Kronecker delta)} \quad (\mu, \nu = 0, 1, \ldots, k_m) .$$

Then, the "extended" linear space Φ_m is the infinite dimensional space of "hyperpolynomials" whose coefficients are functions of the other $N-1$ variables complementary to the independent variable x_m:

$$(22) \qquad \Phi_m = \{f = \sum_{\mu=0}^{k_m} A_\mu x_m^\mu \mid A_\mu = A_\mu(x_1, \ldots, x_{m-1}, x_{m+1}, \ldots, x_N)\} ;$$

and, the extended projector $\mathcal{P}_m$ is just the projector $\tilde{\mathcal{P}}_m$ but with the λ_μ^m now considered to be "parametric functionals" which treat a multivariate function $F(x_1, \ldots, x_N)$ as a function of only the variable x_m — the other $N-1$ variables being treated as parameters†.

The case of bivariate functions $F(x_1, x_2)$ and two projectors $\mathcal{P}_1$ and $\mathcal{P}_2$ of the hyperpolynomial form has been studied extensively in the European literature by D. Mangeron ([13], [14]), A. Marchaud [15], L. Neder [16], M. Picone [17], T. Popoviciu [18], and D. D. Stancu [20].‡ However, until the recent rediscovery of these and other related formulae ([2], [3], [5], [8], [9], [10]), there seems to have been no English language account of the important results of these European mathematicians.

In [20], D. D. Stancu has studied the algebraically maximal Taylor type (i.e. "initial value") expansion $\pi = (\mathcal{P}_1 \oplus \mathcal{P}_2 \oplus \ldots \oplus \mathcal{P}_N)[F]$ and the associated remainder $F - \pi$. However, since Stancu's results were published in Russian and have apparently not been reported in the English literature, and since they provide one of the simplest realizations of the lattice structure Ψ, we shall herein use the algebraic methods of Sections I and II to sketch a derivation of the results of [20]. The corresponding minimal formula of the Taylor type (i.e., the well-known truncated Taylor formula (34) below) and the associated remainder were thoroughly discussed by Stancu [21] for the case of two variables.

Let F be a "sufficiently differentiable" function of the N independent variables $x_1, x_2, \ldots, x_N$ and take the number of projectors M to be equal to the number of independent variables N. Let the projector $\mathcal{P}_m$ be the linear operator

$$(23) \qquad \mathcal{P}_m = \sum_{\mu=0}^{k_m} \frac{x_m^\mu}{\mu!} \frac{\partial^\mu}{\partial x_m^\mu}\bigg|_{x_m=0}$$

† See Appendix B.

‡ The author is indebted to Professor D. D. Stancu of the Faculty of Mathematics, University of Cluj, Rumania for these references.

in which $\partial^\mu/\partial x_m^\mu\Big|_{x_m=0}$ is the linear differential operator ("parametric functional") which takes the μth partial derivative with respect to x_m of a function $F(x_1, x_2, \ldots, x_M)$ and evaluates it on the hyperplane $x_m = 0$. The image π_m of F under $\wp_m$ is the function

$$(24) \qquad \pi_m(x_1, \ldots, x_N) = \sum_{\mu=0}^{k_m} \frac{x_m^\mu}{\mu!} F^{(0,\ldots,\mu,\ldots,0)}(x_1, \ldots, 0, \ldots, x_N)$$

where we use the notation

$$(25) \qquad F^{(0,\ldots,\mu,\ldots,0)}(x_1, \ldots, 0, \ldots, x_N) \equiv \frac{\partial^\mu}{\partial x_m^\mu} F(x_1, \ldots, x_m, \ldots, x_N)\Big|_{x_m=0} .$$

Note especially that the $k_m + 1$ functions $F^{(0,\ldots,\mu,\ldots,0)}(x_1, \ldots, 0, \ldots, x_N)$ $(\mu = 0, 1, \ldots, k_m)$ are functions of $N-1$ variables. The function π_m can be interprested as being the $k_m + 1$ term truncated <u>parametric</u> Taylor expansion, the parameters being the $N-1$ variables $x_1, \ldots, x_{m-1}, x_{m+1}, \ldots, x_N$. Clearly, the μth partial derivatives w.r.t. x_m of π_m and of F coincide on the hyperplane $x_m = 0$:

$$(26) \qquad \pi_m^{(0,\ldots,\mu,\ldots 0)}(x_1, \ldots, 0, \ldots, x_N) = F^{(0,\ldots,\mu,\ldots,0)}(x_1, \ldots, 0, \ldots, x_N) \; (0 \le \mu \le k_m) .$$

Now, for our "sufficiently differentiable" function F, we have that

$$(27) \qquad \wp_\ell \wp_m[F] = \wp_m \wp_\ell[F] \qquad (\ell, m = 1, 2, \ldots, N) ,$$

since all that is required for (27) is that

$$(28)\quad \frac{\partial^{\nu}}{\partial x_{\ell}^{\nu}}\left[\frac{\partial^{\mu}}{\partial x_m^{\mu}} F\Big|_{x_m=0}\right]\Bigg|_{x_\ell=0} = \frac{\partial^{\mu}}{\partial x_m^{\mu}}\left[\frac{\partial^{\nu}}{\partial x_{\ell}^{\nu}} F\Big|_{x_\ell=0}\right]\Bigg|_{x_m=0} \quad \begin{pmatrix} 0\le\mu\le k_m \\ 0\le\nu\le k_\ell \end{pmatrix}$$

In other words, the N projectors of the set $\{\mathcal{P}_m\}$ commute if (28) holds for all ℓ, $m = 1, 2, \ldots, N$ $(\ell \neq m)$. These observations lead to the following results first obtained by Stancu in [20]. They are stated here as corollaries to Theorem 4.

Corollary 1: If $\mathfrak{F}$ is a space of N-variate functions F for which (28) holds, then the set of projectors $\{\mathcal{P}_m\}_{m=1}^{N}$ generate a lattice Ψ. The maximal projector $\mathfrak{M} = \mathcal{P}_1 \oplus \mathcal{P}_2 \oplus \cdots \oplus \mathcal{P}_N$ has the properties that the approximation $\pi = \mathfrak{M}[F]$ interpolates to F and its normal derivatives on the N hyperplanes $x_m = 0$ $(m = 1, 2, \ldots, N)$:

$$(29)\qquad \pi^{(0,\ldots,\mu_m,\ldots,0)}(x_1,\ldots,0,\ldots,x_N) = F^{(0,\ldots,\mu_m,\ldots,0)}(x_1,\ldots,0,\ldots,x_N)$$

for all $0 \le \mu_m \le k_m$ and $1 \le m \le N$. And, the function $\pi' = \mathfrak{L}[F]$, where $\mathfrak{L} = \mathcal{P}_1\mathcal{P}_2\cdots\mathcal{P}_N$ is the minimal projector, interpolates to F and its mixed partial derivatives $F^{(\mu_1,\mu_2,\ldots,\mu_N)}$ at the single point $(x_1, x_2, \ldots, x_N) = (0, 0, \ldots, 0)$:

$$(29')\qquad \pi'^{(\mu_1,\mu_2,\ldots,\mu_N)}(0,0,\ldots,0) = F^{(\mu_1,\mu_2,\ldots,\mu_N)}(0,0,\ldots,0) \qquad (0\le\mu_m\le k_m)\,.$$

The remainder operator $\mathcal{R}_m \equiv I - \mathcal{P}_m$ is the parametric integral operator

$$(30)\qquad \mathcal{R}_m[F] = \int_0^{x_m} \kappa_m(x_m;\xi_m)\,\frac{\partial^{k_m+1}}{\partial \xi_m^{k_m+1}} F(x_1,\ldots,\xi_m,\ldots,x_N)\,d\xi_m$$

where κ_m is the Taylor kernel

$$(31)\qquad \kappa_m(x_m;\xi_m) = (x_m-\xi_m)^{k_m}/k_m! \ .$$

This expression for $\mathfrak{R}_m$ together with the algebraic error formulas in Theorem 4 imply the following.

Corollary 2: The errors $F-\pi$ and $F-\pi'$ associated with the maximal and minimal projectors $\mathfrak{M}$ and $\mathfrak{L}$ are, respectively

$$(32)\qquad \begin{aligned} R(x_1,\ldots,x_N) &= \mathfrak{R}_1\mathfrak{R}_2\ldots\mathfrak{R}_N[F] \\ &= \int_0^{x_1}\ldots\int_0^{x_N}\kappa_1\ldots\kappa_N\, F^{(k_1+1,\ldots,k_N+1)}(\xi_1,\ldots,\xi_N)\,d\xi_1\ldots d\xi_N \end{aligned}$$

$$\begin{aligned} R'(x_1,\ldots,x_N) &= (\mathfrak{R}_1\oplus\mathfrak{R}_2\oplus\ldots\oplus\mathfrak{R}_N)[F] \\ &= \sum_{m=1}^{N}\mathfrak{R}_m[F] - \sum_{m=1}^{N}\sum_{\substack{n=1\\ m\neq n}}^{N}\mathfrak{R}_m\mathfrak{R}_n[F] + \ldots + \\ &\qquad (-1)^{N-1}\mathfrak{R}_1\mathfrak{R}_2\ldots\mathfrak{R}_N[F] \ . \end{aligned}$$

By observing that the Taylor kernels κ_m are positive (or negative) semi-definite functions of ξ_m for $\xi_m \leq x(\geq x)$, we obtain the following order of magnitude estimates for the errors:

$$(33)\qquad R(x_1,\ldots,x_N) = O(x_1^{k_1+1}\, x_2^{k_2+1}\ldots x_N^{k_N+1})$$

$$(33')\qquad R'(x_1,\ldots,x_N) = O(x_1^{k_1+1}) + O(x_2^{k_2+1}) + \ldots + O(x_N^{k_N+1}) \ .$$

Comparison of (33) and (33') shows that the function π converges to F whenever $x_m \to 0$ for <u>any</u> variable x_m; whereas,

π' converges only if $x_m \to 0$ for all $m = 1, 2, \ldots, N$. This simply means that π is a good approximation to F in the vicinity of any of the hyperplanes $x_m = 0$; but, π' is only good in a small neighborhood of the origin $(0, 0, \ldots, 0)$. Expressions (33) and (33') are indicative of the general relationship between maximal and minimal approximations.

We conclude the analysis of generalized Taylor interpolation with the following observations:

a. The function π', which is the truncated Taylor series

$$(34) \quad \pi'(x_1, \ldots, x_N) = \sum_{\mu_1=0}^{k_1} \cdots \sum_{\mu_N=0}^{k_N} \frac{x_1^{\mu_1} \cdots x_N^{\mu_N}}{\mu_1! \cdots \mu_N!} F^{(\mu_1, \ldots, \mu_N)}(0, \ldots, 0) ,$$

represents one of two standard truncations. The other is

$$(35) \quad \pi''(x_1, \ldots, x_N) = \sum\nolimits^* \frac{x_1^{\mu_1} \cdots x_N^{\mu_N}}{\mu_1! \cdots \mu_N!} F^{(\mu_1, \ldots, \mu_N)}(0, \ldots, 0).$$

where $\sum^*$ denotes summation over all values of $\mu_1, \ldots, \mu_N$ such that $\mu_1 + \mu_2 + \ldots + \mu_N \leq k_1 + k_2 + \ldots + k_N$. The function π'' can also be derived operationally from the lattice theoretic formulas of the previous sections. However, because the projector in (35) cannot be factored into simple projectors $\mathcal{P}_1, \mathcal{P}_2, \ldots, \mathcal{P}_N$ as can (34), the construction leading to (35) is slightly more complicated. In this connection, see Sections 2 and 5 of [21].

b. In this example, the algebraic ordering relation has a specific interpolatory meaning. For instance, the relation $\mathfrak{L} \leq \mathfrak{M}$ means that the set of interpolatory properties of π' is a subset of the properties of π. This is clear, since interpolation on the hyperplanes $x_m = 0$ certainly implies interpolation at the single point $(0, 0, \ldots, 0)$.

c. The range of the projector $\mathcal{P}_m$ is the space Φ_m of hyperpolynomials in the variable x_m, cf. expression (22). The function π is in $\bigcup_{m=1}^{N} \Phi_m$ and π' is in the finite dimensional space $\bigcap_{m=1}^{N} \Phi_m$ whose dimensionality is $\prod_{m=1}^{N} (k_m + 1)$. Note that the space $\bigcap_{m=1}^{N} \Phi_m$ is precisely the space of multivariate polynomials $p(x_1, \ldots, x_N)$ (i.e., the tensor product polynomials):

$$(36)\quad p(x_1, \ldots, x_N) = \sum_{\mu_1=0}^{k_1} \cdots \sum_{\mu_N=0}^{k_N} a_{\mu_1, \ldots, \mu_N} x_1^{\mu_1} \cdots x_N^{\mu_N}$$

where $\{a_{\mu_1, \ldots, \mu_N}\}$ is a set of $\prod_{m=1}^{N} (k_m + 1)$ constants.

d. The function π is the unique solution to the partial differential equation

$$(37)\quad \frac{\partial^{k_1 + \ldots + k_N + N}}{\partial x_1^{k_1+1} \cdots \partial x_N^{k_N+1}} \pi(x_1, \ldots, x_N) = 0$$

subject to the boundary conditions (29); and π' is the unique solution to the system of N partial differential equations

$$(38)\quad \frac{\partial^{k_m+1}}{\partial x_m^{k_m+1}} \pi'(x_1, \ldots, x_N) = 0 \qquad (m = 1, 2, \ldots, N)$$

subject to the conditions (29').

The same sort of analysis that has just been carried out showing the contrast between the maximal and minimal Taylor expansions can also be applied to the multivariate analogues of Lagrange polynomial interpolation and to orthogonal projections in a Hilbert space (generalized Fourier expansions). Stated for the bivariate case, the following two corollaries to Theorem 4 again bring out the contrast between maximal and minimal projectors. The first of these results is attributable to L. Neder [16], and is discussed in [9], [18] and [20]. Corollary 4, on the other hand, seems to be a new result. Its proof is given in [9].

Corollary 3: Let $F(x, y)$ be a continuous function defined over the rectangle $[a,b] \times [c,d]$ and let Δ_1: $a = x_1 < x_2 < \ldots < x_{K_1} = b$ and Δ_2: $c = y_1 < y_2 < \ldots < y_{K_2} = d$ be partitions of the intervals $[a,b]$ and $[c,d]$. Let

(39)
$$\mathcal{P}_1[F] = \sum_{i=1}^{K_1} \varphi_i^1(x)\, F(x_i, y)$$
$$\mathcal{P}_2[F] = \sum_{j=1}^{K_2} \varphi_j^2(y)\, F(x, y_j)$$

where $\{\varphi_i^1(x)\}$ and $\{\varphi_j^2(y)\}$ are the polynomials for Lagrange interpolation at the points of Δ_1 and Δ_2, respectively:

(40)
$$\varphi_i^1(x) = \prod_{k \neq i}^{K_1} \left(\frac{x - x_k}{x_i - x_k} \right) \qquad (i = 1, 2, \ldots, K_1)$$
$$\varphi_j^2(y) = \prod_{\ell \neq j}^{K_2} \left(\frac{y - y_\ell}{y_j - y_\ell} \right) \qquad (j = 1, 2, \ldots, K_2) \; .$$

Then, with $\mathfrak{M} = \mathcal{P}_1 + \mathcal{P}_2 - \mathcal{P}_1\mathcal{P}_2$, the maximal approximation $\pi = \mathfrak{M}[F]$ satisfies

(41)
$$\pi(x_i, y) = F(x_i, y) \qquad (i = 1, 2, \ldots, K_1)$$
$$\pi(x, y_j) = F(x, y_j) \qquad (j = 1, 2, \ldots, K_2) \; ,$$

i.e., π interpolates F along the $K_1 + K_2$ mesh-lines of the Cartesian product partition $\Delta_1 \times \Delta_2$. And, the minimal approximation $\pi' = \mathcal{L}[F] = \mathcal{P}_1\mathcal{P}_2[F]$ interpolates F at the $K_1 \cdot K_2$ mesh-points (x_i, y_j):

$$(42) \qquad \pi'(x_i, y_j) = F(x_i, y_j) \quad (i = 1, 2, \ldots, K_1;\ j = 1, 2, \ldots, K_2).$$

Corollary 4: Let $F(x, y)$ be in the Hilbert space $\mathfrak{F} = L^2([a,b] \times [c,d])$ of square integrable functions over the rectangle $[a,b] \times [c,d]$. Let $\{\varphi_i^1(x)\}$ and $\{\varphi_j^2(y)\}$ be orthonormal sets of functions over $[a,b]$ and $[c,d]$ respectively:

$$(43) \qquad \begin{aligned} &\int_a^b \varphi_i^1(x)\, \varphi_k^1(x)\, dx = \delta_{ik} && (i, k = 1, 2, \ldots, K_1) \\ &\int_c^d \varphi_j^2(y)\, \varphi_\ell^2(y)\, dy = \delta_{j\ell} && (j, \ell = 1, 2, \ldots, K_2). \end{aligned}$$

Take

$$(44) \qquad \begin{aligned} \Phi_1 &= \{f = \sum_{i=1}^{K_1} A_i(y)\, \varphi_i^1(x) \mid A_i(y) \in L^2([c,d])\} \\ \Phi_2 &= \{f = \sum_{j=1}^{K_2} B_j(x)\, \varphi_j^2(y) \mid B_j(x) \in L^2([a,b])\} \end{aligned}$$

and let $\mathcal{P}_1$ and $\mathcal{P}_2$ be the orthogonal projectors from $L^2([a,b] \times [c,d])$ onto Φ_1 and Φ_2 defined, respectively, by

$$(45) \qquad \begin{aligned} \mathcal{P}_1[F] &= \sum_{i=1}^{K_1} \varphi_i^1(x) \left(\int_a^b F(\xi, y)\, \varphi_i^1(\xi)\, d\xi \right) \\ \mathcal{P}_2[F] &= \sum_{j=1}^{K_2} \varphi_j^2(y) \left(\int_c^d F(x, \eta)\, \varphi_j^2(\eta)\, d\eta \right). \end{aligned}$$

Then, the maximal approximation $\pi = \mathcal{P}_1[F] + \mathcal{P}_2[F] - \mathcal{P}_1\mathcal{P}_2[F]$ satisfies

$$\int_a^b \pi(\xi, y)\varphi_i^1(\xi)\,d\xi = \int_a^b F(\xi, y)\varphi_i^1(\xi)\,d\xi \qquad (i = 1, 2, \ldots, K_1)$$

(46)

$$\int_c^d \pi(x, \eta)\varphi_j^2(\eta)\,d\eta = \int_c^d F(x, \eta)\varphi_j^2(\eta)\,d\eta \qquad (j = 1, 2, \ldots, K_2)$$

and, among all functions in the space $\Phi_1 \cup \Phi_2$, it is the unique function which minimizes the L^2 norm of the remainder:

$$\|F - \pi\|^2 = \int_a^b \int_c^d (F(x, y) - \pi(x, y))^2\,dxdy\ . \tag{47}$$

The minimal approximation $\pi' = \mathcal{P}_1\mathcal{P}_2[F]$ satisfies:

$$\int_a^b \int_c^d \pi'(\xi, \eta)\varphi_i^1(\xi)\varphi_j^2(\eta)\,d\xi d\eta = \int_a^b \int_c^d F(\xi, \eta)\varphi_i^1(\xi)\varphi_j^2(\eta)\,d\xi d\eta$$

(48)

$$(i = 1, 2, \ldots, K_1;\ j = 1, 2, \ldots, K_2)$$

and is the unique function $f(x, y)$ in $\Phi_1 \cap \Phi_2$ which minimizes $\|F - f\|$.

Generalizations of the class of projectors considered in this example are numerous; and some of these have been dealt with in detail in the recent papers [9] and [10]. In [10], we have considered spline projectors and have established "minimum norm" and "best approximation" (in the sense of the associated pseudo-norm) properties for the maximal spline projector $\mathfrak{M}$. The minimal spline projectors $\mathcal{L} = \mathcal{P}_1\mathcal{P}_2\ldots\mathcal{P}_N$ are the projectors associated with deBoor's scheme of bicubic spline interpolation ([1], [7]) and its N-variate tensor product extensions and generalizations. The maximal spline projectors give a new class of interpolation schemes for solving, for example, the problem of passing a "smooth" surface (bivariate function) through an arbitrary rectangular network of intersecting plane curves (univariate functions), i.e. the conditions of (41). See also the appendix to this paper where we use such a scheme based on cubic splines.

In [9], we have considered a very broad class of multivariate projectors of the form (20) where the λ_μ^m are arbitrary "parametric linear functionals" and the set of functions $\{\varphi_\mu^m\}$ is a basis for the null space of an arbitrary linear ordinary differential operator D_m. We have shown there that the function $\pi = \mathfrak{M}[F]$ is the unique solution to the partial differential equation $\mathcal{D}_1\mathcal{D}_2 \ldots \mathcal{D}_M[\pi] = 0$ subject to the "boundary" conditions $\lambda_\mu^m[\pi] = \lambda_\mu^m[F]$ for all $\mu = 1, 2, \ldots, k_m$ and all $m = 1, 2, \ldots, M$, where $\mathcal{D}_m$ is just the partial differential extension of the operator D_m. On the other hand, the image π' of F under the minimal projector $\mathfrak{L}$ is the unique solution to the system of partial differential equations $\mathcal{D}_1[\pi] = 0, \ldots, \mathcal{D}_M[\pi'] = 0$ subject to the conditions that $\lambda^1_{\mu_1} \ldots \lambda^M_{\mu_M}[\pi'] = \lambda^1_{\mu_1} \ldots \lambda^M_{\mu_M}[F]$ for all $\mu_m = 1, 2, \ldots, k_m$. (Compare the remarks accompanying equations (37) and (38) above.) Note that the intersection of the null spaces of $\mathcal{D}_1, \ldots, \mathcal{D}_M$ is a proper subspace of the null space of the product operator $\mathcal{D}_1 \ldots \mathcal{D}_M$. Results of this nature are precisely what one would expect on the basis of the algebraic results of Sections I and II.

Example 2: An Intrinsically Bivariate Projector

This example is unlike the first in the essential respect that the projector $\mathcal{P}_2$ below is an intrinsically bivariate operator. It cannot, as can the $\mathcal{P}_m$ of Example 1, be obtained as a simple parametric extension of a univariate projector. We will consider a function F of three variables, since this is sufficient to illustrate the main idea. Instead of x_1, x_2, x_3, we will denote the independent variables as r, θ and z.

Let D_1 and D_2 be the unit discs in the planes $z = 0$ and $z = 1$, respectively:

$$D_1 = \{(r,\theta,z) \mid 0 \le r \le 1,\ \ 0 \le \theta \le 2\pi,\ \ z = 0\}$$

(49)

$$D_2 = \{(r,\theta,z) \mid 0 \le r \le 1,\ \ 0 \le \theta \le 2\pi,\ \ z = 1\};$$

and, let Σ be the vertical surface of the truncated right circular cylinder of unit height whose axis is the z-axis:

$$\Sigma = \{(r,\theta,z) \mid r = 1,\ \ 0 \le \theta \le 2\pi,\ \ 0 \le z \le 1\} \ . \tag{50}$$

We shall construct a function $\pi(r,\theta,z)$ which interpolates to an arbitrary continuous trivariate function $F(r,\theta,z)$ on the two ends D_1 and D_2 and on the wall Σ of this unit cylinder

First, consider the simple projector $\mathcal{P}_1$ defined by:

$$\mathcal{P}_1[F] = (1-z)\, F(r,\theta,0) + zF(r,\theta,1) \ . \tag{51}$$

This is a very special case of the polynomial projectors considered in Corollary 3, Example 1. The image function $\pi_1 = \mathcal{P}_1[F]$ has the properties that it interpolates F on the planes $z = 0$ and $z = 1$; or, more specifically, on the discs D_1 and D_2:

$$\pi_1(r,\theta,0) = F(r,\theta,0) \tag{52a}$$

$$\pi_1(r,\theta,1) = F(r,\theta,1) \ . \tag{52b}$$

Next, consider the linear operator $\mathcal{P}_2$ defined by:

$$\mathcal{P}_2[F] = \int_0^{2\pi} G(r,\theta;\theta')\, F(1,\theta',z)\, d\theta' \tag{53}$$

where the boundary Green's function G is the Poisson kernel

$$G(r,\theta;\theta') = \frac{1}{2\pi}\left[\frac{1-r^2}{1-2r\cos(\theta-\theta')+r^2}\right] \ . \tag{54}$$

Note that the variable z is simply a parameter in the integral on the right of (53). For all values of z, the image function

$\pi_2 = \mathcal{P}_2[F]$ satisfies Laplace's equation $\nabla^2 \pi_2 = 0$ † and satisfies the boundary conditions

(55) $\quad \pi_2(1,\theta,z) = F(1,\theta,z) \quad (0 \le \theta \le 2\pi,\ 0 \le z \le 1)$.

In other words, the bivariate function $\pi_2(r,\theta,z^*)$ obtained by slicing the function π_2 with any plane $z = z^*$ is the unique solution of the Dirichlet problem in the unit disc $D^* = \{(r,\theta,z) \mid 0 \le r \le 1,\ 0 \le \theta \le 2\pi,\ z = z^*\}$.

<u>Lemma 8:</u> The operator $\mathcal{P}_2$ is a projector.

Proof: $\mathcal{P}_2$ is clearly a linear operator. To see that it is idempotent, observe that the integrand involves only the values of the argument function $F(r,\theta,z)$ on the circumference Σ of the cylinder, $r = 1$. And, as we noted in (55), π_2 coincides with F on Σ. Hence, $\mathcal{P}_2[\pi_2] = \mathcal{P}_2[\mathcal{P}_2[F]] = \mathcal{P}_2[F]$, so that $\mathcal{P}_2$ is idempotent. Q.E.D.

The two projectors $\mathcal{P}_1$ and $\mathcal{P}_2$ so defined do satisfy the crucial commutativity conditions:

$$\begin{aligned}\mathcal{P}_1\mathcal{P}_2[F] &= \mathcal{P}_2\mathcal{P}_1[F] \\ &= (1-z)\int_0^{2\pi} G(r,\theta;\theta')\, F(1,\theta',0)\, d\theta' \\ &\quad + z\int_0^{2\pi} G(r,\theta;\theta')\, F(1,\theta',1)\, d\theta' \ .\end{aligned}$$

We state our conclusions for this example as a corollary to Theorem 4.

<u>Corollary 5:</u> Let $\mathcal{P}_1$ and $\mathcal{P}_2$ be as in (51) and (53) and let $\mathfrak{m}$ be the maximal projector of the lattice Ψ generated by $\mathcal{P}_1$ and $\mathcal{P}_2$:

---

† $\nabla^2 = \frac{1}{r}\frac{\partial}{\partial r}\left(r\frac{\partial}{\partial r}\right) + \frac{1}{r^2}\frac{\partial^2}{\partial \theta^2}$, i.e. ∇^2 is in polar coordinates.

$$\mathfrak{M} = \mathcal{P}_1 \oplus \mathcal{P}_2 \tag{57}$$
$$= \mathcal{P}_1 + \mathcal{P}_2 - \mathcal{P}_1\mathcal{P}_2 \ .$$

For any continuous trivariate function $F(r,\theta,z)$, the function $\pi = \mathfrak{M}[F]$ coincides with F on the three bounding surfaces D_1, D_2 and Σ of the unit cylinder; i.e., π satisfies all three of the conditions (52a), (52b) and (55). Moreover, it is the unique function which satisfies those boundary conditions and satisfies the partial differential equation $\frac{\partial^2}{\partial z^2}\nabla^2[\pi] = 0$. On the other hand, the image π' of F under the minimal projector $\mathfrak{L} = \mathcal{P}_1\mathcal{P}_2$ interpolates F only on the circumferences of the two discs D_1 and D_2 :

$$\pi'(1,\theta,0) = F(1,\theta,0) \tag{58a}$$
$$(0 \leq \theta \leq 2\pi) \ .$$
$$\pi'(1,\theta,1) = F(1,\theta,1) \tag{58b}$$

and, π' satisfies the pair of partial differential equations $\nabla^2[\pi'] = 0$ and $\partial^2/\partial z^2[\pi'] = 0$. The remainders associated with the approximations π and π' are, respectively:

$$F - \pi = \mathcal{R}_1\mathcal{R}_2[F] \tag{59}$$

$$F - \pi' = \mathcal{R}_1[F] + \mathcal{R}_2[F] - \mathcal{R}_1\mathcal{R}_2[F] \tag{59'}$$

where

$$\mathcal{R}_1[F] = \int_0^1 \kappa_1(z;z')\, F^{(0,0,2)}(r,\theta,z')\,dz' \tag{60}$$

$$\mathcal{R}_2[F] = \int_0^{2\pi}\int_0^1 \kappa_2(r,\theta;r',\theta')\, F(r',\theta',z)\,dr'd\theta' \tag{61}$$

in which $\kappa_1(z;z')$ and $\kappa_2(r,\theta;r',\theta')$ are well-known Green's functions for the differential operators d^2/dz^2 and ∇^2, respectively, subject to homogeneous boundary conditions (cf. [6, p. 371, p. 377]).

This second example is clearly amenable to generalization in a number of interesting directions. For instance, if G is the boundary Green's function for any well-posed linear boundary value problem over an arbitrary domain $\mathcal{D}$, then the operator

$$(62)\qquad \mathcal{P}[F] = \int_{\partial\mathcal{D}} G(x_1,\ldots,x_N;\xi_1,\ldots,\xi_N)F(\xi_1,\ldots,\xi_N)\, d\xi_1 \ldots d\xi_N$$

is a projector and can, therefore, be combined with any other collection of projectors to form a lattice, provided only that they commute.

Example 3: Interpolation on the Perimeter of a Triangle

As our third and final example, we consider a method for interpolating to an arbitrary continuous bivariate function $F(x,y)$ on the edges of the triangle T whose vertices are at $(0,0)$, $(1,0)$, and $(1,1)$. The three projectors $\mathcal{P}_1, \mathcal{P}_2$ and $\mathcal{P}_3$ below were derived on an ad hoc basis by purely geometrical considerations, but guided by the requirements of commutativity. Note that the number of projectors, $M = 3$, is greater than the number, $N = 2$, of independent variables.

The three projectors which we use are:

$$(63a)\qquad \mathcal{P}_1[F] = \left(\frac{x-y}{x}\right) F(x,0)$$

$$(63b)\qquad \mathcal{P}_2[F] = \left(\frac{x-y}{1-y}\right) F(1,y)$$

$$(63c)\qquad \mathcal{P}_3[F] = \left(\frac{1-x}{1-y}\right)\left(\frac{y}{x}\right) F(x,x) .$$

We observe that $\pi_1 = \mathcal{P}_1[F]$ and $\pi_3 = \mathcal{P}_3[F]$ are undefined along the line $x = 0$, and that $\pi_2 = \mathcal{P}_2[F]$ and π_3 are undefined along the line $y = 1$. But, since the domain of interest is the closure of the domain bounded by the triangle T, we need be concerned about these singularities only at the two vertices (0,0) and (1,1), for these are the only points at which the lines $x = 0$ and $y = 1$ intersect our domain. Hence, let us simply agree to exclude the points (0,0) and (1,1) from the domain and proceed.

The image functions π_m assume the following values on the edges of the triangle:

$$\text{(64a)} \qquad \pi_1(x,y) = \begin{cases} F(x,0) & \text{on } y = 0 \\ (1-y)\,F(1,0) & \text{on } x = 1 \\ 0 & \text{on } y = x \end{cases}$$

$$\text{(64b)} \qquad \pi_2(x,y) = \begin{cases} x\,F(1,0) & \text{on } y = 0 \\ F(1,y) & \text{on } x = 1 \\ 0 & \text{on } y = x \end{cases}$$

$$\text{(64c)} \qquad \pi_3(x,y) = \begin{cases} 0 & \text{on } y = 0 \\ 0 & \text{on } x = 1 \\ F(x,x) & \text{on } y = x\,. \end{cases}$$

The projectors do commute:

$$\text{(65)} \qquad \begin{aligned} \mathcal{P}_1\mathcal{P}_2[F] &= \mathcal{P}_2\mathcal{P}_1[F] = (x-y)\,F(1,0) \\ \mathcal{P}_1\mathcal{P}_3[F] &= \mathcal{P}_3\mathcal{P}_1[F] = 0 \\ \mathcal{P}_2\mathcal{P}_3[F] &= \mathcal{P}_3\mathcal{P}_2[F] = 0\,. \end{aligned}$$

Therefore, the three projectors generate a lattice in which the maximal projector $\mathcal{P}_1 \oplus \mathcal{P}_2 \oplus \mathcal{P}_3$ is

$$\text{(66)} \qquad \mathfrak{M} = \mathcal{P}_1 + \mathcal{P}_2 + \mathcal{P}_3 - \mathcal{P}_1\mathcal{P}_2 - \mathcal{P}_2\mathcal{P}_3 - \mathcal{P}_1\mathcal{P}_3 + \mathcal{P}_1\mathcal{P}_2\mathcal{P}_3$$

and the minimal projector $\mathcal{P}_1\mathcal{P}_2\mathcal{P}_3$ is $\mathcal{L} = 0$.

The image of F under $\mathfrak{M}$ is π :

$$\pi(x,y) = \left(\frac{x-y}{x}\right) F(x,0) + \left(\frac{x-y}{1-y}\right) F(1,y) \tag{67}$$
$$- (x-y) F(1,0) + \left(\frac{1-x}{1-y}\right)\left(\frac{y}{x}\right) F(x,x)$$

which, it is easy to verify, interpolates F on the boundary of T, excluding (0,0) and (1,1):

$$\pi(x,0) = F(x,0) \tag{68a}$$

$$\pi(1,y) = F(1,y) \tag{68b}$$

$$\pi(x,x) = F(x,x) \quad . \tag{68c}$$

APPENDIX A: NUMERICAL APPROXIMATION

In Example 1 of Section III, we saw how an arbitrary function $F(x_1, \ldots, x_N)$ of N independent variables can be approximated by simply weighted combinations of functions of $N-1$ and fewer variables; i.e., $F \approx (\mathcal{P}_1 \oplus \mathcal{P}_2 \oplus \ldots \oplus \mathcal{P}_N)[F]$ where $\mathcal{P}_m$ is the "parametric extension" of a univariate projector in the independent variable x_m . In this appendix, we shall show how such decomposition techniques can be applied recursively so as to obtain an approximation in terms of numerical data (i.e., scalar values) extracted from the function F . Also, we shall show that, under fairly mild and realistic assumptions, the amount of numerical data required to achieve a specified accuracy of approximation is substantially less with the decomposition methods than with the usual methods of tensor product approximation (alias cross product, Kronecker product, Cartesian product).

The results of our error analyses for methods based upon cubic splines are summarized in Tables I-VII, at the end of this appendix. These theoretical estimates have been supported by a number of actual computational examples involving related techniques of approximate bivariate and trivariate integration (cf. [9] and [10]).

The details of successive decomposition will be illustrated by considering a bivariate function. The two variable case is, in most respects, typical of the general situation. However, it is not entirely representative, since for higher dimensional functions certain combinatorial effects, which are insignificant for a small number of variables, begin to play an important role. Thus, the approximate error analyses and point counts below are not asymptotically valid as $N \rightarrow \infty$.

We first outline the essential idea of the successive decomposition method (cf. Figure 1). In Corollary 3 of Exampl we saw that a bivariate function (i.e., a surface) $F(x, y)$ car be approximated by weighted sums of univariate functions (i.e. plane curves) $F(x_i, y)$ and $F(x, y_j)$ with an error of $\mathcal{R}_1\mathcal{R}_2[F]$ Since the approximation $\pi = (\mathcal{P}_1 \oplus \mathcal{P}_2)[F]$ interpolates F on the lines $x = x_i$ and $y = y_j$, the error is obviously zero alonc these two sets of lines and, for sufficiently smooth $F(x, y)$ ar sufficiently fine partitions of the x and y intervals, the erro is also small in the interior of each subrectangle $R_{ij} = [x_i, x_{i+}$ $\times [y_j, y_{j+1}]$, its maximum being achieved at approximately the center of the subrectangle. In other words, if we were given a sufficient number of univariate samples $F(x_i, y)$ and $F(x, y_j)$ then we would expect to obtain a very good approximation to th bivariate function $F(x, y)$ by simply interpolating, via Coroll a smooth surface through the given network of curves.

In numerical approximation, however, the goal is to ap proximate F in terms of some finite collection of scalar paran eters — not collections of univariate functions. To this end, next consider the individual functions $F(x_i, y)$ and $F(x, y_j)$ which comprise the curve network. By some univariate schem each of these functions can be approximated to any desired ac curacy in terms of a finite number of numerical values. Let th functions $g_i(y)$ and $h_j(x)$ be finite parameter approximatior to $F(x_i, y)$ and $F(x, y_j)$, respectively.

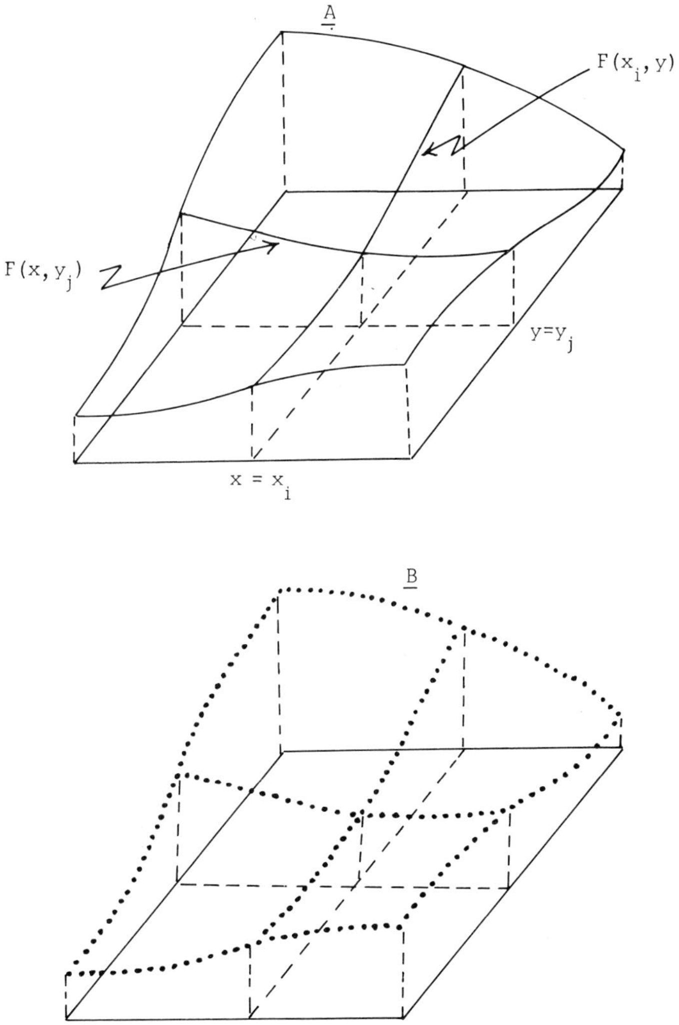

Figure 1: A simple illustration of the decomposition of a bivariate function $F(x, y)$ into (A) a collection of univariate functions and (B) subsequent approximation of the univariate functions in terms of data points.

Suppose that the errors in these approximations are all bounded (in norm) by some ϵ; i.e., $\|F(x_i, y) - g_i(y)\| \leq \epsilon$ and $\|F(x, y_j) - h_j(x)\| \leq \epsilon$. Then, for a sufficiently small ϵ, the error made by replacing the functions $F(x_i, y)$ and $F(x, y_j)$ by their approximations $g_i(y)$ and $h_j(x)$ will be smaller than $\|\mathcal{R}_1\mathcal{R}_2[F]\|$; and therefore the resulting bivariate approximation formula, which now involves only scalar data, is of <u>the same order of accuracy</u> as the original approximation $F \approx (\mathcal{P}_1 \oplus \mathcal{P}_2)[F]$ Moreover, since the total error cannot be made less than the error $\|\mathcal{R}_1\mathcal{R}_2[F]\|$, it is futile to approximate the functions $F(x_i, y)$ and $F(x, y_j)$ to any higher accuracy than is warranted by the initial approximation. In other words, the two levels of decomposition — first into univariate functions and then into scalar parameters — should be <u>consistent</u> in terms of the error criterion.

In contradistinction to the successive decomposition scheme just described are the usual tensor product methods which, as we saw in Example 1, Section III, are based upon the algebraically minimal approximation $F \approx \mathcal{P}_1\mathcal{P}_2[F]$ and involve function values at the mesh-<u>points</u> (in contrast to along mesh-<u>lines</u> as in $(\mathcal{P}_1 \oplus \mathcal{P}_2)[F]$) of a Cartesian product partition

Since, in the general case of decomposition, we must consider N <u>collections</u> of projectors, we shall use superscripts on projectors to denote the associated level. Thus, instead of simply $\mathcal{P}_1, \mathcal{P}_2, \ldots, \mathcal{P}_N$, we shall designate the N projectors associated with the first level — the decomposition of F into $N-1$ variate functions — as $\mathcal{P}^1_1, \mathcal{P}^1_2, \ldots, \mathcal{P}^1_N$. And, in general $\{\mathcal{P}^n_m\}_{m=1}^N$ is the set of projectors used at the nth level, at which one deals with functions of $N-n$ and fewer variables.

<u>Lemma A1</u>: Let $F(x, y)$ be a bivariate function for which the projectors $\mathcal{P}^1_1$ and $\mathcal{P}^1_2$ commute; and let $\mathcal{P}^2_1$ and $\mathcal{P}^2_2$ be two projectors such that $\mathcal{P}^2_1\mathcal{P}^1_2[F] = \mathcal{P}^1_2\mathcal{P}^2_1[F]$ and $\mathcal{P}^2_2\mathcal{P}^1_1[F] = \mathcal{P}^1_1\mathcal{P}^2_2[F]$. Then, with $\mathcal{R}^i_j = I - \mathcal{P}^i_j$ $(i, j = 1, 2)$, the bivariate identity operator I has the decomposition

$$\text{(1A)} \quad I = \mathcal{P}^2_2\mathcal{P}^1_1 + \mathcal{P}^2_1\mathcal{P}^1_2 - \mathcal{P}^1_1\mathcal{P}^1_2 + (\mathcal{R}^2_2\mathcal{P}^1_1 + \mathcal{R}^2_1\mathcal{P}^1_2) + \mathcal{R}^1_1\mathcal{R}^1_2 .$$

Proof: Since $\mathcal{P}_1^1$ and $\mathcal{P}_2^1$ commute, from Section II we have,

$$
\begin{aligned}
\text{(2A)} \qquad I &= \mathcal{P}_1^1 \oplus \mathcal{P}_2^1 + \mathcal{R}_1^1\mathcal{R}_2^1 \\
&= \mathcal{P}_1^1 + \mathcal{P}_2^1 - \mathcal{P}_1^1\mathcal{P}_2^1 + \mathcal{R}_1^1\mathcal{R}_2^1 \ .
\end{aligned}
$$

Since $I = \mathcal{P}_1^2 + \mathcal{R}_1^2$ and $I = \mathcal{P}_2^2 + \mathcal{R}_2^2$ we can write

$$
\begin{aligned}
I &= I\,\mathcal{P}_1^1 + I\,\mathcal{P}_2^1 - \mathcal{P}_1^1\mathcal{P}_2^1 + \mathcal{R}_1^1\mathcal{R}_2^1 \\
&= (\mathcal{P}_2^2 + \mathcal{R}_2^2)\mathcal{P}_1^1 + (\mathcal{P}_1^2 + \mathcal{R}_1^2)\mathcal{P}_2^1 - \mathcal{P}_1^1\mathcal{P}_2^1 + \mathcal{R}_1^1\mathcal{R}_2^1 \\
&= \mathcal{P}_2^2\mathcal{P}_1^1 + \mathcal{P}_1^2\mathcal{P}_2^1 - \mathcal{P}_1^1\mathcal{P}_2^1 + (\mathcal{R}_2^2\mathcal{P}_1^1 + \mathcal{R}_1^2\mathcal{P}_2^1) + \mathcal{R}_1^1\mathcal{R}_2^1 \\
&= \mathcal{P}_2^2\mathcal{P}_1^1 + \mathcal{P}_1^2\mathcal{P}_2^1 - \mathcal{P}_1^1\mathcal{P}_2^1 + (\mathcal{P}_1^1\mathcal{R}_2^2 + \mathcal{P}_2^1\mathcal{R}_1^2) + \mathcal{R}_1^1\mathcal{R}_2^1 \ .
\end{aligned}
$$

Lemma A2: If $F(x, y)$ belongs to the normed linear space $\mathcal{F}$, then with $\mathcal{G} = \mathcal{P}_2^2\mathcal{P}_1^1 + \mathcal{P}_1^2\mathcal{P}_2^1 - \mathcal{P}_1^1\mathcal{P}_2^1$ and the usual definition of operator norm, the error operator $I - \mathcal{G} = \mathcal{P}_1^1\mathcal{R}_2^2 + \mathcal{P}_2^1\mathcal{R}_1^2 + \mathcal{R}_1^1\mathcal{R}_2^1$ is bounded in norm by

$$
\begin{aligned}
\|I - \mathcal{G}\| &\le \|\mathcal{P}_1^1\mathcal{R}_2^2\| + \|\mathcal{P}_2^1\mathcal{R}_1^2\| + \|\mathcal{R}_1^1\mathcal{R}_2^1\| \\
\text{(4A)} \qquad &\le \|\mathcal{P}_1^1\|\,\|\mathcal{R}_2^2\| + \|\mathcal{P}_2^1\|\,\|\mathcal{R}_1^2\| + \|\mathcal{R}_1^1\|\,\|\mathcal{R}_2^1\| \ .
\end{aligned}
$$

If the linear space $\mathcal{F}$ is actually a Hilbert space, and if $\mathcal{P}$ and $\mathcal{Q}$ are two orthogonal projectors, (which is not the case in our example below) then the algebraic ordering relation is equivalent to an ordering in norm, i.e. $\mathcal{P} \le \mathcal{Q}$ (where "$\le$" is the algebraic ordering relation) is equivalent to $\|\mathcal{P}\| \le \|\mathcal{Q}\|$. In such instances, the following lemma is applicable.

Lemma A3: If $\mathcal{F}$ is a Hilbert space and if $\mathcal{P}_j^i \, (i, j = 1, 2)$ are orthogonal projectors, then

(5A) $$\|I - \mathcal{Q}\| \leq \|\mathcal{R}_2^2\| + \|\mathcal{R}_1^2\| + \|\mathcal{R}_1^1 \mathcal{R}_2^1\| \ .$$

Proof: This bound follows directly from (4A) since $\mathcal{P}_1^1 \leq I$ and $\mathcal{P}_2^1 \leq I$ (in the algebraic sense) and hence $\|\mathcal{P}_1^1\| \leq 1$ and $\|\mathcal{P}_2^1\| \leq 1$.

For concreteness, let us consider a specific example of bivariate approximation by successive decomposition. Since the central theme of this symposium has been approximation with spline functions, our example will be based upon cubic splines. For simplicity, we will treat the variables x and y symmetrically in the sense that the projectors $\mathcal{P}_1^1$ and $\mathcal{P}_2^1$ are actually the same projector with the roles of x and y interchanged. Also, the projectors $\mathcal{P}_1^2$ and $\mathcal{P}_2^2$ which are used in the subsequent approximation of the univariate functions $F(x_i, y)$ and $F(x, y_j)$ are identical under the interchange of x and y . The reader will note that the decomposition scheme described heuristically above is considerably more general than the symmetric scheme which we treat here. The details of the general case are, however, easy to infer.

Let $F(x,y) \in C^{4,4}([0,1] \times [0,1])$ and let Δ_1 be the M_1 point uniform partition of the unit interval: $\Delta_1 = \{0, h_1, 2h_1, \ldots, (M_1-1)h_1\}$ where $h_1 = 1/(M_1-1)$. Let $\{\varphi_i(\xi)\}_{i=1}^{M_1}$ be the set of natural† cubic splines with joints (knots) at the mesh-points of Δ_1 and which satisfy the cardinality conditions

(6A) $$\varphi_i(\xi_k) = \delta_{ik} \quad \text{for } \xi_k \in \Delta_1 \ .$$

Consider the "parametric cubic spline projectors" $\mathcal{P}_1^1$ and $\mathcal{P}_2^1$ defined by:

† By "natural", we mean they satisfy the natural boundary conditions $\varphi_i''(0) = \varphi_i''(1) = 0$.

(7A)
$$\mathcal{P}_1^1[F] = \sum_{i=1}^{M_1} \varphi_i(x)\, F(x_i, y) \qquad (x_i \in \Delta_1)$$
$$\mathcal{P}_2^1[F] = \sum_{j=1}^{M_1} \varphi_j(y)\, F(x, y_j) \qquad (y_j \in \Delta_1)\ .$$

With the sup norm

(8A)
$$\|f(x,y)\| = \sup_{(x,y)\,\in\,[0,1]\times[0,1]} |f(x,y)| \quad ,$$

we have the following bounds, which are obvious from the univariate results [1, p. 96] for cubic spline interpolation to $f(x) \in C^4[0,1]$

(9A)
$$\|\mathcal{R}_1^1[F]\| \leq \alpha \|F^{(4,0)}\|\, (h_1)^4$$
$$\|\mathcal{R}_2^1[F]\| \leq \alpha \|F^{(0,4)}\|\, (h_1)^4$$

for some constant α independent of F.

The following result is essentially contained in [10]. It can be established as a corollary to Theorem 4, Section II of the present paper. In [10], we refer to the functions $\varphi_i(\xi)$ as "blending-functions" since they play the role of blending the curves $F(x_i, y)$ and $F(x, y_j)$ into a "smooth surface". Hence, the terminology of the following lemma.

<u>Lemma A4</u>: The <u>spline-blended</u> interpolant $S(x,y)$ given by

$$S(x,y) = (\mathcal{P}_1^1 \oplus \mathcal{P}_2^1)[F]$$

(10A)
$$= \mathcal{P}_1^1[F] + \mathcal{P}_2^1[F] - \mathcal{P}_1^1\mathcal{P}_2^1[F]$$
$$= \sum_{i=1}^{M_1} \varphi_i(x)\, F(x_i, y) + \sum_{j=1}^{M_1} \varphi_j(y) F(x, y_j) - \sum_{i=1}^{M_1} \sum_{j=1}^{M_1} \varphi_i(x)\varphi_j(y) F(x_i, y_j)$$

interpolates F along the $2M_1$ lines $x = x_i$ $(i = 1, 2, \ldots, M_1)$ and $y = y_j$ $(j = 1, 2, \ldots, M_1)$ of the Cartesian product partition $\Delta_1 \times \Delta_1$ of $[0,1] \times [0,1]$. The error $\mathcal{R}_1^1\mathcal{R}_2^1[F] = F - (\mathcal{P}_1^1[F] + \mathcal{P}_2^1[F] - \mathcal{P}_1^1\mathcal{P}_2^1[F])$ has the bound

(11A) $$\|\mathcal{R}_1^1\mathcal{R}_2^1[F]\| \leq A\|F^{(4,4)}\| (h_1)^8$$

for some constant A independent of F.

The important thing to note in Lemma A4 is the $O(h_1^8)$ convergence of S to F. Another, and perhaps more illuminating interpretation of (11A) is that $\|\mathcal{R}_1^1\mathcal{R}_2^1[F]\|$ is proportional to the fourth power of the "areal mesh" $h_1 \cdot h_1$. In the general case of N variables, the error $\|\mathcal{R}_1^1\mathcal{R}_2^1 \ldots \mathcal{R}_N^1[F]\|$ in cubic spline decomposition is proportional to V^4 where $V = h_1^N$ is the volume of the mesh element in the N dimensional Cartesian product partition $\Delta_1 \times \Delta_1 \times \ldots \times \Delta_1$.

We now consider the second level of decomposition, where we approximate the univariate functions $F(x_i, y)$ and $F(x, y_j)$ in (10A) by univariate cubic splines. To carry out these approximations, we consider a second partition Δ_2 of the unit interval. The partition Δ_2 has M_2 uniformly spaced mesh-points $0, h_2, 2h_2, \ldots, (M_2-1)h_2$ where $h_2 = 1/(M_2-1)$ (Normally, $M_2 >> M_1$. That is, the distance h_2 between the mesh points of Δ_2 is much less than the distance h_1 between the adjacent curves $F(x_i, y)$, $F(x_{i+1}, y)$.) We construct the set of natural cubic splines $\{\psi_k(\xi)\}_{k=1}^{M_2}$ such that

(12A) $$\psi_k(\xi_\ell) = \delta_{k\ell} \quad \text{for} \quad \xi_\ell \in \Delta_2 ;$$

and we consider the spline projectors $\mathcal{P}_1^2$ and $\mathcal{P}_2^2$ defined by

(13A) $$\mathcal{P}_1^2[F] = \sum_{k=1}^{M_2} \psi_k(x) F(x_k, y) \qquad (x_k \in \Delta_2)$$
$$\mathcal{P}_2^2[F] = \sum_{\ell=1}^{M_2} \psi_\ell(y) F(x, y_\ell) \qquad (y_\ell \in \Delta_2) ,$$

and having remainders $\mathcal{R}_1^2$ and $\mathcal{R}_2^2$ such that

$$\|\mathcal{R}_1^2[F]\| \leq \beta \|F^{(4,0)}\| \, (h_2)^4 \tag{14A}$$
$$\|\mathcal{R}_2^2[F]\| \leq \beta \|F^{(0,4)}\| \, (h_2)^4$$

for some constant β .

By using Lemmas A1 and A2, we obtain the numerical approximation $S^* \approx F$ of the following lemma.

Lemma A5: The function S^* given by

$$S^*(x,y) = \sum_{i=1}^{M_1} \varphi_i(x) \sum_{k=1}^{M_2} \psi_k(y) F(x_i, y_k)$$
$$+ \sum_{j=1}^{M_1} \varphi_j(y) \sum_{\ell=1}^{M_2} \psi_\ell(x) F(x_\ell, y_j) \tag{15A}$$
$$- \sum_{i=1}^{M_1} \sum_{j=1}^{M_1} \varphi_i(x)\, \varphi_j(y)\, F(x_i, y_j)$$

approximates $F(x,y)$ with an error $F - S^*$ bounded as

$$\|F - S^*\| \leq \|\mathcal{P}_1^1 \mathcal{R}_2^2[F]\| + \|\mathcal{P}_2^1 \mathcal{R}_1^2[F]\| + \|\mathcal{R}_1^1 \mathcal{R}_2^1[F]\|$$
$$\leq \|\mathcal{P}_1^1\| \, \|\mathcal{R}_2^2[F]\| + \|\mathcal{P}_2^1\| \, \|\mathcal{R}_1^2[F]\| + \|\mathcal{R}_1^1 \mathcal{R}_2^1[F]\|$$
$$\leq \beta \|\mathcal{P}_1^1\| \, \|F^{(0,4)}\| \, (h_2)^4 + \beta \|\mathcal{P}_2^1\| \, \|F^{(4,0)}\| \, (h_2)^4 + A\|F^{(4,4)}\| \, (h_1)^8 \tag{16A}$$
$$\leq \kappa[(h_2)^4 + (h_1)^8]$$

where the constant κ is

$$\kappa = \mathrm{Max}\{\beta(\|\mathcal{P}_1^1\| \|F^{(0,4)}\| + \|\mathcal{P}_2^1\| \|F^{(4,0)}\|) , \qquad \text{(17A)}$$
$$A\|F^{(4,4)}\|\} .$$

(Remark: Note that the formula for the function S^* is actuall just the sum of the two tensor product approximations $\mathcal{P}_1^2\mathcal{P}_2^1[F]$ and $\mathcal{P}_1^2\mathcal{P}_2^1[F]$ minus the tensor product function $\mathcal{P}_1^1\mathcal{P}_2^1[F]$. These are bicubic spline functions over, respectively, the Cartesian product meshes $\Delta_1 \times \Delta_2$, $\Delta_2 \times \Delta_1$ and $\Delta_1 \times \Delta_1$ (cf Figure 2) . In fact, the results of Lemma A5 could be derived by forming the sum of the first two of the following expression minus the third

$$I = \mathcal{P}_2^2\mathcal{P}_1^1 + \mathcal{R}_2^2 + \mathcal{R}_1^1 - \mathcal{R}_2^2\mathcal{R}_1^1$$

$$I = \mathcal{P}_1^2\mathcal{P}_2^1 + \mathcal{R}_1^2 + \mathcal{R}_2^1 - \mathcal{R}_1^2\mathcal{R}_2^1 \qquad \text{(18A)}$$

$$I = \mathcal{P}_1^1\mathcal{P}_2^1 + \mathcal{R}_1^1 + \mathcal{R}_2^1 - \mathcal{R}_1^1\mathcal{R}_2^1 ,$$

which are, respectively, the algebraically minimal decompositions of I in the Boolean algebras generated by the pairs $\{\mathcal{P}_2^2, \mathcal{P}_1^1\}$, $\{\mathcal{P}_1^2, \mathcal{P}_2^1\}$ and $\{\mathcal{P}_1^1, \mathcal{P}_2^1\}$. This derivation gives the remainder operator $\mathcal{P}_1^1\mathcal{R}_2^2 + \mathcal{P}_2^1\mathcal{R}_1^2 + \mathcal{R}_1^1\mathcal{R}_2^2$ in the alternative form

$$\mathcal{R}_1^2 + \mathcal{R}_2^2 - \mathcal{R}_1^2\mathcal{R}_2^1 - \mathcal{R}_2^2\mathcal{R}_1^1 + \mathcal{R}_1^1\mathcal{R}_2^1 \qquad \text{(19A)}$$

in which terms involving $\mathcal{R}_1^1$ and $\mathcal{R}_2^1$ alone, which are the dom nant error operators in expressions (18A), do not appear. It is because of this error cancelation that the approximation S^* turns out to be far more accurate than are any of the individual bicubic spline approximations $\mathcal{P}_2^2\mathcal{P}_1^1$, $\mathcal{P}_1^2\mathcal{P}_2^1$ or $\mathcal{P}_1^1\mathcal{P}_2^1$. These observations are, of course, generally true — not just for the cubic spline projectors of our example.)

Consider now the problem of approximating $F(x,y) \in C^{4,4}$ by a function of the form (15A) to an accuracy of 2ϵ by using a minimal number of point values $F(x_\mu, y_\nu)$. From (16A) we see that we should have

$$\kappa[(h_2)^4 + (h_1)^8] = 2\epsilon \ , \tag{20A}$$

which will be achieved by taking

$$\kappa(h_2)^4 = \kappa(h_1)^8 = \epsilon \ . \tag{21A}$$

This implies that the partitions Δ_1 and Δ_2 should have mesh lengths h_1 and h_2, respectively, of the size

$$\begin{aligned} h_1 &= (\epsilon/\kappa)^{1/8} \\ h_2 &= (\epsilon/\kappa)^{1/4} \end{aligned} \tag{22A}$$

from which we draw the interesting conclusion that

$$\boxed{h_2 = (h_1)^2} \ , \tag{23A}$$

i.e. Δ_2 is a much finer partition of $[0,1]$ than is the partition Δ_1.

In general, we do not assume that the partition Δ_2 is a refinement of Δ_1. That is, the mesh-points of Δ_1 are not generally a subset of those of Δ_2. Hence, the total number of distinct points at which F is to be assumed known is equal to the number of distinct points in the union of the three sets of mesh-points represented by $\Delta_1 \times \Delta_2$, $\Delta_2 \times \Delta_1$ and $\Delta_1 \times \Delta_1$, and is easily computed to be (cf. Figure 2).

$$n_D = M_1 M_2 + M_2 M_1 + M_1 M_1 - 4M_1 \tag{24A}$$

or, since $M_1 << M_2$

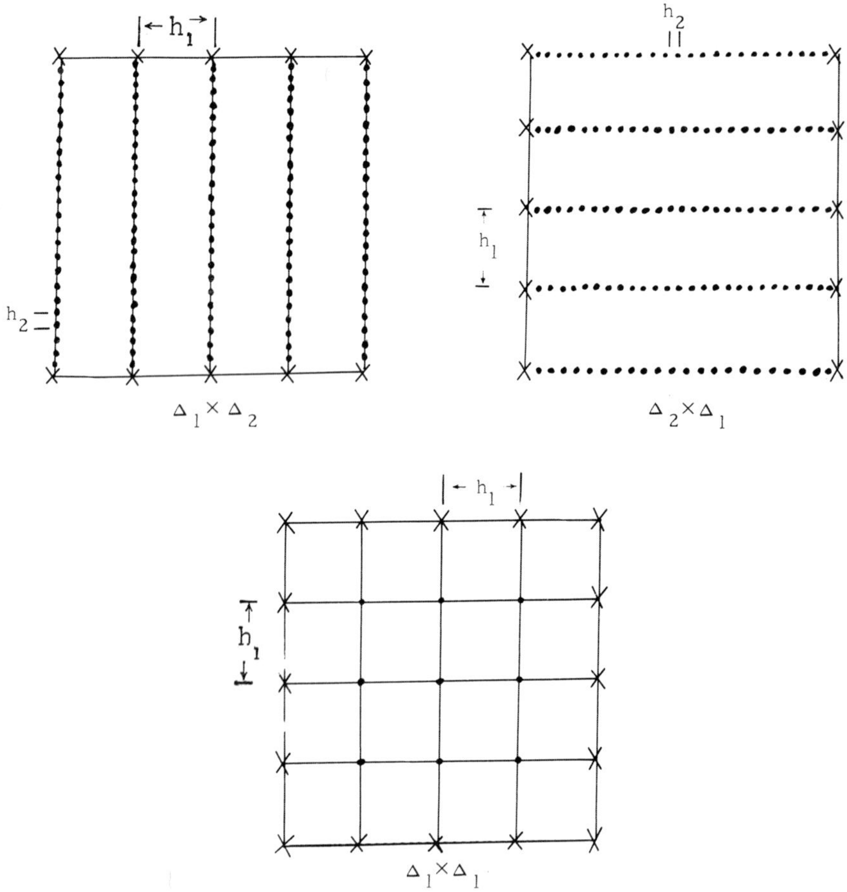

Figure 2: The three Cartesian product meshes involved in the decomposition approximation $S^*(x, y) \approx F(x, y)$ where S^* is given by (15A). Boundary mesh-points (X) are common to $\Delta_1 \times \Delta_2$ or $\Delta_2 \times \Delta_1$. Corner points are common to all three product meshes. The complete set of points involved in the formula for S^* is the union $(\Delta_1 \times \Delta_2) \cup (\Delta_2 \times \Delta_1) \cup (\Delta_1 \times \Delta_1)$. In the above illustration, $M_1 = 5$ and $M_2 = 25$.

(25A) $$n_D \approx 2M_1M_2 .$$

(If Δ_2 is, in fact, a refinement of Δ_1, then $n_D = 2M_1M_2 - M_1M_1$ since all mesh-points in $\Delta_1 \times \Delta_1$ are also in both $\Delta_1 \times \Delta_2$ and $\Delta_2 \times \Delta_1$. In any case, (25A) is valid for $M_1 << M_2$.) By using the relations $M_1 = (h_1)^{-1} + 1$, $M_2 = (h_2)^{-1} + 1$, (25A) takes the more interesting form

(26A) $$n_D \approx 2\left(\frac{1+h_1}{h_1}\right)\left(\frac{1+h_2}{h_2}\right) \approx 2(h_1h_2)^{-1} ;$$

which, using (23A), becomes

(27A) $$n_D \approx 2(h_1)^{-3} = 2(\kappa/\epsilon)^{3/8} .$$

We shall return to these formulas shortly, but let us first consider the obvious tensor product alternative to the cubic spline decomposition, namely, simple bicubic spline interpolation. We can derive the necessary results for this alternative scheme quite readily by specializing all of the above analysis to the case where we take $\mathcal{P}_1^2 = \mathcal{P}_1^1$ and $\mathcal{P}_2^2 = \mathcal{P}_2^1$. In this instance, there is total symmetry of the problem, since the partitions Δ_1 and Δ_2 coincide. Under these circumstances we can drop the superscripts on projectors and the subscripts on the parameters related to the interval partitions. To avoid possible notational confusion between the decomposition formulae and the tensor product formulae we shall use a tilde to designate parameters associated with the latter. Thus, for example, $\tilde{\mathcal{P}}$ is the parametric natural cubic spline projector over the $\tilde{M}$ point uniform partition $\tilde{\Delta}$ with mesh-spacing $\tilde{h} = 1/(\tilde{M}-1)$.

Lemma A6: The bicubic spline $T(x,y)$ given by

$$T(x,y) = \tilde{\mathcal{P}}_1\tilde{\mathcal{P}}_2[F]$$

(28A)
$$= \sum_{\mu=1}^{M}\sum_{\nu=1}^{M} \tilde{\varphi}_\mu(x)\,\tilde{\varphi}_\nu(y)\,F(x_\mu, y_\nu)$$

interpolates F at the $\tilde{M}\cdot\tilde{M}$ mesh-points of $\tilde{\Delta}\times\tilde{\Delta}$ and approximates F with an error $F - T$ bounded as in the following:

$$\|F-T\| = \|(\tilde{\mathcal{R}}_1 + \tilde{\mathcal{R}}_2 - \tilde{\mathcal{R}}_1\tilde{\mathcal{R}}_2)[F]\|$$

$$\leq \|\tilde{\mathcal{R}}_1[F]\| + \|\tilde{\mathcal{R}}_2[F]\| + \|\tilde{\mathcal{R}}_1\tilde{\mathcal{R}}_2[F]\|$$

(29A)
$$\leq \tilde{\alpha}\|F^{(4,0)}\|\,(\tilde{h})^4 + \tilde{\alpha}\|F^{(0,4)}\|\,(\tilde{h})^4 + \tilde{A}\|F^{(4,4)}\|\,(\tilde{h})^8$$

$$\leq \tilde{\kappa}[(\tilde{h})^4 + (\tilde{h})^8]$$

where the constant $\tilde{\kappa}$ is

(30A) $$\tilde{\kappa} = \mathrm{Max}\{\tilde{\alpha}(\|F^{(4,0)}\| + \|F^{(0,4)}\|),\ \tilde{A}\|F^{(4,4)}\|\}\ .$$

The important point to note regarding (29A) is that for $\tilde{h} \ll 1$:

(31A)
$$(\tilde{h})^4 + (\tilde{h})^8 = (\tilde{h})^4[1+O(\tilde{h})^4)] \approx (\tilde{h})^4$$

to a very high degree of accuracy. Thus, we have the useful (approximate) bound

(32A) $$\|F-T\| \leq \tilde{\kappa}(\tilde{h})^4\ .$$

Consider again the problem of obtaining a numerical approximation to F, but using this time a bicubic spline (28A)

To have $\|F - T\| \leq 2\epsilon$, as before, requires that we select a mesh $\tilde{\Delta}$ such that

$$\tilde{h} = (\epsilon/2\tilde{\kappa})^{1/4} . \tag{33A}$$

Hence, the number of points at which F must be known is seen from (28A) to be

$$\begin{aligned} n_T &= \tilde{M}^2 \\ &= \left(\frac{1+\tilde{h}}{\tilde{h}}\right)^2 , \end{aligned} \tag{34A}$$

or, to a good approximation

$$n_T \approx \sqrt{2}\ (\tilde{\kappa}/\epsilon)^{1/2} . \tag{35A}$$

Now, recall formula (27A) which gives the number of points required to achieve an accuracy of 2ϵ by approximating $F(x, y)$ with a function of the form (15A). A measure of the relative efficiency of the two competing methods is the ratio $\rho \equiv n_D/n_T$:

$$\rho = \sqrt{2} \left(\frac{\kappa^{3/8}}{\tilde{\kappa}^{1/2}}\right) \epsilon^{1/8} . \tag{36A}$$

In order to get a feeling for the behavior of n_D, n_T and ρ as functions of the parameters $\kappa, \tilde{\kappa}$ and (most importantly) ϵ, we have computed these quantities for various ranges of the parameters and the results are tabulated in Tables I–VI below. It should be observed that, in lieu of any supplementary information regarding the behavior of $F(x, y)$, the most objective comparisons are those of Tables I–IV, where one assumes that κ and $\tilde{\kappa}$ are of the same magnitude. On the whole, these first four tables indicate that the decomposition method is remarkably superior to the bicubic spline competitor. The values of $\epsilon = 10^{-6}$, 10^{-8}, 10^{-10} and 10^{-12} are in the range of practical interest.

Tables V and VI are biased in favor of, respectively, the decomposition scheme and the tensor product (bicubic spline) method. Functions for which the constants κ and $\tilde{\kappa}$ are related as they are supposed to be in V and VI are, to some degree, exceptional. Thus, unlike I-IV which are intended to be objective, Tables V and VI represent atypical functions $F(x, y)$.

To conclude this appendix, we shall look briefly at functions of three and more variables. Here, the comparison between decomposition techniques and tensor product methods is more dramatic than in the bivariate case.

For a trivariate function $F(x, y, z)$, the analog of (1A) in Lemma A1 is the following formula in which again superscripts denote the level of decomposition, i.e. first, second, or third:

$$
\begin{aligned}
I &= \mathcal{P}_1^1 \oplus \mathcal{P}_2^1 \oplus \mathcal{P}_3^1 + \mathcal{R}_1^1 \mathcal{R}_2^1 \mathcal{R}_3^1 \\
&= \mathcal{P}_1^1 + \mathcal{P}_2^1 + \mathcal{P}_3^1 - \mathcal{P}_1^1\mathcal{P}_2^1 - \mathcal{P}_2^1\mathcal{P}_3^1 - \mathcal{P}_3^1\mathcal{P}_1^1 + \mathcal{P}_1^1\mathcal{P}_2^1\mathcal{P}_3^1 + \\
&\qquad \mathcal{R}_1^1 \mathcal{R}_2^1 \mathcal{R}_3^1 \\
&= (\mathcal{P}_2^2 \oplus \mathcal{P}_3^2)\,\mathcal{P}_1^1 + (\mathcal{P}_1^2 \oplus \mathcal{P}_3^2)\mathcal{P}_2^1 + (\mathcal{P}_1^2 \oplus \mathcal{P}_2^2)\mathcal{P}_3^1 \\
&\qquad - \mathcal{P}_1^1\mathcal{P}_2^1 - \mathcal{P}_2^1\mathcal{P}_3^1 - \mathcal{P}_3^1\mathcal{P}_1^1 + \mathcal{P}_1^1\mathcal{P}_2^1\mathcal{P}_3^1 + [\mathcal{R}_2^2\mathcal{R}_3^2\mathcal{P}_1^1 + \\
&\qquad \mathcal{R}_1^2\mathcal{R}_3^2\mathcal{P}_2^1 + \mathcal{R}_1^2\mathcal{R}_2^2\mathcal{P}_3^1] + \mathcal{R}_1^1 \mathcal{R}_2^1 \mathcal{R}_3^1 \qquad \text{(37A)} \\
&= \ldots \\
&= \mathcal{P}_3^3\mathcal{P}_2^2\mathcal{P}_1^1 + \mathcal{P}^3\mathcal{P}^2\mathcal{P}_1^1 - \mathcal{P}_2^2\mathcal{P}_3^2\mathcal{P}_1^1 + \mathcal{P}_3^3\mathcal{P}_1^2\mathcal{P}_2^1 + \mathcal{P}_1^3\mathcal{P}_3^2\mathcal{P}_2^1 \\
&\quad - \mathcal{P}_1^2\mathcal{P}_3^2\mathcal{P}_2^1 + \mathcal{P}_2^3\mathcal{P}_1^2\mathcal{P}_3^1 + \mathcal{P}_1^3\mathcal{P}_2^2\mathcal{P}_3^1 - \mathcal{P}_1^2\mathcal{P}_2^2\mathcal{P}_3^1 - \mathcal{P}_3^3\mathcal{P}_1^1\mathcal{P}_2^1 - \mathcal{P}_1^3\mathcal{P}_2^1\mathcal{P}_3^1 \\
&\quad - \mathcal{P}_2^3\mathcal{P}_3^1\mathcal{P}_1^1 + \mathcal{P}_1^1\mathcal{P}_2^1\mathcal{P}_3^1 \\
&\quad + \{\mathcal{R}_3^3\mathcal{P}_2^2\mathcal{P}_1^1 + \mathcal{R}_2^3\mathcal{P}_3^2\mathcal{P}_1^1 + \mathcal{R}_3^3\mathcal{P}_1^2\mathcal{P}_2^1 + \mathcal{R}_1^3\mathcal{P}_3^2\mathcal{P}_2^1
\end{aligned}
$$

$$+ \mathcal{R}^3_2 \mathcal{P}^2_1 \mathcal{P}^1_3 + \mathcal{R}^3_1 \mathcal{P}^2_2 \mathcal{P}^1_3 - \mathcal{R}^3_3 \mathcal{P}^1_1 \mathcal{P}^1_2 - \mathcal{R}^3_1 \mathcal{P}^1_2 \mathcal{P}^1_3 - \mathcal{R}^3_2 \mathcal{P}^1_3 \mathcal{P}^1_1\}$$

$$+ [\mathcal{R}^2_2 \mathcal{R}^2_3 \mathcal{P}^1_1 + \mathcal{R}^2_1 \mathcal{R}^2_3 \mathcal{P}^1_2 + \mathcal{R}^2_1 \mathcal{R}^2_2 \mathcal{P}^1_3]$$

$$+ \mathcal{R}^1_1 \mathcal{R}^1_2 \mathcal{R}^1_3 \quad .$$

By carrying through the details, we obtain the trivariate analog of Lemma A5 with the error bound

$$\| F - S^* \| \leq \kappa[(h_3)^4 + (h_2)^8 + (h_1)^{12}] \tag{38A}$$

where h_1 is the distance between plane sectional samples of F ("mesh-planes"), h_2 is the distance between adjacent mesh-lines on those planes, and h_3 is the distance between the mesh-points on those lines. For a consistent decomposition, the relation between these parameters is readily seen to be (compare (23A)):

$$h_3 = (h_2)^2 = (h_1)^3 \tag{39A}$$

where

$$h_1 = (\epsilon/\kappa)^{1/12} \tag{40A}$$

and the number of points required for 3ϵ accuracy of the approximation is

$$\begin{aligned} n_D &\approx (3/h_1)(2/h_2)(1/h_3) \\ &= 6(h_1)^{-11/2} \qquad \text{(41A)} \\ &= 6(\kappa/\epsilon)^{11/24} \end{aligned}$$

The tensor product (tricubic spline) approximation $T = \tilde{\mathcal{P}}_1 \tilde{\mathcal{P}}_2 \tilde{\mathcal{P}}_3 [F]$, on the other hand, has a Cartesian product set of mesh-points $\tilde{\Delta} \times \tilde{\Delta} \times \tilde{\Delta}$ with a linear point spacing of $\tilde{h}$:

(42A) $$\tilde{h} = (\epsilon/3\tilde{\kappa})^{1/4}$$

and hence requires knowledge of F at n_T points:

(43A) $$\begin{aligned} n_T &\approx (\tilde{h})^{-3} \\ &= (3\tilde{\kappa}/\epsilon)^{3/4} \\ &\approx 2.27(\tilde{\kappa}/\epsilon)^{3/4} . \end{aligned}$$

The ratio $\rho \equiv n_D/n_T$ is

(44A) $$\rho \approx 2.6 \left(\frac{\kappa^{11/24}}{\kappa^{3/4}} \right) \epsilon^{7/24} .$$

For $\epsilon = 10^{-8}$ and for several values of $\kappa = \tilde{\kappa}$, Table VII shows the striking comparison between the numbers of points n_D and n_T.

The pattern for the case of N variables is now easy to infer. The number of points required in cubic spline decomposition of the function $F(x_1, \ldots, x_N)$ is

(45A) $$n_D \approx N! \, (\kappa/\epsilon)^{1/4+1/8+\ldots+1/4N} ;$$

and, the number of tensor product points is

(46A) $$n_T \approx (N\tilde{\kappa}/\epsilon)^{N/4} .$$

Thus, the ratio $\rho = n_D/n_T$ is

(47A) $$\rho \approx C \, \epsilon^{1/4[N-1-1/2-1/3-\ldots-1/N]}$$

for some C depending on κ and $\tilde{\kappa}$. Since certain combinatorial approximations have been made in their derivation, formulas (45A) - (47A) are not asymptotically valid for $N \to \infty$

TABLES

Tables I–VI contrast the approximate numbers of points n_D and n_T required to approximate a bivariate function $F(x, y)$ to an accuracy of 2ϵ (i.e. $\|F(x, y) - f(x, y)\| \le 2\epsilon$ where f is the approximation) using the method of successive cubic spline decomposition (i.e. $f = S^*$ of equation (15A)) versus standard bicubic spline interpolation (i.e. $f = T$ of equation (28A)). The parameters κ and $\tilde{\kappa}$ are the constants defined by equations (17A) and (30A), respectively. They depend, of course, upon the function $F(x, y)$. Table VII represents a similar comparison for the approximation of a trivariate function, $F(x, y, z)$. (See paragraphs following (36A).)

Table I: $\epsilon = 10^{-6}$, $\kappa = \tilde{\kappa} = .01, .1, 1.0, 10.0, 100.0$

$\kappa = \tilde{\kappa}$	n_D	n_T	$\rho \equiv n_D/n_T$
0.01	63	140	.45
0.1	150	442	.34
1.0	356	1,400	.25
10.0	845	4,420	.19
100.0	2,020	14,000	.14

Table II: $\epsilon = 10^{-8}$, $\kappa = \tilde{\kappa} = .01, .1, 1.0, 10.0, 100.0$

$\kappa = \tilde{\kappa}$	n_D	n_T	$\rho \equiv n_D/n_T$
0.01	350	1,400	.25
0.1	840	4,400	.20
1.0	2,000	14,000	.14
10.0	4,740	44,200	.11
100.0	11,300	140,000	.081

Table III: $\epsilon = 10^{-10}$, $\kappa = \tilde{\kappa} = .01, .1, 1.0, 10.0, 100.0$

$\kappa = \tilde{\kappa}$	n_D	n_T	$\rho \equiv n_D/n_T$
0.01	2,000	14,000	.14
0.1	4,800	44,200	.11
1.0	11,300	140,000	.081
10.0	26,800	442,000	.061
100.0	64,200	1,400,000	.046

Table IV: $\epsilon = 10^{-12}$, $\kappa = \tilde{\kappa} = .01, .1, 1.0, 10.0, 100.0$

$\kappa = \tilde{\kappa}$	n_D	n_T	$\rho \equiv n_D/n_T$
0.01	1.1×10^4	1.4×10^5	.081
0.1	2.7×10^4	4.4×10^5	.059
1.0	6.3×10^4	1.4×10^6	.044
10.0	1.5×10^5	4.4×10^6	.033
100.0	3.6×10^5	1.4×10^7	.025

Table V: $\epsilon = 10^{-8}$, $\kappa = 1 \leq \tilde{\kappa}$

κ	$\tilde{\kappa}$	n_D	n_T	$\rho \equiv n_D/n_T$
1.0	1.0	2,000	14,000	.14
1.0	10.0	2,000	44,200	.045
1.0	100.0	2,000	140,000	.014
1.0	1,000.0	2,000	442,000	.0045
1.0	10,000.0	2,000	1,400,000	.0014

Table VI: $\epsilon = 10^{-8}$, $\kappa \geq \tilde{\kappa} = 1$

κ	$\tilde{\kappa}$	n_D	n_T	$\rho \equiv n_D/n_T$
1.0	1.0	2,000	14,000	.14
10.0	1.0	4,700	14,000	.34
100.0	1.0	11,300	14,000	.81
1,000.0	1.0	26,000	14,000	1.85
10,000.0	1.0	63,000	14,000	4.5

Table VII: Trivariate approximation, $\epsilon = 10^{-8}$, $\kappa = \tilde{\kappa}$

$\kappa = \tilde{\kappa}$	n_D	n_T	$\rho \equiv n_D/n_T$
0.01	3.4×10^3	7.2×10^4	.048
0.1	1.0×10^4	4.1×10^5	.025
1.0	2.7×10^4	2.3×10^6	.012
10.0	8.0×10^4	1.3×10^7	.0063
100.0	2.3×10^5	7.2×10^7	.0032

APPENDIX B: A SPECIAL FORM OF THEOREM 4

In Example 1 of Section III, we actually make use of a special form of Theorem 4 in which the generators $\{\mathcal{P}_m\}_{m=1}^M$ are themselves the maximal elements of some collection of distributive lattices ψ_m $(m = 1, 2, \ldots, M)$ which are generated by M collections of "micro"-projectors $\{p_1^m, p_2^m, \ldots, p_{k_m}^m\}_{m=1}^M$. In this appendix we shall briefly sketch the details of the construction of the "super"-lattice Ψ beginning with M collections of "micro"-projectors.

Lemma 4': Let $\{p_\mu^m\}_{\mu=1}^{k_m}$ be a commutative collection of projectors. The maximal element of the distributive lattice ψ_m generated by this collection is

$$\mathcal{P}_m = p_1^m \oplus p_2^m \oplus \cdots \oplus p_{k_m}^m \tag{1B}$$

and $\mathcal{P}_m$ has the properties that

$$p_\mu^m \mathcal{P}_m = p_\mu^m \qquad (\mu = 1, 2, \ldots, k_m) \; . \tag{2B}$$

Theorem 4': Let $\{p_{\mu_1}^1\}_{\mu_1=1}^{k_1}$, $\{p_{\mu_2}^2\}_{\mu_2=1}^{k_2}$, $\ldots$, $\{p_{\mu_M}^M\}_{\mu_M=1}^{k_M}$ be M sets of projectors which satisfy the hypotheses of Lemma 4'; and, let $\{\mathcal{P}_m\}_{m=1}^M$ be the associated set of maximal projectors in the lattices $\{\psi_m\}_{m=1}^M$, i.e. $\mathcal{P}_m$ is given by equation (1B) For all m and n with $m \neq n$, assume that the "micro"-projectors $p_{\mu_m}^m$ and $p_{\nu_n}^n$ commute:

$$p_{\mu_m}^m p_{\nu_n}^n = p_{\nu_n}^n p_{\mu_m}^m \qquad \begin{pmatrix} \mu_m = 1, 2, \ldots, k_m \\ \nu_n = 1, 2, \ldots, k_n \end{pmatrix} \; . \tag{3B}$$

Then, the projectors $\mathcal{P}_m$ and $\mathcal{P}_n$ commute

$$\mathcal{P}_m \mathcal{P}_n = \mathcal{P}_n \mathcal{P}_m \qquad (m, n = 1, 2, \ldots, M) \tag{4B}$$

and therefore $\{\mathcal{P}_m\}_{m=1}^{M}$ generates a distributive lattice Ψ with maximal element $\mathfrak{m}$

$$\begin{aligned}\mathfrak{m} &= \mathcal{P}_1 \oplus \mathcal{P}_2 \oplus \ldots \oplus \mathcal{P}_M \\ &= (p_1^1 \oplus p_2^1 \oplus \ldots \oplus p_{k_1}^1) \oplus (p_1^2 \oplus p_2^2 \oplus \ldots \oplus p_{k_2}^2) \oplus \\ &\quad \ldots \oplus (p_1^M \oplus p_2^M \oplus \ldots \oplus p_{k_M}^M)\end{aligned} \tag{5B}$$

and minimal element $\mathfrak{L}$

$$\begin{aligned}\mathfrak{L} &= \mathcal{P}_1 \mathcal{P}_2 \ldots \mathcal{P}_M \\ &= \prod_{m=1}^{M} (p_1^m \oplus p_2^m \oplus \ldots \oplus p_{k_m}^m) \ .\end{aligned} \tag{6B}$$

$\mathfrak{m}$ has the properties that

$$p_{\mu_m}^m \mathfrak{m} = p_{\mu_m}^m \qquad \text{for all} \quad \mu_m = 1, 2, \ldots, k_m ; \quad m = 1, 2, \ldots, M \ . \tag{7B}$$

and $\mathfrak{L}$ satisfies

$$p_{\mu_1}^1 p_{\mu_2}^2 \ldots p_{\mu_M}^M \mathfrak{L} = p_{\mu_1}^1 p_{\mu_2}^2 \ldots p_{\mu_M}^M \ . \tag{8B}$$

The special form of Theorem 4' which we use in Example 1, Section III has p_μ^m of the form (cf. Equation (20))

$$p_\mu^m = \varphi_\mu^m (x_m) \lambda_\mu^m \tag{9B}$$

where λ_μ^m is a "parametrically extended" linear functional. Moreover, in addition to the hypotheses of Theorem 4', the "micro"-projectors of that example also satisfy the orthogonality conditions

(10B) $$p_i^m p_j^m = 0 \qquad \text{for} \quad i \neq j = 1, 2, \ldots, k_m ,$$

i.e. for each fixed $m = 1, 2, \ldots, M$, the projectors $p_1^m, p_2^m, \ldots, p_{k_m}^m$ are mutually orthogonal. Hence, (10B) implies

(11B) $$p_1^m \oplus p_2^m \oplus \ldots \oplus p_{k_m}^m = p_1^m + p_2^m + \ldots + p_{k_m}^m .$$

(In connection with this last relation, see [11, p. 148].) Therefore, the maximal and minimal elements $\mathfrak{M}$ and $\mathfrak{L}$ of the lattice Ψ are seen from (5B), (6B) and (11B) to assume the simpler form

(12B) $$\mathfrak{M} = \left(\sum_{\mu_1=1}^{k_1} p_{\mu_1}^1\right) \oplus \left(\sum_{\mu_2=1}^{k_2} p_{\mu_2}^2\right) \oplus \ldots \oplus \left(\sum_{\mu_M=1}^{k_M} p_{\mu_M}^M\right)$$

and

(13B) $$\mathfrak{L} = \prod_{m=1}^{M} \left(\sum_{\mu_m=1}^{k_m} p_{\mu_m}^m\right) .$$

Acknowledgements

During the course of the research reported here, the author has benefited from many valuable discussions with his colleagues Messrs. A. V. Butterworth, W. W. Meyer, and D. H. Thomas of the General Motors Research Laboratories, with Professor Robert Barnhill of the University of Utah and with Professor Garrett Birkhoff of Harvard University. We especially want to acknowledge the important contribution of Professor Birkhoff in pointing out that the algebraic structure of our results was isomorphic to that of a distributive lattice.

REFERENCES

[1] J. H. Ahlberg, E. N. Nilson, J. L. Walsh, The Theory of Splines and Their Applications (Academic Press, New York, 1967).

[2] D. V. Ahuja, S. A. Coons, "Geometry for construction and display," IBM Systems Journal 7 (1968), Nos. 3-4, 188-205.

[3] G. Birkhoff, W. J. Gordon, "The Draftsman's and related equations," Journal of Approximation Theory 1 (1968), 199-208.

[4] G. Birkhoff, S. MacLane, A Survey of Modern Algebra (Macmillan, New York, 1953).

[5] S. A. Coons, "Surfaces for the computer aided design of space forms," Project MAC, MIT. Revised to Mac-TR-41, June 1967. Available through CFSTI, Sills Building, 5285 Port Royal Road, Springfield, Va. 22151.

[6] R. Courant, D. Hilbert, Methods of Mathematical Physics Vol. I (Interscience, New York, 1953).

[7] C. deBoor, "Bicubic spline interpolation," J. Math. Physics 41 (1962), 212-218.

[8] A. R. Forrest, "Coons surfaces and multivariable functional interpolation," To appear in ACM Communications.

[9] W. J. Gordon, "Blending-function methods of bivariate and multivariate interpolation and approximation," Report GMR-834, October 1968. To appear in SIAM Journal on Numer. Anal., Fall 1969.

[10] W. J. Gordon, "Spline-blended surface interpolation through curve networks," Report GMR-799, July 1968. To appear in J. Math. Mech., Summer 1969.

[11] P. R. Halmos, Finite-Dimensional Vector Spaces (Van Nostrand, Princeton, 1958).

[12] S. MacLane, G. Birkhoff, Algebra (Macmillan, New York, 1967).

[13] D. Mangeron, "Introduzione nello studio dei sistemi polivibrante con rimanenza ed argomenti retardati," Accad. Nazionale dei Lincei (1965), Series VIII, Vol. XXXIX, fasc. 1-2, p. 22-28. (See also the numerous other papers by Mangeron, Picone, et. al. referenced in the bibliography of that paper.)

[14] D. Mangeron, "Sopra un problema al contorno..," Rend Accad. Sci. Fis. Mat. Napoli 2 (1932), 28-40.

[15] A. Marchaud, "Sur les dérivées et sur la différence des fonctions de variables réeles," J. de Math Pures et Appl. 6 (1927), 332-425.

[16] L. Neder, "Interpolationsformeln für Funktionen mehrerer Argumente," Scandinavisk Aktuarietdskrift 9 (1926), 59-69.

[17] M. Picone, "Vedute generali sull'interpolazione e qualche loro consequenza," Ann. Scuola Norm. Sup. Pisa (1951), Series 3, Vol. 5, fasc. 3-4.

[18] T. Popoviciu, "Sur les solutions bornées et les solutions measurables de certaines équations fonctionnelles Mathematica 14 (1938), 47-106.

[19] A. Sard, "Optimal approximation," J. of Functional Analysis 1, No. 2 (Aug. 1967), 222-243.

[20] D. D. Stancu, "One some Taylor expansions for functions of many variables," Revue de Math. Pures Appl. (1959), 249-265 (In Russian).

21] D. D. Stancu, "The remainder of certain linear approximation formulas in two variables," SIAM J. Numer. Anal. Ser. B. Vol. 1, (1964), 137-163.

Multivariate Spline Functions and Elliptic Problems

MARTIN H. SCHULTZ

§1. Introduction

In the past few years there has been renewed interest in the Rayleigh-Ritz-Galerkin method for approximating the solutions of well-posed boundary value problems for linear and nonlinear partial differential equations. For classical accounts of this method the reader is referred to [23], [24], [30], [41], and [47] and for modern accounts the reader is referred to [3], [14], [15], [22], [56], [57], [58], and [71]. For a bibliography of the extensive Russian work on this method, the reader is referred to [48].

The incentive for this renewed interest has been the realization beginning in the mathematical literature in [7] and [80], that by using finite dimensional spaces of spline type functions in this method one obtains sparse systems of nonlinear algebraic equations which can be solved very efficiently by a modern computer. Moreover, the resulting approximations are very accurate. We remark that this combination is called the finite element method in the engineering literature, cf. [82].

In this paper, we describe and theoretically analyze this combination for a large class of nonlinear elliptic partial differential equations and linear elliptic eigenvalue problems. The results of this paper directly generalize, extend, and improve the corresponding results of [7]-[9], [16]-[22], [26], [28], [37], [40], [55], [64]-[74], [80], and [81]. For a discussion of this combination for other types of boundary value problems, the reader is referred to [25], [76], [77], and [78].

In particular, we develop the approximation theory of multivariate spline subspaces of Sobolev spaces in sections 2-5. In section 6, we discuss nonlinear two-point boundary value problems and in section 7 we discuss nonlinear elliptic partial differential equations. We give the analysis for Neuma problems in section 8, for linear elliptic eigenvalue problems in section 9, and for a class of singular problems in section 10 Finally, in section 11, we illustrate how the theoretical error bounds obtained in sections 6-9 may be extended to bounds in higher-order Sobolev norms and to the uniform norm.

We remark that many people are currently working on multivariate spline approximation theory and its applications to the study of the Rayleigh-Ritz-Galerkin method with spline subspaces. In particular, we refer to [4], [13], and [27], all of whom use Fourier analysis techniques.

§2. One Dimensional Interpolation Theory

In this section we discuss computable upper and lower bounds for the error in interpolating a class of functions of on real variable belonging to a Sobolev space by means of functic belonging to a finite dimensional spline subspace. These results extend and improve the corresponding results of [1], [6] [40], and [64]. The full details of the proofs are to appear i [73].

Let $-\infty < a < b < \infty$ be fixed and for each nonnegative integer, q, and each $1 \leq r \leq \infty$ $H^{q,r}(a,b)$ denote the compl tion of the set

$$\{f \in C^{\infty}(a,b) \mid f \text{ is real-valued and } \|f\|_{H^{q,r}(a,b)} \equiv$$

$$\left(\int_a^b \sum_{j=0}^{q} [D^j f(x)]^r dx\right)^{1/r} < \infty$$

with respect to the norm $\|\cdot\|_{H^{q,r}(a,b)}$ where $Df \equiv \frac{df}{dx}$ denotes the derivative of f. We remark that it is easy to verify that $H^{q,r}(a,b)$ consists of all those real-valued functions on $[a,b]$ which have $q-1$ absolutely continuous derivatives and whose q-th derivative exists almost everywhere in $[a,b]$ and is in $L^r(a,b)$. For the case of $r = 2$, we have a special convention that $H^{q,2}(a,b) \equiv H^q(a,b)$.

For each nonnegative integer, M, let $\mathcal{P}_M(a,b)$ denote the set of partitions, Δ, of $[a,b]$ of the form

$$(2.1) \qquad \Delta: a = x^0 < x^1 < \ldots < x^M < x^{M+1} = b$$

and let $\mathcal{P}(a,b) \equiv \bigcup_{M=0}^{\infty} \mathcal{P}_M(a,b)$. If $\Delta \in \mathcal{P}_M(a,b)$, d is a positive integer, and z is an integer such that $-1 \leq z \leq d-1$, we define the spline space, $S(d, \Delta, z)$, to be the set of all real-valued functions $s(x) \in C^z[a,b]$, such that on each subinterval (x^i, x^{i+1}), $0 \leq i \leq M$, $s(x)$ is a polynomial of degree d, where by $C^{-1}[a,b]$ we mean those functions which have a simple jump discontinuity at each partition point, x^i, $1 \leq i \leq M$. Clearly $S(d, \Delta, z) \subset H^{z+1,\infty}(a,b)$.

It is easy to verify and important to note that all the results of this paper remain essentially unchanged if one allows the number z to depend on the partition points, x^i, $1 \leq i \leq M$, in such a way that $-1 \leq z(x^i) \leq d-1$ for all $1 \leq i \leq M$ or if one considers appropriate spaces of "polynomial g-splines", cf. [40] and [64]. Thus, for example it is possible to obtain direct generalizations and improvements of Theorems 3.1-3.6 of [40] and Theorems 18-21 of [64]. The details are left to the reader.

In this section, we assume that $d \equiv 2m-1$ for some positive integer m and $\frac{d-1}{2} \equiv m-1 \leq z \leq 2m-2 = d-1$. Following [64], we define the interpolation mapping $\mathcal{J}: C^{m-1}[a,b] \to S(2m-1, \Delta, z)$ by $\mathcal{J}f \equiv s$, where

$$(2.2) \qquad D^k s(x^i) = D^k f(x^i), \quad \begin{cases} 0 \leq k \leq 2m-2-z, & 1 \leq i \leq M, \\ 0 \leq k \leq m-1, & i = 0 \text{ and } M+1. \end{cases}$$

We remark that the preceding interpolation mapping corresponds to the Type I interpolation of [64]. It is easy to modify the results of this section for the cases in which the interpolation mapping corresponds to Types II, III, and IV interpolation of [64]. The details we left to the reader.

To be specific, in this section we give explicit upper and in some cases lower bounds for the quantities

$$\wedge(p,q;S(2m-1,\Delta,z);j,r) \equiv \sup\{\|D^j(f-\mathcal{J}f)\|_{H^{0,r}(a,b)} / \tag{2.3}$$
$$\|D^p f\|_{H^{0,q}(a,b)} \mid f \in H^{p,q}(a,b),\ \|D^p f\|_{H^{0,q}(a,b)} \neq 0\},$$

where $1 \leq m$, $m \leq p \leq 2m$, $m-1 \leq z \leq 2m-2$, $0 \leq j \leq p$, $\Delta \in \mathcal{P}(a,b)$, and $\mathcal{J}$ denotes the interpolation mapping defined in (2.2).

Now we recall some basic results from [64] and introduce some additional notation.

Theorem 2.1. The interpolation mapping given by (2.2) is well defined for all $\Delta \in \mathcal{P}(a,b)$, $1 \leq m$, and $m-1 \leq z \leq 2m-2$.

Theorem 2.2. (First Integral Relation). If $f \in H^m(a,b)$, $1 \leq m$, $\Delta \in \mathcal{P}(a,b)$, and $m-1 \leq z \leq 2m-2$,

$$\|D^m f\|^2_{H^0(a,b)} = \|D^m(f-\mathcal{J}f)\|^2_{H^0(a,b)} + \|D^m \mathcal{J}f\|^2_{H^0(a,b)} . \tag{2.4}$$

Theorem 2.3. (Second Integral Relation). If $f \in H^{2m}(a,b)$ $1 \leq m$, $\Delta \in \mathcal{P}(a,b)$, and $m-1 \leq z \leq 2m-2$,

$$\|D^m(f-\mathcal{J}f)\|^2_{H^0(a,b)} = \int_a^b (f-\mathcal{J}f) D^{2m} f\, dx . \tag{2.5}$$

Finally, following Kolmogorov, cf. [46, pg. 146], if and k are positive integers, let $\lambda_k(t)$ denote the k-th eigenvalue of the boundary value problem,

(2.6) $$(-1)^t D^{2t} y(x) = \lambda y(x), \quad a < x < b,$$

(2.7) $$D^k y(a) = D^k y(b) = 0, \quad t \leq k \leq 2t-1,$$

where the λ_k are arranged in order of increasing magnitude and repeated according to their multiplicity. We remark that the problem (2.6)-(2.7) has a countably infinite number of eigenvalues, all of which are nonnegative and it may be shown that $\lambda_k = (\frac{\pi}{b-a})^{2t} k^{2t}[1+O(k^{-1})]$, as $t < k \to \infty$.

Letting $\overline{\Delta} \equiv \max_{0 \leq i \leq M} (x^{i+1} - x^i)$ and $\underline{\Delta} \equiv \min_{0 \leq i \leq M} (x^{i+1} - x^i)$ for any $\Delta \in \mathcal{P}_M(a,b)$, we have the following results.

Theorem 2.4.

(2.8) $$\lambda_k^{-\frac{1}{2}(m-j)} \leq \wedge(m,2;\, S(2m-1,\, \Delta,\, z);j,2) \leq K_{m,m,z,j}(\overline{\Delta})^{m-j},$$

where

(2.9) $$k \equiv (M+1)(2m-z+1) + z - j + 2$$

and

(2.10) $$K_{m,m,z,j} \equiv \begin{cases} 1, & \text{if } m-1 \leq z \leq 2m-2, \quad j = m, \\ (\frac{1}{\pi})^{m-j}, & \text{if } m-1 = z, \quad 0 \leq j \leq m-1, \\ \frac{(z-2-m)!}{\pi^{m-j}}, & \text{if } m-1 \leq z \leq 2m-2, \\ \frac{(z-2-m)!}{j!\,\pi^{m-j}}, & \text{if } m-1 \leq z \leq 2m-2,\ 2m-2-z \leq j \leq m-1, \end{cases}$$

for all $1 \leq m$, $0 \leq M$, $\Delta \in \mathcal{P}_M(a,b)$, $m-1 \leq z \leq 2m-2$, and $0 \leq j \leq m$.

We remark that in this case it is easy to verify that there exists a positive constant, K, such that

$$\lambda_k^{-\frac{1}{2}} \geq \left(\frac{b-a}{\pi}\right)^{m-j} \frac{1}{(M+1)^{m-j}} \frac{1}{S^{m-j}} \frac{1}{1+KS^{-1}(M+1)^{-1}} \geq \frac{1}{\pi^{m-j}} \frac{1}{S^{m-j}} \cdot$$

$$\cdot \frac{1}{1+KS^{-1}(M+1)^{-1}} (\underline{\Delta})^{m-j}, \quad \text{where } S \equiv \left(2m-z+1+\frac{z-j+2}{M+1}\right) .$$

Theorem 2.5

$$(2.11) \quad \lambda_k^{-\frac{1}{2}}(2m-j) \leq \wedge(2m, 2; S(2m-1, \Delta, z); j, 2) \leq K_{m, 2m, z, j}($$

where

$$(2.12) \qquad k \equiv (M+1)(2m-z+1)+z-j+2$$

and

$$(2.13) \qquad K_{m, 2m, z, j} \equiv (K_{m, m, z, j})(K_{m, m, z, 0}) ,$$

for all $1 \leq m$, $0 \leq M$, $\Delta \in \mathcal{P}_M(a, b)$, $m-1 \leq z \leq 2m-2$, and $0 \leq j \leq m$.

Theorem 2.6

$$(2.14) \quad \lambda_k^{-\frac{1}{2}}(p-j) \leq \wedge(p, 2; S(2m-1, \Delta, z); j, 2) \leq K_{m, p, z, j}(\overline{\Delta})^{p-}$$

where

$$(2.15) \qquad k \equiv (M+1)(2m-z+1)+z-j+2 \quad ,$$

$$(2.16) \qquad K_{m, p, z, j} \equiv K_{p, p, 2m-1, j} + K_{m, 2m, z, j} \cdot$$

$$\cdot\; 2^{\frac{1}{2}(2m-p)} \left[\frac{p!}{(2p-2m)!}\right]^2 (\overline{\Delta}/\underline{\Delta})^{2m-p} \quad ,$$

for all $1 \leq m$, $0 \leq M$, $\Delta \in \mathcal{P}_M(a,b)$, $m < p < 2m$, $4m-2p-1 \leq z \leq 2m-2$, and $0 \leq j \leq m$.

In the special cases of $m < p < 2m$ and $m < j < p$, $\mathcal{J}f$ is not necessarily in $H^j(a,b)$ if $z+1 < j \leq p$, and we modify the definition of $\wedge$ by defining

$$\|D^j(f-\mathcal{J}f)\|_{H^0(a,b)} \equiv \left(\sum_{i=0}^{M} \|D^j(f-\mathcal{J}f)\|^2_{H^0(x^i,\, x^{i+1})} \right)^{\frac{1}{2}} .$$

Theorem 2.7

$$(2.17) \qquad \wedge(p,2;\, S(2m-1),\Delta,z);\, j,2) \leq K_{m,p,z,j} (\overline{\Delta})^{p-j} ,$$

where

$$(2.18)\ K_{m,p,z,j} \equiv K_{p,p,p,j} + (K_{m,p,z,m} + K_{p,p,p,m})\, 2^{\frac{1}{2}(j-m)} \left[\frac{(2p+m)!}{(2p-j)!}\right]^2 (\overline{\Delta}/\underline{\Delta})^{m-j} ,$$

for all $1 \leq m$, $0 \leq M$, $\Delta \in \mathcal{P}_M(a,b)$, $m < p \leq 2m$, $4m-2p-1 \leq z \leq 2m-2$, and $m < j \leq p$.

Theorem 2.8

$$(2.19) \qquad \wedge(m,2;\, S(2m-1,\Delta,z);\, j,\infty) \leq K^{\infty}_{m,m,z,j} (\overline{\Delta})^{m-j-\frac{1}{2}} ,$$

where

$$(2.20)\quad K^{\infty}_{m,m,z,j} =$$

$$\begin{cases} K_{m,m,z,j+1} & , \text{ if } m-1 = z,\ 0 \le j \le m-1 , \\ K_{m,m,z,j+1}, & \text{if } m-1 < z \le 2m-2,\ 0 \le j \le 2m-2-z \\ (j-2m+3+z)^{\frac{1}{2}} K_{m,m,z,j+1}, & \text{if } m-1 < z \le 2m-2 , \\ & 2m-2-z < j \le m-1 , \end{cases}$$

for all $1 \le m$, $0 \le M$, $\Delta \in \mathcal{P}_M(a,b)$, $m-1 \le z \le 2m-2$, and $0 \le j \le m-1$.

Theorem 2.9

$$(2.21)\quad \wedge(2m, 2; S(2m-1,\Delta,z); j, \infty) \le K^{\infty}_{m,2m,z,j} (\overline{\Delta})^{2m-j-\frac{1}{2}} ,$$

where

$$(2.22)\quad K^{\infty}_{m,2m,z,j+1} =$$

$$\begin{cases} K_{m,2m,z,j+1} & , \text{ if } m-1 = z,\ 0 < j \le m \\ K_{m,2m,z,j+1}, & \text{if } m-1 < z \le 2m-2,\ 0 \le j \le 2m-2 \\ (j-2m+3+z)^{\frac{1}{2}} K_{m,2m,z,j+1}, & \text{if } m-1 < z \le 2m-2 , \\ & 2m-2-z < j \le m-1 , \end{cases}$$

for all $1 < m$, $0 \le M$, $\Delta \in \mathcal{P}_M(a,b)$, $m-1 \le z \le 2m-2$, and $0 \le j \le m-1$.

Theorem 2.10

$$(2.23)\quad \wedge(p, 2; S(2m-1,\Delta,z); j, \infty) \le K^{\infty}_{m,p,z,j} (\overline{\Delta})^{p-j-\frac{1}{2}} ,$$

where

$$(2.24)\qquad K^{\infty}_{m,p,z,j} \equiv K^{\infty}_{p,p,2m-1,j} + K^{\infty}_{m,2m,z,j} \cdot$$

$$\cdot\ 2^{\frac{1}{2}(2m-p)}\left[\frac{p!}{(2p-2m)!}\right]^2 (\bar{\Delta}/\underline{\Delta})^{2m-p} ,$$

for all $1 \leq m$, $0 \leq M$, $\Delta \in \mathcal{P}_M(a,b)$, $m < p < 2m$, $4m-2p-1 \leq z \leq 2m-2$, and $0 \leq j \leq m-1$.

In the special cases of $m < p \leq 2m$ and $m < j \leq p$, $\mathcal{J}f$ is not necessarily in $H^{j,\infty}(a,b)$ if $z+1 < j \leq p$ and we modify the definition of $\wedge$ by defining

$$\|D^j(f-\mathcal{J}f)\|_{H^{0,\infty}(a,b)} \equiv \left(\sum_{i=0}^{M} \|D^j(f-\mathcal{J}f)\|^2_{H^{0,\infty}(x^i,x^{i+1})}\right)^{\frac{1}{2}} .$$

Theorem 2.11

$$(2.25)\qquad \wedge(p,2;S(2m-1,\Delta,z);j,\infty) \leq K^{\infty}_{m,p,z,j}(\bar{\Delta})^{p-j-\frac{1}{2}}$$

where

$$(2.26)\qquad K^{\infty}_{m,p,z,j} \equiv K^{\infty}_{p,p,p,j} + (K^{\infty}_{m,p,z,j} + K^{\infty}_{p,p,p,j})\, 2^{j-m+1} .$$

$$\cdot\ \left[\frac{(2p-m)!}{(2p-j-1)!}\right]^2 (\bar{\Delta}/\underline{\Delta})^{j-m+1} ,$$

for all $1 \leq m$, $0 \leq M$, $\Delta \in \mathcal{P}_M(a,b)$, $m < p \leq 2m$, $4m-2p-1 < z \leq 2m-2$ and $m \leq j \leq p-1$.

In some special cases, eg. subspaces of piecewise Hermite polynomials, cf. [9], and subspaces of cubic splines, cf. [6] and [77], the previous results may be extended to other values of q and r. For completeness we state two special results for the important case of $q = r = \infty$.

From [9], we have

Theorem 2.12

$$(2.27)\qquad \wedge(p,\infty; S(2m-1,\Delta,m-1); j,\infty) \leq C_{j,m,p,\infty,\infty}(\bar{\Delta})^{p-j},$$

where $C_{j,m,p,\infty,\infty}$ is the constant given in [8], for all $1 \leq m$ $0 \leq M$, $\Delta \in \mathcal{P}_M(a,b)$, $m \leq p \leq 2m$, and $0 \leq j \leq m$. While from [6] and [77] we have

Theorem 2.13. There exists a nonnegative, continuous functio μ, on $[0,\infty)$ such that

$$(2.28)\qquad \wedge(p,\infty; S(3,\Delta,2); j,\infty) \leq \mu(\bar{\Delta}/\underline{\Delta})(\bar{\Delta})^{p-j},$$

for all $0 \leq M$, $\Delta \in \mathcal{P}_M(a,b)$, $2 \leq p \leq 4$, and $0 \leq j \leq 3$.

§3. One Dimensional Approximation Theory

In this section we discuss lower and upper bounds for the error in approximating a class of functions of one real-variable belonging to a Sobolev space by functions belonging to a one dimensional spline subspace. We consider both sub-spaces of splines of odd degree and subspaces of splines of even degree and the only restriction we place on z is that $-1 \leq z \leq d-1$. The results of this section extend, generalize, and improve the corresponding results of [1], [40], and [64]. The full details of the proofs are to appear in [73].

As in section 2, let $-\infty < a < b < \infty$ be fixed. For eac pair of integers $0 \leq n \leq q$ let $H_n^q(a,b)$ denote the completio of the set $\{f \in C^\infty(a,b) \mid f$ is real-valued, $\|f\|_{H^q(a,b)} < \infty$, and $D^k f(a) = D^k f(b) = 0$, $0 \leq k \leq n-1\}$ with respect to the norm $\|\cdot\|_H$

The symbol K will be used repeatedly throughout this paper t denote a positive constant, not necessarily the same at each occurrence.

Given two nonnegative integers $p \leq r$, S a finite dimensional subspace of $H^p(a,b)$, and S_p a finite dimensional subspace of $H_p^p(a,b)$, we are interested in upper and lower bounds for the quantities

$$(3.1)\quad E(r,p,S) \equiv \sup\{\inf_{s \in S}(\|f-s\|_{H^p(a,b)} / \|f\|_{H^r(a,b)}) \mid f \in H^r(a,b),\ f \not\equiv 0\}$$

and

$$(3.2)\quad E(r,p,S_p) \equiv \sup\{\inf(\|f-s\|_{H^p(a,b)} / \|f\|_{H^r(a,b)}) \mid f \in H_p^r(a,b),\ f \not\equiv 0\}.$$

Specifically we are interested in the choice of $S = S(d,\Delta,z) \subset H^{z+1}(a,b)$ and if $0 \leq p \leq z+1$, $S_p \equiv S_p(d,\Delta,z) \equiv S(d,\Delta,z) \cap H_p^p(a,b)$.

Using the results of section 2, we can prove the following results.

Theorem 3.1. If $d \equiv 2m-1$, where m is a positive integer, and $k \equiv (M+1)(2m-z-1) + z - p + 2$, there exists a positive constant, K, such that

$$(3.3)\quad \lambda_k^{-\frac{1}{2}}(r-p) \leq E(r,p,S(2m-1,\Delta,z)) \leq K(\bar{\Delta})^{r-p},$$

for all $1 \leq m$, $0 \leq M$, $\Delta \in \mathcal{P}_M(a,b)$, $-1 \leq z \leq 2m-2$, $0 \leq r \leq 2m$, and $0 \leq p \leq \min(z+1,r)$,

and

$$(3.4)\quad \lambda_k^{-\frac{1}{2}}(r-p) \leq E(r,p,S_p(2m-1,\Delta,z)) \leq K(\bar{\Delta})^{r-p},$$

for all $1 \leq m$, $0 \leq M$, $\Delta \in \mathcal{P}_M(a,b)$, $-1 \leq z \leq 2m-2$, $0 \leq r \leq 2m$, and $0 \leq p \leq \min(z+1,r,m)$.

From Theorem 3.1, we have the following analogous result for even order spline subspaces.

Theorem 3.2. If $d \equiv 2m$, where m is a positive integer, and $k \equiv (M+1)(2m-z+2) + z - p + 2$, there exists a positive constant, K, such that

(3.5) $$\lambda_k^{-\frac{1}{2}}(r-p) \leq E(r,p,S(2m,\Delta,z)) \leq K(\bar{\Delta})^{r-p},$$

for all $1 \leq m$, $0 \leq M$, $\Delta \in \mathcal{P}_M(a,b)$, $-1 \leq z \leq 2m-1$, $0 \leq r \leq 2m+$ and $0 \leq p \leq \min(z+1,r)$,

and

(3.6) $$\lambda_k^{-\frac{1}{2}}(r-p) \leq E(r,p,S_p(2m,\Delta,z)) \leq K(\bar{\Delta})^{r-p},$$

for all $1 \leq m$, $0 \leq M$, $\Delta \in \mathcal{P}_M(a,b)$, $-1 \leq z \leq 2m-1$, $0 \leq r \leq 2m+$ and $0 \leq p \leq \min(z+1,r,m)$.

§4. Multivariate Splines Defined in Rectangular Parallelepipe

In this section, we discuss lower and upper bounds fo the error in approximating a class of functions, defined on a rectangular parallelepiped and belonging to a Sobolev space, H^q, by functions belonging to a multivariate spline subspace. These results extend, generalize, and improve the correspond- ing results of [7], [9], [66], and [69]. The full details of t proofs are to appear in [68]. For analogous results in $H^{q,\infty}$ cf. [65]. For results on multivariate interpolation, the reade is referred to [7], [9], and [72].

Let N be a fixed positive integer. For each $1 \leq i \leq N$ let

$$-\infty < a_i < b_i < \infty, \; R \equiv \mathop{X}_{i=1}^{N} [a_i, b_i], \text{ and } \mathcal{P}(R) \equiv \{\rho \equiv \mathop{X}_{i=1}^{N} \Delta_i \,|\, \Delta_i \in \mathcal{P}(a_i b_i)$$

$1 \leq i \leq N\}$. Let $\bar{\rho} \equiv \max_{1 \leq i \leq N} \bar{\Delta}_i$, $\underline{\rho} \equiv \min_{1 \leq i \leq N} \underline{\Delta}_i$ for all $\rho \in \mathcal{P}(R)$, define a set, $C \subset \mathcal{P}(R)$, of partitions of R to be quasi uniform if and only if there exists a positive constant, K, such that $(\bar{\rho}/\underline{\rho}) \leq K$ for all $\rho \in C$. For each nonnegative integer, q, let $H^q(R)$ denote the completion of the set $\{f \in C^\infty(R) \mid f$ is real-valued and $\|f\|_{H^q(R)} \equiv (\sum_{|\alpha| \leq q} \int_R |D^\alpha f(x)|^2 dx)^{\frac{1}{2}} < \infty\}$ with respect to the Sobolev norm $\|\cdot\|_{H^q(R)}$, where $\alpha \equiv (\alpha_1, \ldots, \alpha_N)$, α_i nonnegative integers, $1 \leq i \leq N$, $|\alpha| \equiv \sum_{i=1}^N \alpha_i$, and

$$D^\alpha \equiv \frac{\partial^{\alpha_N}}{\partial x_N^{\alpha_N}}.$$

Furthermore, for each pair of integers $0 \leq n \leq q$, let $H_n^q(R)$ denote the completion of the set $\{f \in C^\infty(R) \mid f$ is real-valued, $\|f\|_{H^q(R)} < \infty$, and $D^\alpha f(x) = 0$ for all $x \in \partial R$, $|\alpha| \leq n-1\}$ with respect to the norm $\|\cdot\|_{H^q(R)}$.

Given two nonnegative integers $p \leq r$, S a finite dimensional subspace of $H^p(R)$, and S_p a finite dimensional subspace of $H_p^p(R)$, we are interested in upper and lower bounds for the quantities.

$$(4.1)\quad E(r,p,S) \equiv \sup\{\inf_{s \in S} (\|f-s\|_{H^p(R)} / \|f\|_{H^r(R)}) \mid f \in H^r(R),\ f \neq 0\}$$

and

$$(4.2)\quad E(r,p,S_p) \equiv \sup\{\inf_{s \in S_p} (\|f-s\|_{H^p(R)} / \|f\|_{H^r(R)}) \mid f \in H_p^r(R),\ f \neq 0\}.$$

Following Jerome, cf. [38] and [39], if k and j are positive integers, let $\lambda_k(j)$ denote the k-th eigenvalue of the boundary value problem,

$$(4.3) \qquad \int_R \sum_{|\alpha| \leq j} D^{\alpha} y(x) D^{\alpha} \varphi(x)\, dx = \lambda \int_R y(x) \varphi(x)\, dx$$

for all $\varphi \in H^j(R)$,

where the λ_k are arranged in order of increasing magnitude and repeated according to their multiplicity. We remark that it may be shown that $\lambda_k(j) \sim k^{2j/N}$, as $j < k \to \infty$, cf. [39].

Letting $\bigotimes_{i=1}^{N} S(d, \Delta_i, z_i)$ denote the tensor product of the spaces $S(d, \Delta_i, z_i)$, we have the following two results.

Theorem 4.1. Let $d \equiv 2m - 1$, where m is a positive integer, $k \equiv 1 + \prod_{i=1}^{N} \{(M_i+1)(2m-z_i+1) + z_i + 1\}$, and $C \subset \mathcal{P}(R)$ be a quasi-uniform set of partitions of R. There exists a positive constant, K, such that

$$(4.4) \qquad \lambda_k^{-\frac{1}{2}}(r-p) \leq E(r, p, \bigotimes_{i=1}^{N} S(2m-1, \Delta_i, z_i)) \leq K(\bar{\rho})^{r-p} ,$$

for all $1 \leq m$, $\rho \in C$, $-1 \leq z_i \leq 2m-2$, $0 \leq r \leq 2m$, and $0 \leq p \leq \min_{1 \leq i \leq N} \min(z_i+1, r)$,

$$(4.5) \qquad \lambda_k^{-\frac{1}{2}}(r-p) \leq E(r, p, \bigotimes_{i=1}^{N} S_p(2m-1, \Delta_i, z_i)) \leq K(\bar{\rho})^{r-p} ,$$

for all $1 \leq m$, $\rho \in C$, $-1 \leq z_i \leq 2m-2$, $0 \leq r \leq 2m$, and $0 \leq p \leq \min_{1 \leq i < N} \min(z_i+1, r, m)$.

Theorem 4.2. Let $d \equiv 2m$, where m is a positive integer, $k \equiv 1 + \prod_{i=1}^{N} \{(M_i+1)(2m-z_i+2) + z_i + 1\}$, and $C \subset \mathcal{P}(R)$ be a quasi-uniform set of partitions of R. There exists a positive constant, K, such that

$$(4.6)\qquad \lambda_k^{-\frac{1}{2}}(r-p) \le E(r,p, \bigotimes_{i=1}^{N} S(2m,\Delta_i,z_i)) \le K(\bar{\rho})^{r-p} \quad ,$$

for all $1 \le m$, $\rho \in C$, $-1 \le z_i \le 2m-1$, $0 \le r \le 2m+1$, and $0 \le p \le \min_{1\le i\le N} \min(z_i+1, r)$,

and

$$(4.7)\qquad \lambda_k^{-\frac{1}{2}}(r-1) \le E(r,p, \bigotimes_{i=1}^{N} S_p(2m,\Delta,z_i)) \le K(\bar{\rho})^{r-p} \quad ,$$

for all $1 \le m$, $\rho \in C$, $-1 \le z_i \le 2m-1$, $0 \le r \le 2m+1$, and $0 \le p \le \min_{1\le i\le N} (z_i+1, r, m)$.

§5. Multivariate Spline Approximation in General Domains

In this section, we first discuss lower and upper bounds for the error in approximating a class of functions, defined on a domain Ω, of the type first considered by Harrick, cf. [34] and [35], and belonging to a Sobolev space, $H_p^p(\Omega)$, by functions belonging to a "finite dimensional weighted multivariate spline subspace". Second, we discuss lower and upper bounds for the error in approximating a class of functions, defined on a Lipschitz domain, Ω, cf. [50], and belonging to a Sobolev space $H^p(\Omega)$, by restrictions of functions belonging to a finite dimensional spline space. The results extend, generalize, and improve the corresponding results of [67]. The details of the proofs are to appear in [68].

If Ω is a bounded, measururable subset of R^N and p is a positive integer, let $H^p(\Omega)$ denote the closure of the set $\{f \in C^\infty(\Omega) |$ f is a real-valued function and $\|f\|_{H^p(\Omega)} \equiv (\int_\Omega \sum_{|\alpha|\le p} |D^\alpha u(x)|^2 dx)^{\frac{1}{2}} < \infty\}$ with respect to the Sobolev norm $\|\cdot\|_{H^p(\Omega)}$, and $H_n^p(\Omega)$, $0 \le n \le p$, denote the closure of the set $\{f \in C^{\infty}(\Omega) | f$ is a real-valued function, $\|f\|_{H^p(\Omega)} < \infty$, and $D^{\alpha} f(x) = 0$ for all $x \in \partial\Omega$ and $0 \leq |\alpha| \leq n-1\}$ with respect to the Sobolev norm $\|\cdot\|_{H^p(\Omega)}$.

Given two nonnegative integers $p \leq r$, S a finite dimensional subspace of $H^p(\Omega)$, and S_p a finite dimensional subspace of $H^p_p(\Omega)$, we are interested in upper and lower bounds for the quantities

$$(5.1) \quad E(r,p,S) \equiv \sup\{\inf_{s \in S} (\|f-s\|_{H^p(\Omega)} / \|f\|_{H^r(\Omega)}) | f \in H^r(\Omega),\ f \neq 0\}$$

and

$$(5.2) \quad E(r,p,S_p) \equiv \sup\{\inf_{s \in S_p} (\|f-s\|_{H^p(\Omega)} / \|f\|_{H^r(\Omega)}) | f \in H^r_p(\Omega),\ f \neq 0\} .$$

Finally, following [38] and [39], if k and j are positive integers, let $\lambda_k(j)$ denote the k-th eigenvalue of the boundary value problem,

$$(5.3) \qquad \int_{\Omega} \sum_{|\alpha| \leq j} D^{\alpha} y(x) D^{\alpha} \varphi(x)\, dx = \lambda \int_{\Omega} y(x) \varphi(x)\, dx$$

for all $\varphi \in H^j(\Omega)$,

where the $\lambda_k(j)$ are arranged in order of increasing magnitude and repeated according to their multiplicity. We remark that it may be shown that $\lambda_k(j) \sim k^{2j/N}$, as $j < k \to \infty$, cf. [39].

To begin we state some denseness results. The proofs are to appear in [68].

Theorem 5.1. Let $d \equiv 2m-1$, where m is a positive integer, $\Omega \subset R \equiv \mathop{X}_{i=1}^{N} [a_i, b_i]$ be a closed, bounded subset of R^N, and

$C \equiv \{\rho_k\}_{k=1}^{\infty} \subset \mathcal{P}(R)$ be a quasi-uniform sequence of partitions of R such that $\bar{\rho}_k \to 0$ as $k \to \infty$. If there exists $\gamma \in C^p(\Omega)$ with $\gamma(x) > 0$ for all $x \in \text{int}\,\Omega$ and $D^{\alpha}\gamma(x) = 0$, $0 \leq |\alpha| \leq p-1$, for all $x \in \partial\Omega$, $-1 \leq z_i \leq 2m-2$, and $0 \leq p \leq \min_{1 \leq i \leq N} (z_i+1)$, then $\bigcup_{\rho \in C} \gamma(\bigotimes_{i=1}^{N} S(2m-1, \Delta_i, z_i))$ is dense in $H_p^p(\Omega)$.

<u>Theorem 5.2.</u> Let $d \equiv 2m$, where m is a positive integer, $\Omega \subset R \equiv \mathop{X}_{i=1}^{N} [a_i, b_i]$ be a closed, bounded subset of R^N, and $C \equiv \{\rho_k\}_{k=1}^{\infty} \subset \mathcal{P}(R)$ be a quasi-uniform sequence of partitions of R such that $\bar{\rho}_k \to 0$ as $k \to \infty$. If there exists $\gamma \in C^p(\Omega)$ with $\gamma(x) > 0$ for all $x \in \text{int}\,\Omega$ and $D^{\alpha}\gamma(x) = 0$, $0 \leq |\alpha| \leq p-1$, for all $x \in \partial\Omega$, $-1 \leq z_i \leq 2m-1$, and $0 \leq p \leq \min_{1 \leq i \leq N} (z_i+1)$, then $\bigcup_{\rho \in C} \gamma(\bigotimes_{i=1}^{N} S(2m, \Delta_i, z_i))$ is dense in $H_p^p(\Omega)$.

We must comment on the method of constructing the function $\gamma(x)$. Clearly, it suffices to find a function, denoted by $\theta(x)$, such that $\theta(x) \in C^p(\Omega)$, $\theta(x) > 0$ for all $x \in \text{int}\,\Omega$, and $\theta(x) = 0$ for all $x \in \partial\Omega$, since then we can use $\gamma(x) \equiv \theta^p(x)$ in the preceding theorem.

Following [41] and [67], we now indicate some specific cases in which we can <u>explicitly</u> construct such a function, $\theta(x)$. If the boundary of the region Ω has an equation of the form $F(x) = 0$, where $F \in C^p(\Omega)$, then we can use $\theta(x) = \pm F(x)$, the sign being chosen once so as to make $\theta(x)$ positive in the interior of Ω. For example, if Ω is the closed unit ball, then we can choose $\theta(x) \equiv 1 - \sum_{i=1}^{N} x_i^2$. For the case of a convex polygon, the equations of whose sides are $a_1^i x_1 + a_2^i x_2 + \dots + a_N^i x_N = c_i$, $1 \leq i \leq m$, we can use $\theta(x) \equiv \pm \prod_{i=1} (a_1^i x_1 + a_2^i x_2 + \dots + a_N^i x_N - c_i)$, the sign being chosen so as to make $\theta(x)$ positive in the interior of Ω.

For the case of a region bounded by several smooth curves a similar result holds. For example, for the domain Ω of Figure 1, if $F_i(x) \in C^p(\Omega)$, $1 \le i \le 3$,

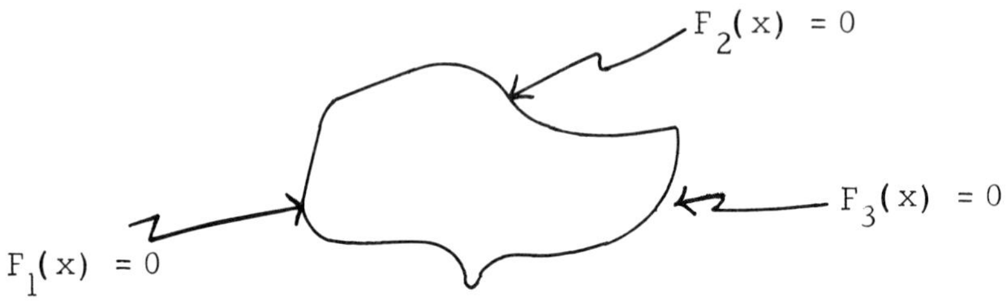

Figure 1

then we may use $\theta(x) \equiv \pm F_1(x) F_2(x) F_3(x)$.

For the case of a region, Ω, with holes, if $\theta_e(x)$ is already defined as above for the exterior boundary of Ω and the holes are given by $F_i(x) = 0$, $1 \le i \le m$, then we can use $\theta(x) \equiv \pm\theta_e(x) \prod_{i=1}^{m} F_i(x)$. For example, in the plane consider the region, Ω, which is inside the unit disk, but outside the disk of radius $\frac{1}{2}$ centered at the point with coordinates $(\frac{1}{2}, 0)$. We can use the function $\theta(x_1, x_2) \equiv (1 - x_1^2 - x_2^2)(x_1^2 + x_2^2 - x_1)$.

For a special, but very important class of domains, Ω, we can actually bound the approximation error for finite dimensional spaces of "weighted spline functions" as introduced in Theorems 5.1 and 5.2. Following [67] and [68] we have

Theorem 5.3. Let $d \equiv 2m-1$, where m is a positive integer, $\Omega \subset R \equiv \mathop{X}_{i=1}^{N} [a_i, b_i]$ be a closed, bounded subset of R^N, $\theta \in C^1(R$ be such that $\partial\Omega \equiv \{x \in R^N \mid \theta(x) = 0\}$, $\theta(x) > 0$ for all $x \in$ int and $\sum_{i=1} |D_i \theta(x)| \neq 0$ for all $x \in \partial\Omega$. If $C \subset \mathcal{P}(R)$ is a quasi-uniform collection of partitions of R and $\theta \in C^r(\Omega)$, there exists a positive constant, K, such that

$$(5.4)\quad \lambda_k^{-\frac{1}{2}}(r-p) \leq E(r,p,\theta^p(\bigotimes_{i=1}^{N} S(2m-1,\Delta_i,z_i)) \leq K(\bar{\rho})^{r-2p},$$

where $k \equiv 1 + \dim(D^\alpha \theta^p(\bigotimes_{i=1}^{N} S(2m-1,\Delta_i,z_i)))$, $|\alpha| = p$, for all $1 \leq m$, $\rho \in C$, $-1 \leq z_i \leq 2m-2$, $2p \leq r \leq 2m+p$, and $0 \leq p \leq \min_{1\leq i\leq N} (z_i+1)$.

Theorem 5.4. Let $d \equiv 2m$, where m is a positive integer, $\Omega \subset R \equiv \mathop{X}_{i=1}^{N} [a_i, b_i]$, be a closed, bounded subset of R^N, and $\theta \in C^1(R^N)$ be such that $\partial\Omega \equiv \{x \in R^N | \theta(x) = 0\}$, $\theta(x) > 0$ for all $x \in \text{int}\,\Omega$ and $\sum_{i}^{N} |D_i \theta(x)| \neq 0$ for all $x \in \partial\Omega$. If $C \subset \mathcal{P}(R)$ is a quasi-uniform collection of partitions of R and $\theta \in C^r(\Omega)$, there exists a positive constant, K, such that

$$(5.5)\quad \lambda_k^{-\frac{1}{2}}(r-p) \leq E(r,p,\theta^p(\bigotimes_{i=1}^{N} S(2m,\Delta_i,z_i))) \leq K(\bar{\rho})^{r-2p},$$

where $k \equiv 1 + \dim(D^\alpha \theta^p(\bigotimes_{i=1}^{N} S(2m,\Delta_i,z_i)))$, $|\alpha| = p$, for all $1 \leq m$, $\rho \in C$, $-1 \leq z_i \leq 2m-1$, $2p \leq r \leq 2m+p+1$, and $0 \leq p \leq \min_{1\leq i\leq N} (z_i+1)$.

We remark that inequalities (5.4) and (5.5) do <u>not</u> show that the subspaces, $\theta^p(\bigotimes_{i=1}^{N} S(d,\Delta_i,z_i))$, are optimal. However, this may be due to our method of proof. We conjecture that using the theory of Besov spaces, we can show that (5.4) holds with $(\bar{\rho})^{r-2p}$ replaced by $(\bar{\rho})^{r-p}|\log \bar{\rho}|$ and $2p \leq r \leq 2m+p$ replaced by $p \leq r \leq 2m$, and (5.5) holds with

$(\bar{\rho})^{r-2p}$ replaced by $(\bar{\rho})^{r-p}$ and $2p \leq r \leq 2m+p+1$ replaced by $p \leq r \leq 2m+1$, cf. [4].

A bounded, open set $\Omega \subset R^N$ is said to be a <u>Lipschitz domain</u> if and only if there exists a finite open covering $\{U_i\}_{i=1}^n$ of the boundary $\partial\Omega$, finitely many cones $\{\gamma_i\}_{i=1}^n$, and a positive number ϵ such that every point of $\partial\Omega$ is the center of a sphere of radius ϵ entirely contained in one of the sets U_i and every point of $U_i \cap \Omega$ is the vertex of a translate of γ_i, $1 \leq i \leq n$, contained entirely in Ω.

Interpreting $\bigotimes_{i=1}^{N} S(2m-1,\Delta_i,z_i)$ and $\bigotimes_{i=1}^{N} S(2m,\Delta_i,z_i)$ as the restriction to Ω of the appropriate spline functions defined on R, we have

<u>Theorem 5.5</u>. Let $\Omega \subset R \equiv \mathop{X}_{i=1}^{N} [a_i, b_i]$ be a Lipschitz domain, m be a positive integer, and C be a quasi-uniform set of partitions of R. There exists a positive constant, K, such that

$$(5.6) \qquad \lambda_k^{-\frac{1}{2}(r-p)} \leq E(r,p,\bigotimes_{i=1}^{N} S(2m-1,\Delta,z_i)) \leq K(\bar{\rho})^{r-p} ,$$

where $k \equiv 1 + \prod_{i=1}^{N} \{(M_i+1)(2m-z_i+1) + z_i+1\}$, for all $1 \leq m$, $\rho \in C$, $-1 \leq z_i \leq 2m-2$, $0 \leq r \leq 2m$, $0 \leq p \leq \min_{1 \leq i \leq N} \min(z_i+1, r)$, and

$$(5.7) \qquad \lambda_k^{-\frac{1}{2}(r-p)} \leq E(r,p,\bigotimes_{i=1}^{N} S(2m,\Delta_i,z_i)) \leq K(\bar{\rho})^{r-p} ,$$

where $k \equiv 1 + \prod_{i=1}^{N} \{(M_i+1)(2m-z_i+2) + z_i+1\}$, for all $1 \leq m$, $\rho \in C$, $-1 \leq z_i \leq 2m-1$, $0 \leq r \leq 2m+1$, and $0 \leq p \leq \min_{1 \leq i \leq N} \min(z_i+1, r)$.

§ 6. The Rayleigh-Ritz Method For Nonlinear Two Point Boundary Value Problems

As a particular application of the results of sections 2 and 3, we consider here the Rayleigh-Ritz method for a class of nonlinear two-point boundary value problems, cf. [17], [18], [40], and [64]. Using the notation of [18], we consider the problem

$$L[u(x)] = f(x,u), \quad -\infty < a < x < b < \infty, \tag{6.1}$$

with boundary conditions

$$D^k u(a) = D^k u(b) = 0, \quad 0 \le k \le n-1, \tag{6.2}$$

where the linear differential operator L is defined by

$$L[u(x)] \equiv \sum_{j=0}^{n} (-1)^j D^j [p_j(x) D^j u(x)], \quad n \ge 1. \tag{6.3}$$

We remark that inhomogeneous boundary conditions: $D^k u(a) = \alpha_k$, $D^k u(b) = \beta_k$, $0 < k < n-1$, can always be reduced to those of (6.2) by means of a suitable change of the dependent variable. Moreover, other types of boundary conditions, eg. Neumann, mixed, or periodic, can be handled by the methods described in this section. The reader is referred to [21] for the details concerning periodic boundary conditions and to [19] for the details concerning nonlinear, Neumann, and mixed boundary conditions.

We assume throughout this section that (i) $p_j(x) \in L^\infty(a,b)$, $0 \le j \le n$, and are real-valued, (ii) there exists a positive constant μ such that

$$\mu \|w\|_{H_n^n(a,b)} \le \int_a^b \sum_{i=0}^{n} p_j(x) (D^j w(x))^2 \, dx \tag{6.4}$$

for all $w \in H_n^n(a,b)$, (iii) $f(x,u)$ is a real-valued, measurable function on $[a,b] \times R$ such that there exists a positive number M such that $\left| \frac{f(x,u) - f(x,v)}{u-v} \right| \le M$ for almost all $x \in [a,b]$

and all $-\infty < u, v < \infty$ with $u \neq v$, and (iv) there exists a real constant γ such that

$$(6.5)\quad \frac{f(x,u) - f(x,v)}{u - v} \leq \gamma < \Lambda \equiv \inf_{w \in H_n^n[a,b]} \frac{\int_a^b \sum_{j=0}^n p_j(x)(D^j[w(x)])^2 dx}{\int_a^b [w(x)]^2 dx}$$

for almost all $x \in [a,b]$ and all $-\infty < u, v < \infty$ with $u \neq v$.

We make the important remark that (iii) is essentially equivalent to the other assumptions plus (iii') for each positive real number c, there exists a positive number $M(c)$ such that $\left|\frac{f(x,u) - f(x,v)}{u - v}\right| \leq M(c)$ for almost all $x \in [a,b]$, and all $-\infty < u, v \leq \infty$ with $u \neq v$ and $|u| \leq c$, $|v| \leq c$ and $f(x,u(x)) \in L^2(a,b)$ for all $u \in H_n^n(a,b)$. In fact, under these assumptions, we can find an a priori bound for any generalized solution of (6.1)-(6.2), and hence modify the right-hand side of (6.1) to $\tilde{f}(x,u)$ so that $\tilde{f}$ bounded, (iv) holds for $\tilde{f}$, and $\tilde{f}(x,u) = f(x,u)$ for all $x \in [a,b]$ and all u satisfying the a priori bound.

Following [18] and [45] we define the functional

$$(6.6)\quad F[w] \equiv \int_a^b \{\frac{1}{2} \sum_{j=0}^n P_j(x)(D^j w(x))^2 - \int_0^{w(x)} f(x,\eta)\, d\eta\} dx$$

for all $w(x) \in H_n^n(a,b)$ and following [14] and [51] we say that u is a generalized solution of (6.1)-(6.2) if and only if u is a solution of the optimization problem:

$$(6.7)\quad \inf_{w \in H_n^n(a,b)} F[w].$$

We now state a general result which follows from the theory of monotone operators, cf. [14], [22], and [49].

Theorem 6.1. If assumptions (i)-(iv) hold, then (6.1)-(6.2) has a unique generalized solution.

We consider now any M dimensional subspace S_M of $H_n^n(a, b)$ and let $\{B_i(x)\}_{i=1}^M$ be M linearly independent functions from the subspace. The analogue of Theorem 6.1, concerning the minimization of the functional $F[w]$ over the subspace S_M, is given in

Theorem 6.2. If assumptions (i)-(iv) hold, then there exists a unique function, $u_M(x)$, in the subspace S_M which minimizes the functional $F[w]$ over S_M. Moreover, there exists a positive constant, K, such that

$$(6.8)\quad \|D^j(u-u_M)\|_{L^\infty(a,b)} \le K\|u-u_M\|_{H^n(a,b)} \le K^2 \inf_{y\in S_M} \|u-y\|_{H^n(a,b)},$$

$$0 \le j \le n-1 .$$

For a proof of this result the reader is referred to [22].

To find the unique element $u_M(x) \equiv \sum_{i=1}^{M} \hat{\beta}_i B_i(x)$ in S_M which minimizes $F[w]$ over S_M, we must solve the M nonlinear equations

$$(6.9)\quad \int_a^b \{\sum_{j=0}^n p_j(x)(\sum_{k=1}^M \beta_k D^j B_k(x)) D^j B_i(x)\} =$$

$$\int_a^b f(x, \sum_{k=1}^M \beta_k B_k(x)) B_i(x)\, dx, \qquad 1 \le i \le M,$$

for the M unknowns $\hat{\beta}_1, \hat{\beta}_2, \ldots, \hat{\beta}_M$ which arise from

$$(6.10) \qquad \frac{\partial F[\sum_{i=1}^{M} \beta_i B_i(x)]}{\partial \beta_i} = 0, \quad 1 \le i \le M .$$

Sometimes it is convenient to write the equations in (6.9) in matrix form as

$$(6.10) \qquad A\underline{\beta} = \underline{g}(\underline{\beta}) ,$$

where $A \equiv (a_{ik})$ is a real $M \times M$ matrix and $g(\underline{\beta}) \equiv (g_1(\underline{\beta}), \ldots, g_M(\underline{\beta}))^T$ is a real column vector, both being determined by

$$(6.11) \qquad a_{ik} \equiv \int_a^b \sum_{j=0}^{n} p_j(x) D^j B_i(x) D^j B_k(x)\, dx, \quad 1 \le i,k \le M ,$$

and

$$(6.12) \qquad g_k(\underline{\beta}) \equiv \int_a^b f(x, \sum_{i=1}^{M} \beta_i B_i(x)) B_k(x)\, dx, \quad 1 \le k \le M .$$

We remark that in general these entrees will have to be computed by a quadrature rule. For an analysis of this aspect of the problem cf. [32] and [33].

It is important to remark that as a by-product of the proof of Theorem 6.2, we have that the matrix A is always symmetric and positive definite. Hence, if (6.1) is linear, i.e., $f(x,u)$ is independent of u, the point successive overrelaxation iterative method can be rigorously applied to determine $\hat{\underline{\beta}}$, cf. [79, pg. 59]. Moreover, if (6.1) is nonlinear, the nonlinear Gauss-Seidel iterative method, i.e.,

$$(6.13) \qquad \sum_{j<i} a_{ij}\beta_j^{(r+1)} + \sum_{j>i} a_{ij}\beta_j^{(r)} = g_i(\beta_1^{(r+1)}, \beta_i^{(r+1)}, \beta_{i+1}^{(r)}, \ldots, \beta_M^{(r)})$$

and the nonlinear successive overrelaxation iterative method are known to be convergent, cf. [53], [61], and [62].

However, if we had a band matrix we could use Gaussian elimination, which is quite stable with respect to rounding errors. Moreover, it is clear that the speed of convergence of the above mentioned methods depends in part on the sparseness of the matrix A.

Before coupling the approximation theory results of section 3 with Theorem 6.1, we examine the question of basis functions for the spline subspaces. Indeed, were it not for the fact that sparceness is so important and orthonormalization so difficult, we would always use polynomial type subspaces.

We restrict ourselves to the two extreme cases, i.e., $d = 2m-1$, $z = m-1$ and d arbitrary, $z = d-1$. The other cases may be obtained from the $z = d-1$ case by letting some partition points coalesce. The space $S(2m-1, \Delta, m-1)$ has been called the space of piecewise Hermite polynomials on Δ, cf. [80]. A way of describing an arbitrary element of $S(2m-1, \Delta, m-1)$ is to give m interpolation parameters $d_i^{(k)}$, $0 \leq k \leq m-1$, at each point $x^i \in \Delta \in \mathcal{P}_M(a,b)$, $0 \leq i \leq M+1$. In each subinterval $[x^i, x^{i+1}]$, there is a unique interpolating polynomial $v_i(x)$ of degree $2m-1$ such that

$$(6.14) \qquad D^k v_i(x^i) = d_i^{(k)} \quad \text{and} \quad D^k v_i(x^{i+1}) = d_{i+1}^{(k)}, \quad 0 \leq k \leq m-1 .$$

The associated function $v(x)$, defined on $[a,b]$ by the $v_i(x)$ on each subinterval of Δ, is of class $C^{m-1}[a,b]$, and is thus an element of $S(2m-1, \Delta, m-1)$. As the number of free parameters $d_i^{(k)}$ associated with any element of $S(2m-1, \Delta, m-1)$ is $m(M+2)$, then $S(2m-1, \Delta, m-1)$ is a vector space of dimension $m(M+2)$. As in [18], a convenient basis for $S(2m-1, \Delta, m-1)$ is $\{s_{i,k}(x;m;\Delta)\}_{i=0,\ k=0}^{M+1,\ m-1}$ where the element $s_{i,k}(x;m;\Delta)$ is defined by

$$(6.15) \qquad D^\ell s_{i,k}(x^j;m;\Delta) = \delta_{ij}\delta_{\ell k}, \quad 0 \leq \ell \leq m-1, \ 0 \leq j \leq M+1 ,$$

so that the support of $s_{i,k}(x;m;\Delta)$ is contained in $[x^{i-1}, x^{i+1}]$. For the special case of $m = 1$, the basis function $s_{i,0}(x;1;\Delta)$ is the piecewise linear function

$$s_{i,0}(x;1;\Delta) \equiv \begin{cases} (x - x^{i-1})/(x^i - x^{i-1}), & x^{i-1} \le x \le x^i , \\ (x^{i+1} - x)/(x^{i+1} - x^i), & x^i \le x \le x^{i+1} , \\ 0 & , x \in \{[a,b] - [x^{i-1}, x^{i+1}]\} , \end{cases}$$

$1 \le i \le M$,

$$s_{0,0}(x;1;\Delta) \equiv \begin{cases} (x^1 - x)/(x^1 - x^0), & a \le x \le x^1 , \\ 0 & , \quad x^1 \le x \le b , \end{cases}$$

$$s_{m+1,0}(x;1;\Delta) \equiv \begin{cases} 0 & , \quad a \le x \le x_M , \\ (x - x_M)/(x_{m+1} - x_M), & x_M \le x \le b , \end{cases}$$

pictured in Figure 2 if $1 \le i \le M$.

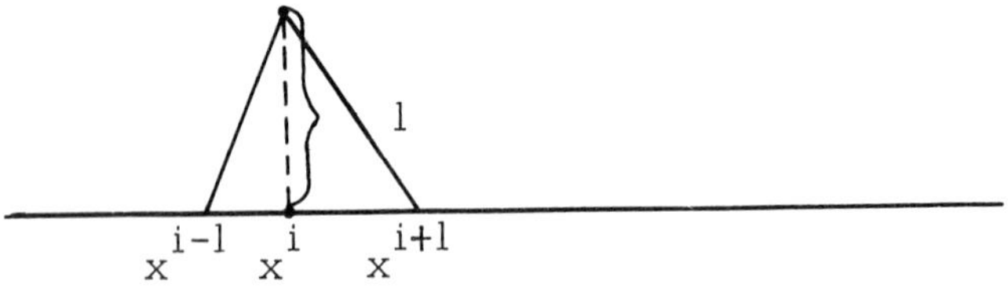

Figure 2

For the special case of $m = 2$, the basis function $s_{i,0}(x;2,\Delta)$ is the piecewise cubic function

$$s_{i,0}(x;2;\Delta) \equiv \begin{cases} -2\left(\frac{x-x^{i-1}}{x^{i}-x^{i-1}}\right)^3 + 3\left(\frac{x-x^{i-1}}{x^{i}-x^{i-1}}\right)^2, & x^{i-1} \le x \le x^{i}, \\ 2\left(\frac{x-x^{i}}{x^{i+1}-x^{i}}\right)^3 - 3\left(\frac{x-x^{i}}{x^{i+1}-x^{i}}\right)^2 + 1, & x^{i} \le x \le x^{i+1}, \\ 0, & x \in \{[a,b] - [x^{i-1}, x^{i+1}]\}, \end{cases}$$

$1 \le i \le M$,

$$s_{0,0}(x;2;\Delta) \equiv \begin{cases} 2\left(\frac{x-x^{0}}{x^{1}-x^{0}}\right)^3 - 3\left(\frac{x-x^{0}}{x^{1}-x^{0}}\right)^2 + 1, & a \le x \le x^{1}, \\ 0, & x^{1} \le x \le b. \end{cases}$$

$$s_{M+1,0}(x;2;\Delta) \equiv \begin{cases} -2\left(\frac{x-x^{M}}{x^{M+1}-x^{M}}\right)^3 + 3\left(\frac{x-x^{M}}{x^{M}-x^{M-1}}\right), & x^{M} \le x \le b, \\ 0, & a \le x \le x^{M} \end{cases}$$

pictured in Figure 3, if $1 \le i \le M$ and

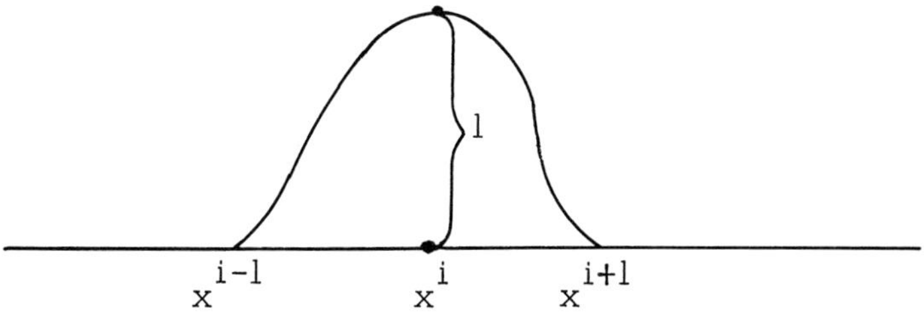

Figure 3

$s_{i,1}(x;2;\Delta)$ is the piecewise cubic function

$$s_{i,1}(x;2;\Delta) \equiv \begin{cases} \left(\frac{x-x^{i-1}}{x^i - x^{i-1}}\right)^2 (x-x^i), & x^{i-1} \le x \le x^i , \\ (x-x^i)\left(\frac{x^{i+1}-x}{x^{i+1}-x^i}\right)^2, & x^i \le x < x^{i+1} , \\ 0 & , \; x \in \{[a,b] - [x^{i-1},x^{i+1}]\} , \end{cases}$$

$1 \le i \le M$,

$$s_{0,1}(x;2;\Delta) \equiv \begin{cases} (x-x^0)\left(\frac{x^1-x}{x^1-x^0}\right)^2, & a \le x \le x^1 , \\ 0 & , \; x^1 \le x \le b , \end{cases}$$

$$s_{M+1,1}(x;2;\Delta) \equiv \begin{cases} \left(\frac{x-x^M}{x^{M+1}-x^M}\right)^2 (x-x^{M+1}), & x^M \le x \le b , \\ 0 & , \; a \le x \le x^1 , \end{cases}$$

pictured in Figure 4 if $1 \le i \le M$.

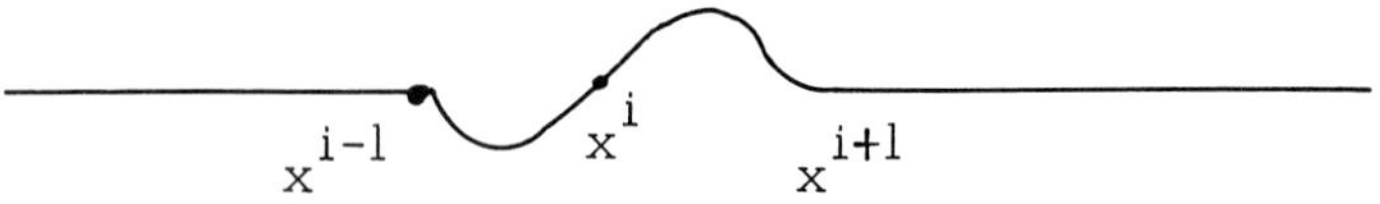

Figure 4

Those elements of $S(2m-1,\Delta,m-1)$, $m \geq n$, which satisfy the boundary conditions of (6.2) form an $m(M+2) - 2n$ dimensional subspace of $H_n^n(a,b)$. This amounts to the restriction that $d_0^{(k)} = d_{M+1}^{(k)} = 0$ for $0 \leq k \leq n-1$.

It is clear from (6.11) and the fact that supp $s_{i,k}(x;m;\Delta)$ is contained in $[x^{i-1}, x^{i+1}]$, $0 \leq k \leq m-1$, that the matrix A is a $3m$ diagonal matrix for the subspace $S_n(2m-1;\Delta;m-1)$ and the given basis functions. Moreover, if we partition A so that the coordinates corresponding to the m basis functions $s_{i,k}(x;m;\Delta)$, $0 \leq k \leq m-1$, are lumped together for each $0 \leq i \leq M+1$, then we obtain a block tridiagonal matrix.

We now explicitly write down the matrix A for the operator $L[u] \equiv -D^2u(x)$, $[a,b] \equiv [0,1]$, and the subspaces $S_1(1,\Delta_h;0)$ and $S_1(3,\Delta_h,1)$, where Δ_h: $0 < h < 2h < \ldots < (M+1)h = 1$, and the above basis functions.

For the case of $S_1(1,\Delta_h,0)$ we have that A is the well-known matrix

$$\frac{1}{h}\begin{pmatrix} 2 & -1 & & \bigcirc \\ -1 & \ddots & \ddots & \\ & \ddots & \ddots & -1 \\ \bigcirc & & -1 & 2 \end{pmatrix}$$

which is irreducibly diagonally dominant, cf. [79], and appears in the "usual" three point central difference approximation to $-D^2$. For the case of $S_1(3,\Delta_h,1)$ we normalize the basis functions $S_{i,1}(x;2;\Delta)$ by dividing by h and we obtain

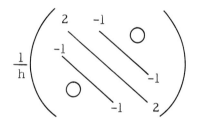

$$A \equiv \frac{1}{h}\begin{pmatrix}
\frac{2}{15} & -\frac{1}{10} & -\frac{1}{30} & 0 & 0 & & & & & \\
-\frac{1}{10} & \frac{12}{5} & 0 & -\frac{6}{5} & \frac{1}{10} & & & & \bigcirc & \\
-\frac{1}{30} & 0 & \frac{4}{15} & -\frac{1}{10} & -\frac{1}{30} & & & & & \\
0 & -\frac{6}{5} & -\frac{1}{10} & \ddots & & \ddots & & & & \\
0 & \frac{1}{10} & -\frac{1}{30} & & \ddots & & \ddots & & & \\
 & & \ddots & & & \ddots & -\frac{6}{5} & \frac{1}{10} & 0 \\
 & & & \ddots & & & -\frac{1}{10} & -\frac{1}{30} & 0 \\
 & \bigcirc & & -\frac{6}{5} & -\frac{1}{10} & \frac{12}{5} & 0 & \frac{1}{10} \\
 & & & \frac{1}{10} & -\frac{1}{30} & 0 & \frac{4}{15} & -\frac{1}{30} \\
 & & & 0 & 0 & \frac{1}{10} & -\frac{1}{30} & \frac{2}{15}
\end{pmatrix}$$

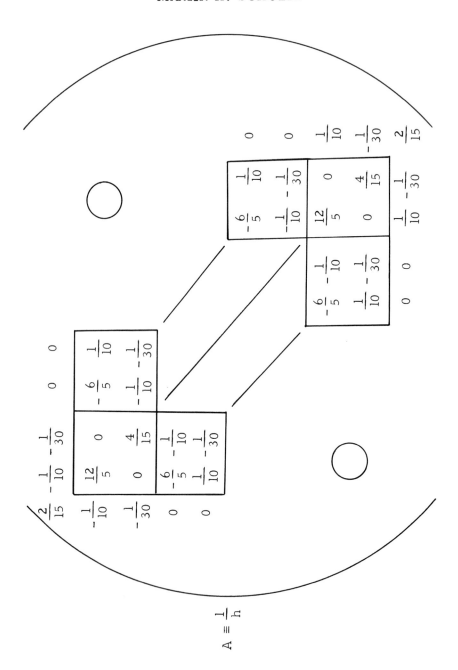

which is irreducibly diagonally dominant.

We turn now to a discussion of the space $S(d,\Delta,d-1)$. Unfortunately the most obvious basis functions for $S(d,\Delta,d-1)$ for the special cases of $d \equiv 2m-1$, the cardinal spline functions, cf. [1], do not give rise to a sparse matrix and hence are not appropriate for the Raleigh-Ritz method. To define suitable basis functions, we follow [18] and [63] and augment the partition $\Delta: a = x^0 < x^1 < \ldots < x^{M+1} = b$ with the points $x^{-d} < x^{-d+1} < \ldots < x^{-1} < x^0$ and $x^{M+1} < x^{M+1+1} < \ldots < x^{M+1+d}$ to form a new partition $\tilde{\Delta}: x^{-d} < x^{-d+1} < x^0 < \ldots < x^{M+1} < x^{M+2} < \ldots < x^{M+1+d}$. Letting

$$x_+^d \equiv \begin{cases} x^d, & \text{if } x \geq 0, \\ 0, & \text{if } x < 0, \end{cases} \qquad \text{and} \qquad \omega_i(x) \equiv \prod_{k=0}^{d+1} (x - x^{i+k}) \qquad \text{for}$$

$-d \leq i \leq M$, we define $M_{d,i}(x;\tilde{\Delta}) \equiv \sum_{k=0}^{d+1} (d+1) \dfrac{(x^{i+k} - x)_+^d}{\omega_i'(x^{i+k})}$, for

$-d \leq i \leq M$. As a basis for $S(d,\Delta,d-1)$, we take the restriction of the functions $\{M_{d,i}(x;\tilde{\Delta})\}_{i=-d}^{M}$ to the interval $[a,b]$. In fact, it is easy to verify that the set $\{M_{d,i}(x;\tilde{\Delta})\}_{i=-d}^{M} \subset S(d,\Delta,d-1)$ and is linearly independent. Moreover, the support of $M_{d,i}(x;\tilde{\Delta})$ is contained in the interval $[x_i, x_{i+d+1}]$. For the special case of $d = 1$, $S(1,\Delta,0)$ is also a space of piecewise, linear Hermite polynomials and the basis functions $M_{1,i}(x,\tilde{\Delta})$ are identical with the basis functions $s_{i,0}(x;1;\Delta)$ pictured in Figure 2.

For the special case of $d = 3$, $[a,b] \equiv [0,1]$, and uniform partition $\tilde{\Delta}_h$, i.e., $x^i = ih$, $-3 \leq i \leq M+4$, we have the explicit representation of the basis functions

$$M_{3,i}(x;\tilde{\Delta}_h) \equiv \frac{1}{6h^4}\begin{cases} (x-x^i)^3, & x^i \le x \le x^{i+1} \\ h^3 + 3h^2(x-x^{i+1}) + 3h(x-x^{i+1})^2 - 3(x-x^{i+1})^3, & x^{i+1} \le x \le x^{i+2}, \\ h^3 + 3h^2(x^{i+3}-x) + 3h(x^{i+3}-x)^2 - 3(x^{i+3}-x)^3, & x^{i+2} \le x \le x^{i+3}, \\ (x^{i+4}-x)^3, & x^{i+3} \le x \le x^{i+4}, \\ 0, & \text{all other } x \in [0,1] \end{cases}$$

for $-3 \le i \le M$, which are pictured in Figure 4.

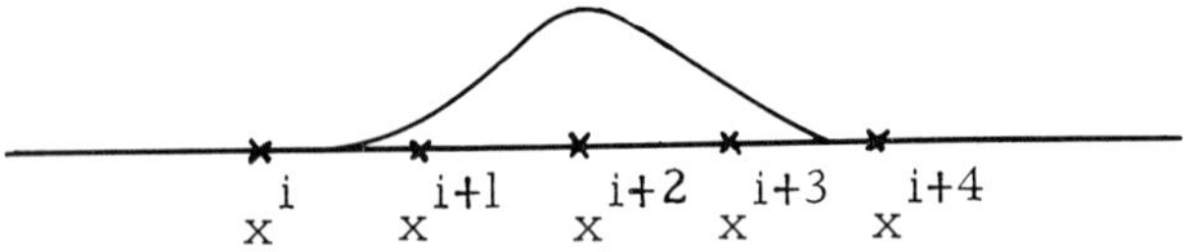

In order to obtain a convenient basis for $S_1(3,\tilde{\Delta}_h,2)$, it is necessary to modify these basis functions somewhat.

Let $\tilde{M}_{3,-2}(x;\tilde{\Delta}_h) \equiv M_{3,-2}(x;\tilde{\Delta}_h) - 4M_{3,-3}(x;\tilde{\Delta}_h)$, $\tilde{M}_{3,-1}(x;\tilde{\Delta}_h) \equiv M_{3,-1}(x;\tilde{\Delta}_h) - M_{3,-3}(x;\tilde{\Delta}_h)$, $\tilde{M}_{3,i}(x;\tilde{\Delta}_h) \equiv M_{3,i}$, ($0 \le i \le M-3$, $\tilde{M}_{3,M-2}(x;\tilde{\Delta}_h) \equiv M_{3,M-2}(x;\tilde{\Delta}_h) - M_{3,M}(x;\tilde{\Delta}_h)$, $\tilde{M}_{3,M-1}(x;\tilde{\Delta}_h) \equiv M_{3,M-1}(x;\tilde{\Delta}_h) - 4M_{3,M}(x;\tilde{\Delta}_h)$. Then $\{\tilde{M}_{3,i}(x;\tilde{\Delta}_h)\}_{i=-2}^{M-1}$ is a basis for $S_1(3,\tilde{\Delta}_h,2)$ and the suppor each $\tilde{M}_{3,i}(x;\tilde{\Delta}_h)$ is contained in at most four adjacent subintervals of $[0,1]$.

In general, the use of the space $S_n(d;\tilde{\Delta}, d-1)$, $2n \leq d+1$, in the Rayleigh-Ritz method yields a band matrix with band width of $2d+1$. In the special case of $d = 2m-1$, this is $4m-1$ and is to be contrasted with the band width of $3m$ in the matrix corresponding to the subspace $S_n(2m-1;\Delta;m-1)$. However, if the former matrix, i.e., the one with the larger band width, is a $k \times k$ matrix, the latter matrix will be a $t \times t$ matrix where t is approximately mk.

We now couple the results of section 3 with Theorem 6.2. Generalizing [17], [18], [40], and [64] we have

Theorem 6.3. If assumptions (i) - (iv) hold and $n \leq \min(z+1, m) \leq 2m-1$, $S_n(2m-1, \Delta, z) \subset H_n^n(a,b)$ for all $\Delta \in \mathcal{P}(a,b)$ and there exists a unique function, $s(x) \in S_n(2m-1, \Delta, z)$, which minimizes the functional $F[w]$ over $S_n(2m-1, \Delta, z)$. Moreover, if $u \in H_n^r(a,b)$ where $n \leq r \leq 2m$, there exists a positive constant, K, such that

$$(6.16)\quad \|D^j(u-s)\|_{L^\infty(a,b)} \leq K\|u-s\|_{H^n(a,b)} \leq K^2(\bar{\Delta})^{r-n}, \quad 0 \leq j \leq n-1,$$

for all $\Delta \in \mathcal{P}(a,b)$.

Theorem 6.4. If assumptions (i) - (iv) hold and $n \leq \min(z+1, m) \leq 2m$, $S_n(2m, \Delta, z) \subset H_n^n(a,b)$ for all $\Delta \in \mathcal{P}(a,b)$ and there exists a unique function, $s(x) \in S_n(2m, \Delta, z)$, which minimizes the functional $F[w]$ over $S_n(2m, \Delta, z)$. Moreover, if $u \in H_n^r(a,b)$ where $n \leq r \leq 2m+1$, there exists a positive constant, K, such that

$$(6.17)\quad \|D^j(u-s)\|_{L^\infty(a,b)} \leq K\|u-s\|_{H^n(a,b)} \leq K^2(\bar{\Delta})^{r-n}, \quad 0 \leq j \leq n-1,$$

for all $\Delta \in \mathcal{P}(a,b)$.

Under special circumstances the result of Theorem 6.3 can be sharpened. As an improvement of the corresponding results of [16], [37], [55], [60], and [74], we have

Theorem 6.5. If assumptions (i) - (iv) hold, $p_n(x) = 1$, $p_j(x) \equiv 0$ for all $1 \leq j \leq n-1$, $m-1 \leq z \leq 2m-2$, and $n \leq \min(z+1,$ $\leq 2m-1$, then $S_n(2m-1, \Delta, z) \subset H^n_n(a,b)$ for all $\Delta \in \mathcal{P}(a,b)$ and there exists a unique function, $s(x) \in S_n(2m-1, \Delta, z)$, which minimizes the functional $F[w]$ over $S_n(2m-1, \Delta, z)$. Moreover, there exists a positive constant, K, such that

$$(6.16) \quad \|D^j(u-s)\|_{L^2(a,b)} \leq K \|D^{j+2m-2n}(y - \mathcal{J}y)\|_{L^2(a,b)}, \quad 0 \leq j \leq m$$

and

$$(6.17) \quad \|D^j(u-s)\|_{L^\infty(a,b)} \leq K \|D^{j+2m-2n}(y - \mathcal{J}y\|_{L^\infty(a,b)},$$

$$0 \leq j \leq m-1,$$

where $y(x) \equiv \int_a^x \int_a^{x_{2m-2n}} \cdots \int_a^{x_2} u(x_1)\, dx_1 dx_2 \ldots dx_{2m-2n} \equiv \mathfrak{m}(u)$ and $\mathcal{J}$ is the interpolation mapping into $S(4m-2n-1, \Delta, z+2m-2n$

Proof. It suffices to prove (6.16) as the rest of the Theorem is identical to Theorem 6.3. If assumptions (i) - (iv) hold and we have the equation (6.1) $-D^{2n}u(x) + p_0(x)u(x) = f(x,u)$, we may rewrite (6.1) by moving the term $p_0(x)u(x)$ to the right side of the equation to obtain

$$(6.1') \quad -D^{2n}u(x) = f(x,u) - p_0(x)u(x) \equiv \tilde{f}(x,u).$$

Moreover, it is easy to directly verify that assumptions (i) - (iv) hold for this rewritten form (6.1'). Hence, in (6.1) it suffices to assume that $p_n(x) \equiv 1$ and $p_j(x) = 0$ for all $0 \leq j \leq n-1$.

We prove only (6.16) as the proof of (6.17) is essentially identical. By the triangle inequality it suffices to show that there exists a positive constant, K, such that

$$\|D^n(s - D^{2m-2n}\mathcal{J}y)\|_{L^2(a,b)} \leq K\|D^{2m-2n}(y - \mathcal{J}y)\|_{L^2(a,b)} .$$

Using the fact, cf. [64], that $\int_a^b D^{2m-n}(y - \mathcal{J}y)\, D^{2m-n} w\, dx = 0$ for all $w \in S(4m-2n-1, \Delta, z+2m-2n)$, we have

$$(1 - \tfrac{\gamma}{\wedge}) \int_a^b D^n(D^{2m-2n}\mathcal{J}y - s)\, D^n(D^{2m-2n}\mathcal{J}y - s)\, dx \leq$$

$$\int_a^b D^n(D^{2m-2n}\mathcal{J}y - s)\, D^n(D^{2m-2n}\mathcal{J}y - s)\, dx - \gamma \int_a^b (D^{2m-2n}\mathcal{J}y - s)^2 dx$$

$$\leq \int_a^b D^n(D^{2m-2n}\mathcal{J}y)\, D^n(D^{2m-2n}\mathcal{J}y - s)\, dx - \int_a^b f(x, D^{2m-2n}\mathcal{J}y)(D^{2m-2n}\mathcal{J}y - s)\, dx$$

$$= \int_a^b (D^{2m-n}\mathcal{J}y)\, D^{2m-n}(\mathcal{J}y - \mathfrak{m}(s))\, dx - \int_a^b f(x, D^{2m-2n}\mathcal{J}y)(D^{2m-2n}\mathcal{J}y - s)\, dx$$

$$= \int_a^b D^n(D^{2m-2n} y)\, D^n(D^{2m-2n}(\mathcal{J}y - \mathfrak{m}(s)))\, dx - \int_a^b f(x, D^{2m-2n}\mathcal{J}y)(D^{2m-2n}\mathcal{J}y - s)\, dx$$

$$= \int_a^b [f(x, D^{2m-2n} y) - f(x, D^{2m-2n}\mathcal{J}y)]\,(D^{2m-2n}\mathcal{J}y - s)\, dx .$$

The result now follows from assumptions (iii) and (iv) . Q.E.D.

From Theorems 2.4, 2.5, 2.6, 2.8, 2.9, and 2.10 we have

Corollary 1. (i) If the hypotheses of Theorem 6.5 hold and $u \in H_n^r(a,b)$, where r is either m or $2m$, then there exists a positive constant K, such that

$$\|D^j(u-s)\|_{L^2(a,b)} \leq K(\bar{\Delta})^{r-j}, \quad 0 \leq j \leq m, \tag{6.18}$$

for all $\Delta \in \mathcal{P}(a,b)$.

(ii) If the hypotheses of Theorem 6.5 hold, $C \subset \mathcal{P}(a,b$ is a quasi-uniform set of partitions of (a,b), $u \in H_n^r(a,b)$, where $m < r < 2m$, and $4m-2r-1 \leq z \leq 2m-2$ then there exists a positive constant, K, such that

$$\|D^j(u-s)\|_{L^2(a,b)} \leq K(\bar{\Delta})^{r-j}, \quad 0 \leq j \leq m, \tag{6.19}$$

for all $\Delta \in C$.

(iii) If the hypotheses of Theorem 6.5 hold and $u \in H_n^r(a,b)$, where r is either m or $2m$, then there exists a positive constant, K, such that

$$\|D^j(u-s)\|_{L^\infty(a,b)} \leq K(\bar{\Delta})^{r-j-\frac{1}{2}}, \quad 0 \leq j \leq m-1, \tag{6.20}$$

for all $\Delta \in \mathcal{P}(a,b)$.

(iv) If the hypotheses of Theorem 6.5 hold, $C \subset \mathcal{P}(a,b$ is a quasi-uniform set of partitions of (a,b), $u \in H_n^r(a,b)$, where $m < r < 2m$, and $4m-2r-1 \leq z \leq 2m-2$, then there exists a positive constant, K, such that

$$\|D^j(u-s)\|_{L^\infty(a,b)} \leq K(\bar{\Delta})^{r-j-\infty}, \quad 0 \leq j \leq m-1, \tag{6.21}$$

for all $\Delta \in C$.

For the special case of piecewise Hermite polynomial subspaces, we may extend the results of Corollary 1 by using Theorem 2.12.

Corollary 2. If the hypotheses of Theorem 6.5 hold, $u \in H^{r,\infty}(a,b)$, where $m \leq r \leq 2m$, and $z = m-1$, there exists a positive constant, K, such that

$$(6.22) \qquad \| D^j(u-s) \|_{L^\infty(a,b)} \leq K(\bar{\Delta})^{r-j}, \quad 0 \leq j \leq m ,$$

for all $\Delta \in C$.

For the special case of C^2-cubic spline subspaces, we may extend the results of Corollary 1 by using Theorem 2.13.

Corollary 3. If the hypothesis of Theorem 6.5 hold, $C \subset \mathcal{P}(a,b)$, is a quasi-uniform set of partitions of (a,b), $1 \leq n \leq m = 2$, $u \in H_n^{r,\infty}(a,b)$, where $2 \leq r \leq 4$, and $z = 2$, there exists a positive constant, K, such that

$$(6.23) \qquad \| D^j(u-s) \|_{L^\infty(a,b)} \leq K(\bar{\Delta})^{r-j}, \quad 0 \leq j \leq r ,$$

for all $\Delta \in C$.

§7. The Galerkin Method for Nonlinear Elliptic Differential Equations

In this section we extend, generalize, and improve the results of [9], [22], [66], [67], and [69], concerning the use of spline functions with the Galerkin method, by invoking the new approximation theory results of sections 4 and 5. First, we study the Galerkin method for approximating the solutions of a class of abstract monotone mapping equations in Hilbert spaces, as originally considered in [14] and [49]. Second, we give some concrete examples of boundary value problems for nonlinear elliptic differential equations which give rise to such operator equations. Third, we apply the approximation theory results of sections 4 and 5 to give error bounds for the Galerkin method for a class of abstract monotone mapping equations in special Sobolev spaces. In particular, new results for the model semi-linear equation, treated in [22], [31], [45], [54], and [71], are obtained and these compare very favorably with results of [42], [43], and]52].

Let H be a Hilbert space over the real field with inner-product $(\cdot, \cdot)$ and $\|\cdot\| \equiv (\cdot, \cdot)^{\frac{1}{2}}$. Following [14] we say that a nonlinear mapping T of H into itself is <u>finitely continuous</u> if and only if T is continuous from finite-dimensional subspaces of H into H with the weak-star topology and <u>strongly monotone</u> if and only if there exists a positive constant, γ, such that

$$(7.1) \qquad \gamma\|u-v\|^2 \leq |(Tu-Tv, u-v)| \quad \text{for all } u, v \in H \; .$$

For a discussion of a weaker form of (7.1) see [14].

In this section, we are particularly interested in the problem of determining those $u \in H$ such that

$$(7.2) \qquad Tu = 0$$

or

$$(7.3) \qquad (Tu, v) = 0 \quad \text{for all } v \in H \; .$$

The Galerkin method for such a problem is to give a finite-dimensional subspace H_k of H and to determine those $u_k \in H$ such that

$$(7.4) \qquad (Tu_k, v_k) = 0 \quad \text{for all } v_k \in H_k \; .$$

We now state the following result which was proved in [14].

<u>Theorem 7.1.</u> If T is finitely continuous and strongly monotone on H, then (7.2) has a unique solution u. Likewise, (7.4) has a unique solution for every finite-dimensional subspace H_k.

To study the convergence properties of the u_k's as $\dim H_k \to \infty$, we need additional hypotheses on the mapping T. We begin with

Theorem 7.2. Let T be a finitely continuous, strongly monotone mapping of H into itself. If T is bounded, i.e., maps bounded subsets of H into bounded subsets, then there exists a positive constant, K, such that

$$c\|u-u_k\|^2 \leq K \inf_{y \in H_k} \|u-y\| , \tag{7.5}$$

for all finite dimensional subspaces H_k of H. Moreover, if T is Lipschitz continuous for bounded arguments, i.e., given $M > 0$, there exists a constant $C(M)$ such that

$$\|Tu-Tv\| \leq C(M)\|u-v\| \tag{7.6}$$

for all $u, v \in H$ with $\|u\|, \|v\| \leq M$, there exists a positive constant, K, such that

$$\|u-u_k\| \leq \inf_{y \in H_k} \|u-y\| , \tag{7.7}$$

for all finite dimensional subspaces H_k of H.

Corollary. Let T be a finitely continuous, bounded, strongly monotone mapping of H into itself. If $\{H_k\}_{k=1}^{\infty}$, a sequence of finite-dimensional subspace of H, is such that $\lim_{k \to \infty} \{\inf_{y \in H_k} \|u-y\|\} = 0$, where u is the unique solution of (7.2), then

$$\lim_{k \to \infty} \|u-u_k\| = 0 , \tag{7.8}$$

where the u_k, $1 \leq k$, are the unique solutions of (7.4).

In this section, we restrict ourselves to a discussion of examples of nonlinear elliptic differential equations satisfying

Dirichlet type boundary conditions. For a discussion of nonlinear elliptic differential equations satisfying other types of boundary conditions see [14], [19], [21], and [22]. In particular, if Ω is a Lipschitz domain in R^N, $N \geq 1$, we consider the problem of finding a solution u of

$$\sum_{0 \leq |\alpha| \leq n} (-1)^{|\alpha|} D^{\alpha}\{A_{\alpha}(x,u,\ldots,D^n u)\} = 0, \quad x \in \Omega , \tag{7.9}$$

$$D^{\beta} u(x) = 0, \quad x \in \partial\Omega, \quad \text{for all } |\beta| \leq n-1 , \tag{7.10}$$

where $A_{\alpha}(x,u,\ldots,D^n u)$ denotes a function which can depend upon x and any $D^{\gamma}u$ with $|\gamma| \leq n$.

For any $u, v \in \overset{\circ}{H}{}^n_n(\Omega)$ we define the form

$$a(u,v) \equiv \sum_{0 \leq |\alpha| \leq n} \int_{\Omega} A_{\alpha}(x,u,\ldots,D^n u) D^{\alpha} v dx , \tag{7.11}$$

and we say that (7.9) - (7.10) has a <u>generalized solution</u>, u, if and only if

$$a(u,v) = 0 \quad \text{for all } v \in \overset{\circ}{H}{}^n_n(\Omega) . \tag{7.12}$$

Following [50], we have from the Sobolev Imbedding Theorem

<u>Theorem 7.3.</u> Let the functions $A_{\alpha}(x,u,\ldots,D^n u)$ appearing in (7.9) be measurable in $x \in \Omega$ and continuous in their other arguments for almost all $x \in \Omega$. Let $1 \leq \rho < \infty$ and $g(r)$ be a nonnegative continuous function on $[0,+\infty)$ such that

$$|A_{\alpha}(x,u,\ldots,D^n u)| \leq \left\{ g\Big(\sum_{|\beta| < n-(N/2)} |D^{\beta} u| \Big) \right\} \cdot \tag{7.13}$$

$$\cdot \left\{ 1 + \sum_{|\beta| = n-(N/2)} |D^{\beta} u|^{\rho} + \sum_{n-(N/2) < |\beta| \leq n} |D^{\beta} u|^{q_\beta / p_\alpha} \right\}$$

for all $|\alpha| \leq m$, almost all $x \in \Omega$, and all $D^\alpha u$, $|\gamma| \leq n$, where $q_\beta = \frac{2N}{N-2n+2|\beta|}$ and

$$\frac{1}{p_\alpha} = \begin{cases} 1 & \text{if} \quad |\alpha| < n-(N/2) \\ \frac{1}{2} + \frac{n-|\alpha|}{N}, & \text{if} \quad |\alpha| \geq n-(N/2) . \end{cases}$$

Then, $a(u,v) = (Tu,v)$ for all $u, v \in H_n^n(\Omega)$ defines a bounded, finitely continuous mapping of $H_n^n(\Omega)$ into itself. Moreover, if there exists a positive constant, γ, such that $|a(u,u-v) - a(v,u-v)| \geq \gamma \|u-v\|^2_{H^n(\Omega)}$ for all $u, v \in H_n^n(\Omega)$, then T is strongly monotone, (7.9) - (7.10) has a unique generalized solution, the Galerkin method yields a unique approximation u_k for each finite dimensional subspace H_k of $H_n^n(\Omega)$, and there exists a positive constant, K, such that

$$(7.14) \qquad \|u-u_k\|^2_{H^n(\Omega)} \leq \inf_{y \in H_k} \|u-y\|_{H^n(\Omega)} .$$

We remark that it is easily seen that the growth conditions imposed upon the coefficient functions $A_\alpha(x,u,\ldots,D^n u)$ by the inequalities of (7.13) are very restrictive. A general trick for avoiding this theoretical, difficulty consists of finding a priori L^∞-bounds for the solution and also in some cases its derivatives and using these bounds to modify (7.9) so that the new coefficient functions $\tilde{A}_\alpha(x,u,\ldots,D^n u)$ satisfy (7.13) in such a way that the unique solution of the modified nonlinear Dirichlet problem satisfies the same a priori bound. We refer to [22] for the details of how this is done in many important special cases. Furthermore, it usually turns out that the strongly monotone mapping, T, associated with the modified problem is uniformly Lipschitz continuous and hence the error estimate (7.7) holds.

Consider the special cases in which the differential equation of (7.9) has the form

$$(7.5) \qquad L[u] - f(x,u) = 0 ,$$

where $L[u] \equiv \sum_{0 \leq |\alpha|, |\beta| \leq n} (-1)^{|\alpha|} D^{\alpha}(a_{\alpha\beta}(x) D^{\beta} u(x))$,

$a_{\alpha\beta}(x) \in L^{\infty}(\Omega)$ for all $0 \leq |\alpha|, |\beta| \leq n$, $a_{\alpha\beta}(x) = a_{\beta\alpha}(x)$ for all $0 \leq |\alpha|, |\beta| \leq n$ and all $x \in \Omega$, $f(x,u)$ is a bounded, measurable function of $x \in \Omega$ and $-\infty < u < \infty$, there exists a positive constant, μ, such that

$$\int_{\Omega} \sum_{0 \leq |\alpha|, |\beta| \leq n} a_{\alpha\beta}(x) D^{\alpha} u(x) D^{\beta} u(x)\, dx \geq \mu \|u\|^2_{H^n(\Omega)} ,$$

for all $u \in H^n_n(\Omega)$, there exists a positive constant, K, such that, $\left|\frac{f(x,u) - f(x,v)}{u - v}\right| \leq K$ for all $x \in \Omega$ and all $u \neq v$, and $\frac{f(x,u) - f(x,v)}{u - v} \leq \gamma < \Lambda \equiv$

$$\inf_{w \in H^n_n(\Omega)} \frac{\int_{\Omega} \sum_{0 \leq |\alpha|, |\beta| \leq n} a_{\alpha\beta}(x) D^{\alpha} w(x) D^{\beta} w(x)\, dx}{\int_{\Omega} w^2(x)\, dx}, \quad \text{for all } x \in \Omega$$

$u \neq v$. These cases are a direct generalization of the one-dimensional problem discussed in section 6 and as such have a variational formulation in terms of the functional

$$\text{(7.16)} \quad F[w] \equiv \int_{\Omega} \frac{1}{2} \sum_{0 \leq |\alpha|, |\beta| \leq n} a_{\alpha\beta}(x) D^{\alpha} w(x) D^{\beta} w(x) - \int_0^{w(x)} f(x, \eta)\, d\eta \quad dx .$$

Moreover, it is easily verified that such problems give rise to a uniformly Lipschitz, strongly monotone mapping in $H^n_n(\Omega)$

and that the algebraic equations given by the Galerkin method for any subspace H_k of $H_n^n(\Omega)$ are identical with those given by the Rayleigh-Ritz method for the subspace. The details are left to the reader who is referred to [22], [32], [67], and [71] for more specific results for this class of problems.

To complete our list of applications of the monotone mapping theory to the study of Galerkin's method for elliptic differential equations we refer to [9] and [15] in which linear, nonselfadjoint Dirichlet problems are discussed, [70] where a class of, nonlinear, nonselfadjoint Dirichlet problems are discussed, and to [22] where general nonlinear Dirichlet problems and systems of nonlinear two-point boundary value problems are discussed. There are also many applications to the study of nonlinear integral equations.

We now combine the results of sections 4 and 5 with Theorem 7.2. For examples of other subspaces the reader is referred to [23], [28], [41], and [83]. From Theorems 4.1 and 4.2, we have

<u>Theorem 7.4.</u> Let $R \subset R^N$ be a rectangular parallelepiped, $C \subset \mathcal{P}(R)$ be a quasi-uniform set of partitions, and T be a strongly monotone mapping of $H_p^p(R)$ into itself which is Lipschitz continuous for bounded arguments. (i) If m is a positive integer, $-1 \le z_i \le 2m-2$, and $0 \le p \le \min_{1 \le i \le N} \min(z_i + 1, m)$, then $\bigotimes_{i=1}^{N} S(2m-1, \Delta_i, z_i) \subset H_p^p(R)$ for all $\rho \in C$. Moreover, if the generalized solution, u, of (7.2) belongs to $H^r(R)$ where $p \le r \le 2m$, then there exists a positive constant, K, such that

$$\|u-s\|_{H^p(R)} \le K(\rho)^{r-p} , \tag{7.17}$$

where s is the Galerkin approximation to u in $\bigotimes_{i=1}^{N} S_p(2m-1,\Delta_i, z_i)$, for all $\rho \in C$.

(ii) If m is a positive integer, $-1 \le z_i \le 2m-1$, and $0 \le p \le \min_{1 \le i \le N} \min(z_i + 1, m)$, then $\bigotimes_{i=1}^{N} S_p(2m, \Delta_i, z_i) \subset H_p^p(R)$ for all

$\rho \in C$. Moreover, if the generalized solution, u, of (7.2) belongs to $H^r(R)$, where $p \leq r \leq 2m+1$, then there exists a positive constant, K, such that

$$\|u-s\|_{H^p(R)} \leq K(\bar{\rho})^{r-p}, \tag{7.18}$$

where s is the Galerkin approximation to u in $\bigotimes_{i=1}^{N} S_p(2m,\Delta_i,z_i)$ for all $\rho \in C$.

We remark that in the special case of (7.15) in which f is independent of u, n = 1, and the hypotheses of Theorem 7.4 hold, if m = 1, $z_i = 0$, and r = 2, we have $\|u-s\|_{H^1(r)} \leq K\bar{\rho}$ for all $\rho \in C$. This is the analogue of results for discrete approximation of u given in [42], [43], and [52].

From Theorem 5.3 and 5.4, we have

Theorem 7.5. Let $\Omega \subset R \equiv \mathop{X}_{i=1}^{N} [a_i, b_i]$, a closed, bounded subset of R^N, and $\theta \in C^1(R^N)$ be such that $\partial\Omega \equiv \{x \in R^N \mid \theta(x) = 0\}$, $\theta(x) > 0$ for all $x \in \operatorname{int}\Omega$, and $\sum_{i=1}^{N} |D_i\theta(x)| \neq 0$ for all $x \in \partial\Omega$. Let $C \subset \mathcal{P}(R)$ be a quasi-uniform set of partitions and T be a strongly monotone mapping of $H_p^p(\Omega)$ into itself which is Lipschitz continuous for bounded arguments.

(i) If m is a positive integer, $-1 \leq z_i \leq 2m-2$, $0 \leq p \leq \min_{1 \leq i \leq N} (z_i+1)$, and $\theta \in C^r(\Omega)$ where $2p \leq r \leq 2m+p$ then $\theta^p(\bigotimes_{i=1}^{N} S(2m-1,\Delta_i,z_i)) \subset H_p^p(\Omega)$ for all $\rho \in C$. Moreover, if the generalized solution, u, of (7.2) belongs to $H^r(\Omega)$, there exists a positive constant, K, such that

$$\|u-s\|_{H^p(\Omega)} \leq K(\bar{\rho})^{r-2p}, \tag{7.19}$$

where s is the Galerkin approximation to u in $\theta^p(\bigotimes_{i=1}^{N} S(2m-1,\Delta_i,z_i))$, for all $\rho \in C$.

(ii) If m is a positive integer, $-1 \leq z_i \leq 2m-1$, $-1 \leq p \leq \min_{1 \leq i \leq N} (z_i+1)$ and

$\theta \in C^r(\Omega)$, where $2p \leq r \leq 2m+p+1$, then $\theta^p(\bigotimes_{i=1}^{N} S(2m, \Delta_i, z_i)) \subset H^p(\Omega)$ for all $\rho \in C$. Moreover, if the generalized solution, u, of (7.2) belongs to $H^r(\Omega)$, there exists a positive constant, K, such that

$$\|u-s\|_{H^p(\Omega)} \leq K(\bar{\rho})^{r-2p}, \tag{7.20}$$

where s is the Galerkin approximation to u in $\theta^p(\bigotimes_{i=1}^{N} S(2m,\Delta_i z_i))$ for all $\rho \in C$.

Several remarks concerning the nonlinear algebraic problem (7.4) for the nonlinear Dirichlet problems, for these subspaces are now in order. First of all if $\{B_k(x_j)\}_{k=1}^{M_j}$ are the one-dimensional basis elements for $S(d, \Delta_j, z_j)$, $1 \leq j \leq N$, described in section 6, then it is easy to verify that the set of natural basis elements defined in $\theta^p(\bigotimes_{i=1}^{N} S(d, \Delta_i, z_i))$ by $\theta^p \mathop{X}_{j=1}^{N} B_{k_j}(x_j)$, $1 \leq k_j \leq M_j$, $1 \leq j \leq N$, form a basis for $\theta^p(\bigotimes_{i=1}^{N} S(d, \Delta_i, z_i))$. Similarly, if $\{B_k(x_j)\}_{k=1}^{M_j}$ are the one-dimensional basis elements for $S_p(d, \Delta_j, z_j)$, $1 \leq j \leq N$, described in section 6, then it is easy to verify that the set of natural basis elements defined in $\bigotimes_{i=1}^{N} S_p(d, \Delta_i, z_i)$ by $\mathop{X}_{j=1}^{N} B_{k_j}(x_j)$, $1 \leq k_j \leq M_j$, $1 \leq j \leq N$, form a basis for $\bigotimes_{i=1}^{N} S_p(d, \Delta_i, z_i)$. Moreover, it is also easy to verify that with this choice of basis elements, the nonlinear algebraic system (7.4) is sparce. In fact, if $d \equiv 2m-1$ and $z_i \equiv m-1$, then each unknown is coupled to at most $3mN-1$ others and if d is either even or odd and $z_i \equiv d-1$, each unknown is coupled to at most $(4m-1)N-1$ others. However, unlike the situation in one dimension described in section 6, an unknown is <u>not</u> in general coupled <u>only</u> to its nearest neighbors, i.e., the nonlinear system (7.4) is <u>not</u> a "band system".

Finally, if the problem (7.9)-(7.10) is self-adjoint then we may rigorously employ all the iterative methods discussed in section 6. Not much is known about iterative methods for non-selfadjoint, nonlinear problems. However, we do know that Picard's iteration method does converge for those problems yielding Lipschitz continuous, strongly monotone mapping, cf. [49].

§8. The Neumann Problem

In this section, we give an example of the application of the approximation theory results of section 5 to the study of the Rayleigh-Ritz-Galerkin method for elliptic differential equations with "natural" or "nonessential" boundary conditions. In particular, we discuss the Neumann Problem for the Poisson equation. For details about more general problems see [19], [28], [65], and [68].

In $\Omega \subset R^N$ is a Lipschitz domain, the problem under consideration is to find a solution $u(x)$ of the equation

$$-\Delta u(x) \equiv \sum_{i=1}^{N} D_i^2 u(x) = f(x), \quad x \in \Omega , \tag{8.1}$$

subject to the boundary condition

$$\frac{\partial u}{\partial n}(x) = g(x), \quad x \in \partial\Omega , \tag{8.2}$$

where n refers to the exterior normal to the boundary $\partial\Omega$ of the domain Ω.

In order that a solution exist it is necessary that the inhomogeneous data, f and g, satisfy the compatibility equation

$$\oint_{\partial\Omega} g ds = \int_{\Omega} f dx . \tag{8.3}$$

Under appropriate smoothness conditions on $\partial\Omega$, f, and g, a solution will exist and be unique to within an additive constant. We make the solution unique by requiring that

$$\int_{\Omega} u dx = 0 . \tag{8.4}$$

Let the functional $H[\theta,\psi]$ be defined by

$$H[\theta,\psi] \equiv \int_\Omega \left\{ \sum_{i=1}^{N} (D_i\theta)(D_i\psi) \right\} dx \quad . \tag{8.5}$$

Then, the solution of (8.1) - (8.2) subject to (8.4) minimizes the functional

$$F[\theta] \equiv \tfrac{1}{2} H[\theta,\theta] + \int_\Omega f\theta\, dx - \oint_{\partial\Omega} g\theta ds \tag{8.6}$$

over the space $\Phi \equiv \{\theta \in H^1(\Omega) | \int_\Omega \theta dx = 0\}$. We remark that we do not require the functions in Φ to satisfy the boundary condition (8.2).

If S_M is any finite dimensional subspace of Φ, we may apply the Rayleigh-Ritz method and minimize F over S_M . Doing this we have the important result of [28].

Theorem 8.1. The Rayleigh-Ritz method yields a unique approximation, u_M, for each finite dimensional subspace S_M of Φ and there exists a positive constant, K, such that

$$\|u-u_M\|_{H^1(\Omega)} \leq K \inf_{y\in S_M} \|u-y\|_{H^1(\Omega)} \quad . \tag{8.7}$$

If $\Omega \subset R \equiv \mathop{X}_{i=1}^{N} [a_i, b_i]$ and $\rho \in \mathcal{P}(R)$, we may choose the restriction to Ω of $\bigotimes_{i=1}^{N} S(d,\Delta_i, z_i)$, $0 \leq z_i \leq d-1$, as our finite dimensional subspace of $H^1(\Omega)$. Furthermore, if $\{B_i\}_{i=1}^{M}$ are any basis of $\bigotimes_{i=1}^{N} S(d,\Delta_i, z_i)$ we define a corresponding subspace of Φ by taking S_M to be the space spanned by the functions $\{\widetilde{B}_i \equiv B_i - \int_\Omega B_i dx\}_{i=1}^{M}$. Clearly the functions $\widetilde{B}_i \in \bigotimes_{i=1}^{N} S(d,\Delta_i, z_i)$.

Combining the results of Theorems 5.5 and 8.1, we have

Theorem 8.2. Let $\Omega \subset R \equiv \mathop{X}\limits_{i=1}^{N} [a_i, b_i]$ be a Lipschitz domain, m be a positive integer, and C be a quasi-uniform set of partitions of R. (i) If $u \in H^r(\Omega)$, $1 \le r \le 2m$, then

$$\|u-u_M\|_{H^1(\Omega)} \le K(\bar{\rho})^{r-1}, \tag{8.8}$$

where u_M is the Rayleigh-Ritz approximation in the subspace S_M corresponding to $\bigotimes_{i=1}^{N} S(2m-1, \Delta_i, z_i)$, for all $1 \le m$, $0 \le z_i \le 2m-2$, and $\rho \in C$.

(ii) If $u \in H^r(\Omega)$, $1 \le r \le 2m+1$, then

$$\|u-u_M\|_{H^1(\Omega)} \le K(\bar{\rho})^{r-1} \tag{8.9}$$

where u_M is the Rayleigh-Ritz approximation in the subspace S_M corresponding to $\bigotimes_{i=1}^{N} S(2m, \Delta_i, z_i)$, for all $1 \le M$, $0 \le z_i \le 2m-1$, and $\rho \in C$.

Proof: We prove only (8.8), as the proof of (8.9), as the proof of (8.9) is essentially identical. Moreover, it suffices to bound the right-hand side of (8.8).

By Theorem 5.5 for each $\bigotimes_{i=1}^{N} S(2m-1, \Delta_i, z_i)$ satisfying the hypotheses, there exists $v \in \bigotimes_{i=1}^{N} S(2m-1, \Delta_i, z_i)$ such that (8.10) $\|u-v\|_{H^1(\Omega)} \le K(\bar{\rho})^{r-1}$. Moreover, from (8.4) and (8.10) we have,

$$\int_\Omega v\,dx = \int_\Omega (v-u)\,dx + \int_\Omega u\,dx \leq (\text{meas}\,\Omega)^{\frac{1}{2}} \|u-v\|_{H^0(\Omega)} \leq$$

$$(\text{meas}\,\Omega)^{\frac{1}{2}} \|u-v\|_{H^1(\Omega)} \leq (\text{meas}\,\Omega)^{\frac{1}{2}} K(\bar{\rho})^{r-1} .$$

Hence

$$\|u - v - \int_\Omega v\,dx\|_{H^1(\Omega)} \leq (1 + (\text{meas}\,\Omega)^{\frac{1}{2}}) K(\bar{\rho})^{r-1} . \qquad \text{Q. E. D.}$$

We remark that if $u \in H^4(\Omega)$, then the above result yields $\|u-u_M\|_{H^1(\Omega)} \leq K(\bar{\rho})^3$ for bicubic spline subspaces and if $u \in H^2(\Omega)$, the result yields $\|u-u_M\|_{H^1(\Omega)} \leq K\bar{\rho}$ for bilinear spline subspaces. Thus, these approximation schemes compare favorably with those of [11] and [12] for the Neumann problem for the Poisson equation. In fact, their finite difference approximation schemes yield an approximate solution, u_h, with the properties that $\|u-u_h\|_\infty = O(h^2|\ln h|)$, as $h \to 0$, if $u \in C^4(\bar{\Omega})$, and $\|u-u_h\|_\infty = O(h|\ln h|)$, as $h \to 0$, if $u \in C^3(\bar{\Omega})$, where $\|\cdot\|_\infty$ denotes the maximum over the mesh points. Furthermore, if the solution u is sufficiently differentiable the Rayleigh-Ritz method will be of "high-order accuracy" if the degree d of the spline subspace is chosen sufficiently large, while no "high-order accurate" finite difference approximation schemes are known.

§9. Eigenvalue Problems

In this section, we consider the Rayleigh-Ritz method for computing the eigenvalues, λ, and eigenfunctions, u, for the following real, self-adjoint boundary value problem:

$$(9.1)\quad \mathcal{L}[u] \equiv \sum_{0\leq|\alpha|\leq n} (-1)^{|\alpha|} D^{\alpha}(A_{\alpha}(x) D^{\alpha}u)(x) = \lambda \sum_{0\leq|\alpha|\leq t} (-1)^{|\alpha|} D^{\alpha}(B_{\alpha}(x)D^{\alpha}u)(x) \equiv \lambda\mathfrak{M}[u]$$

for all $x \in \Omega \subset R^N$, and

$$(9.2)\qquad D^{\alpha}u(x) = 0, \quad \text{for all } 0 \leq |\alpha| \leq n-1 \text{ and } x \in \partial\Omega,$$

where $A_{\alpha}, B_{\alpha} \in C^{|\alpha|}(\Omega)$ and are real-valued, and $0 \leq t < n-1$. In particular, we are interested in spline type spaces and in coupling the approximation theory results of sections 4 and 5 with a theoretical analysis of the Rayleigh-Ritz method. For an analogous analysis of these problems with more general boundary conditions, at least in one dimension, the reader is referred to [8] and [20]. The reader is referred to [23], [30] and [49] for more information about variational methods for (9.1)-(9.2).

If $\mathcal{D}$ denotes the set of all real-valued functions in $C^{2n}(\Omega)$ which satisfy the homogeneous boundary conditions (9.2), then we assume that there exist two positive constant K_1 and K_2 such that

$$(9.3)\qquad (\mathcal{L}u, u)_{L^2(\Omega)} \geq K_1 (\mathfrak{M}u, u)_{L^2(\Omega)} \geq K_2 (u, u)_{L^2(\Omega)}$$

for all $u \in \mathcal{D}$.

To describe the theory for the eigenvalue problem (9.1)-(9.2) we define the following inner products on $\mathcal{D}$:

$$(9.4)\qquad (u, v)_{\mathfrak{M}} \equiv (\mathfrak{M}u, v)_{L^2(\Omega)}, \quad \text{for all } u, v \in \mathcal{D},$$

and

$$(9.5)\qquad (u, v)_{\mathcal{L}} \equiv (\mathcal{L}u, v)_{\mathcal{L}^2(\Omega)}, \quad \text{for all } u, v \in \mathcal{D}.$$

We remark that the fact that (9.4) and (9.5) define legitimate inner-products on $\mathcal{D}$ follows from (9.3) and if we denote by $H_{\mathfrak{M}}$ and $H_{\mathfrak{L}}$ the completions of $\mathcal{D}$ with respect to the norms $\|\cdot\|_{\mathfrak{M}} \equiv (\cdot,\cdot)_{\mathfrak{M}}^{\frac{1}{2}}$ and $\|\cdot\|_{\mathfrak{L}}^{\frac{1}{2}} \subset H_{\mathfrak{M}}$. We now recall a fundamental result, cf. [30].

<u>Theorem 9.1.</u> If $\mathfrak{L}$, $\mathfrak{M}$, and Ω are such that bounded subsets of $H_{\mathfrak{L}}$ are precompact in $H_{\mathfrak{M}}$, then the eigenvalue problem (9.1)-(9.2) has countably many eigenvalues $\{\lambda_j\}_{j=1}^{\infty}$ which are real, have no finite limit point, and can be arranged as

$$(9.6) \qquad 0 < \lambda_1 \leq \lambda_2 < \ldots < \lambda_k \leq \lambda_{k+1} \leq \ldots .$$

There exists a corresponding sequence of eigenfunctions $\{\varphi_j(x)\}_{j=1}^{\infty}$ where $\varphi_j \in H_{\mathfrak{L}}$ and $\mathfrak{L}\varphi_j(x) = \lambda_j \mathfrak{M}\phi_j(x)$, $j \geq 1$. Moreover, these eigenfunctions can be chosen so that

$$(9.7) \qquad (\varphi_i, \varphi_j)_{\mathfrak{M}} = \delta_{ij}, \quad \text{for all } i, j = 1, 2, \ldots ,$$

$$(9.8) \qquad (\varphi_i, \varphi_j)_{\mathfrak{L}} = \lambda_i \delta_{ij}, \quad \text{for all } i, j = 1, 2, \ldots ,$$

$\{\varphi_j\}_{j=1}^{\infty}$ is complete in $H_{\mathfrak{M}}$, and

$$(9.9) \qquad \lambda_k = \inf\{\|w\|_{\mathfrak{L}}^2 / \|w\|_{\mathfrak{M}}^2 \mid w \in H_{\mathfrak{L}}, \text{ such that}$$

$$(w, \varphi_\ell)_{\mathfrak{M}} = 0 \text{ for all } 1 \leq \ell \leq k-1\} = \|\varphi_k\|_{\mathfrak{L}}^2 / \|\varphi_k\|_{\mathfrak{M}}^2, \quad \text{for all } k \geq 1 .$$

Now let S_M be any M-dimensional subspace of $H_{\mathfrak{L}}$. The Rayleigh-Ritz method is to find the extremal points of $R[w] \equiv \|w\|_{\mathfrak{L}}^2 / \|w\|_{\mathfrak{M}}^2$ over the subspace S_M rather than over the whole space $H_{\mathfrak{L}}$. It is relatively easy to verify, cf. [23], [30], and [49], that the Rayleigh-Ritz method for S_M yields M approximate eigenvalues, $0 \leq \lambda_1(S_M) \leq \lambda_2(S_M) \leq \ldots \leq \lambda_M(S_M)$ and M approximate eigenfunctions, $\varphi_1(S_M), \ldots, \varphi_M(S_M), \ldots, \varphi_M(S_M)$.

Following Theorems 2 and 4 of [20], we have

Theorem 9.2. Let the differential operators $\mathcal{L}$ and $\mathfrak{M}$ be such that (9.3) holds and let bounded subsets of $H_{\mathcal{L}}$ be precompact in $H_{\mathfrak{M}}$. If S_M is any M-dimensional subspace of $H_{\mathcal{L}}$ and $\{\tilde{\varphi}_i\}_{i=1}^{k}$ is any "globally approximating set of functions" to the first k eigenfunctions of (9.1)-(9.2), where $M > k$, i.e.,

$$(9.10) \qquad \sum_{i=1}^{k} \|\tilde{\varphi}_i - \varphi_i\|_{\mathfrak{M}}^2 < 1 ,$$

then

$$(9.11) \quad \lambda_j \leq \lambda_j(S_M) \leq \lambda_j + \left(\sum_{i=1}^{j} \|\tilde{\varphi}_i - \varphi_i\|_{\mathcal{L}}^2 \Big/ 1 - \sqrt{\sum_{i=1}^{j} \|\tilde{\varphi}_i - \varphi_i\|_{\mathfrak{M}}} \right)^2$$

for all $1 \leq j \leq k$. Moreover, if in addition $0 < \lambda_1 < \lambda_2 < \ldots < \lambda$ there exists a positive constant, K, such that

$$(9.12) \qquad \|\varphi_k(S_M) - \varphi_k\|_{\mathcal{L}} \leq K \left\{ \sum_{j=1}^{k} (\lambda_j(S_M) - \lambda_j) \right\}^{\frac{1}{2}} .$$

Corollary. Let $\mathcal{L}, \mathfrak{M}$, and Ω be such that (9.3) holds and bounded subsets of $H_{\mathcal{L}}$ are precompact in $H_{\mathfrak{M}}$. If $\{S_{M_\ell}\}_{\ell=1}^{\infty}$ is any sequence of M_ℓ-dimensional subspaces of $H_{\mathcal{L}}$ such that $\lim_{\ell \to \infty} \{ \inf_{y \in S_{M_\ell}} \|\varphi_j - y\|_{\mathcal{L}} \} = 0$ for all $1 \leq j \leq k$, where k is a fixed positive integer, then $\lambda_j(S_{M_\ell}) \downarrow \lambda_j$ as $\ell \to \infty$ for all $1 \leq j \leq k$ and $\|\varphi_j(S_{M_\ell}) - \varphi_j\|_{\mathcal{L}} \to 0$ as $\ell \to \infty$ for all $1 \leq j \leq k$.

For special classes of domains, Ω, we may now use the approximation theory results of sections 4 and 5, eg. Theorems 5.1 and 5.2, to find spline type subspaces of $H_{\mathcal{L}}$. The detail are left to the reader. We now provide an example of asymptot bounds for the errors for various subspaces under varying regularity hypotheses. The following results extend and improve th corresponding results of [8], [20], and [81].

Theorem 9.3. Let $R \subset R^N$ be a rectangular parallelepiped, $C \subset \mathcal{P}(R)$ be a quasi-uniform set of partitions, and $\mathcal{L}$ and $\mathfrak{m}$ be such that (9.3) holds and bounded subsets of $H_{\mathcal{L}}$ are precompact in $H_{\mathfrak{m}}$. (i) If m is a positive integer, $-1 \leq z_i \leq 2m-2$, and $0 \leq n \leq \min_{1 \leq i \leq N} \min(z_i+1, m)$, then $\bigotimes_{i=1}^{N} S_n(2m-1, \Delta_i, z_i) \subset H_n^n(R)$ for all $\rho \in C$. Moreover, if $\varphi_1, \ldots, \varphi_j \in H^r(R)$, $1 \leq j$, where $n \leq r \leq 2m$, then there exists a positive constant, K, such that

$$(9.13) \quad |\lambda_k - \lambda_k(\bigotimes_{i=1}^{N} S_n(2m-1, \Delta_i, z_i))| \leq K(\bar{\rho})^{2r-2n}, \quad 1 \leq k \leq j, \quad \text{for all } \rho \in C,$$

and if $0 < \lambda_1 < \lambda_2 < \ldots < \lambda_j$,

$$(9.14) \quad \|\varphi_k - \varphi_k(\bigotimes_{i=1}^{N} S_n(2m-1, \Delta_i, z_i))\|_{\mathcal{L}} \leq K(\bar{\rho})^{r-n}, \quad 1 \leq k \leq j, \quad \text{for all } \rho \in C.$$

(ii) If m is a positive integer, $-1 \leq z_i \leq 2m-1$, and $0 \leq n \leq \min_{1 \leq i \leq N} \min(z_i+1, m)$, then $\bigotimes_{i=1}^{N} S_n(2m-1, \Delta_i, z_i) \subset H_n^n(R)$ for all $\rho \in C$. Moreover, if $\varphi_1, \ldots, \varphi_j \in H^r(R)$, $1 \leq j$, where $n \leq r \leq 2m+1$, then there exists a positive constant, K, such that

$$(9.15) \quad |\lambda_k - \lambda_k(\bigotimes_{i=1}^{N} S_n(2m, \Delta_i, z_i))| \leq K(\bar{\rho})^{2r-2n}, \quad 1 \leq k \leq j, \quad \text{for all } \rho \in C,$$

and if $0 < \lambda_1 < \lambda_2 < \ldots < \lambda_j$,

$$(9.16) \quad \|\varphi_j - \varphi_j(\bigotimes_{i=1}^{N} S_n(2m, \Delta_i, z_i))\|_{\mathcal{L}} \leq K(\bar{\rho})^{r-n}, \quad \text{for all } \rho \in C.$$

<u>Theorem 9.4.</u> Let $\Omega \subset R \equiv \overset{N}{\underset{i=1}{X}} [a_i, b_i]$, be a closed, bounded subset of R^N, and $\theta \in C^1(R^N)$ be such that $\partial\Omega \equiv \{x \in R^N | \theta(x) = 0$ $\theta(x) > 0$ for all $x \in \text{int}\Omega$, and $\sum_{i=1}^{N} |D_i \theta(x)| \neq 0$ for all $x \in \partial\Omega$. Let C be a quasi-uniform set of partitions of R and $\mathfrak{L}$ and $\mathfrak{M}$ be such that (9.3) holds and bounded subsets of $H_{\mathfrak{L}}$ are precompact in $H_{\mathfrak{M}}$. (i) If m is a positive integer, $-1 \leq z \leq 2m-2$, and $0 \leq n \leq \min_{1 \leq i \leq N} (z_i+1)$, and $\theta \in C^r(\Omega)$, where

$2n \leq r \leq 2m+n$, then $\theta^n(\bigotimes_{i=1}^{N} S(2m-1, \Delta_i, z_i)) \subset H_n^n(\Omega)$ for all $\rho \in C$. Moreover, if $\varphi_1, \ldots, \varphi_j \in H^r(\Omega)$, $1 \leq j$, then there exists a positive constant, K, such that

$$(9.17) \quad |\lambda_k - \lambda_k(\theta^n(\bigotimes_{i=1}^{N} S(2m-1, \Delta_i, z_i)))| \leq K(\bar{\rho})^{2r-2n}, \quad 1 \leq k \leq j,$$

for all $\rho \in C$,

and if $0 < \lambda_1 < \lambda_2 \lambda \ldots \lambda_j$,

$$(9.18) \quad \|\varphi_k - \varphi_k(\theta^n(\bigotimes_{i=1}^{N} S(2m-1, \Delta_i, z_i)))\|_{\mathfrak{L}} \leq K(\bar{\rho})^{r-n}, \quad 1 \leq k \leq j,$$

for all $\rho \in C$.

(ii) If m is a positive integer, $-1 \leq z_i \leq 2m-1$, and $0 \leq n \leq \min_{1 \leq i \leq N} (z_i+1)$, and $\theta \in C^r(\Omega)$, where $2n \leq r \leq 2m+n+1$, then $\theta^n(\bigotimes_{i=1}^{N} S(2m, \Delta_i, z_i)) \subset H_n^n(\Omega)$ for all $\rho \in C$. Moreover if $\varphi_1, \ldots, \varphi_j \in H^r(\Omega)$, $1 \leq j$, then there exists a positive constant, K, such that

$$(9.19) \quad |\lambda_k - \lambda_k(\theta^n(\bigotimes_{i=1}^{N} S(2m, \Delta_i, z_i)))| \leq K(\bar{\rho})^{2r-2n}, \quad 1 \leq k \leq j,$$

for all $\rho \in C$,

and if $0 < \lambda_1 < \lambda_2 < \ldots < \lambda_j$,

$$(9.20)\quad \|\varphi_k - \varphi_k(\bigotimes_{i=1}^{N} S(2m,\Delta_i,z_i)))\|_{\mathcal{L}} \leq K(\bar{\rho})^{r-n}, \quad 1 \leq k \leq j ,$$

for all $\rho \in C$.

§10. Planar Rectangular Domains

Let $R \equiv [a_1,b_1] \times [a_2,b_2] \subset R^2$ be a bounded rectangular domain. The purpose of this section is to show how the spline spaces introduced in §4 may be augmented by appropriate "singular" basis functions so that the Rayleigh-Ritz-Galerkin method for a class of eigenvalue and source problems in R will be of high-order accuracy, even though derivatives of the solutions are singular at the corners. The results of this section improve and extend those of [26] and the reader is referred to [26] for comments on how to extend these results to more general planar, rectangular polygons and for computational results. In particular, we obtain approximation theory results analogous to those of section 4 and leave to the reader the task of applying these to the study of the Rayleigh-Ritz-Galerkin method as was done in section 7 and 9.

We start by considering the real, source problem

(10.1) $-\nabla(p(x,y)\nabla u) + q(x,y)u = f$, for all $(x,y) \in \text{int}\, R$, subject to the boundary condition.

$$(10.2)\qquad u(x,y) = 0, \quad \text{for all } (x,y) \in \partial R ,$$

where $p \in C^1(R)$, $q,f \in C(R)$, $p(x,y) > 0$ for all $(x,y) \in R$, $q(x,y) \geq 0$ for all $(x,y) \in R$, $p, q,$ and f are analytic in neighborhoods of the four corners of R, and $\nabla \equiv \frac{\partial}{\partial x} + \frac{\partial}{\partial y}$. We will need the following fundamental result of [44], which we state.

Theorem 10.1. If u is the solution (10.1) - (10.2) and (ξ,η) is a corner of R, u has the following asymptotic expansion near (ξ,η): if (r,θ) are polar coordinates with origin at (ξ,η), $u(r,\theta) = c_1 r^2\{\ln r \sin 2\theta + [\theta - \frac{\pi}{2}]\cos 2\theta\} + c_2 r^3\{\ln r[\sin 3\theta + \sin\theta] + \theta[\cos 3\theta + \cos\theta]\} + c_3 r^3\{\ln r[\cos 3\theta - \cos\theta] - [\theta - \frac{\pi}{2}][\sin 3\theta - \sin\theta]\} + \{\text{terms analytic in } r \text{ and } \theta\} + O(r^4 \ln r)$.

Given a positive integer d, let n_i, $1 \le i \le 4$, denote the number of terms in the asymptotic expansion of u near (ξ_i, η_i) through terms of order $O(r^d)$. The terms in this expansion can be used to define singular basis functions w_{k_i}, $1 \le k_i \le n_i$, $1 \le i \le 4$, as follows. Let $r_1 \le \min(b_1-a_1, b_2-a_2)$ be any fixed positive number and define w_{k_i} to be equal to the corresponding term in the asymptotic expansion for u in the sector $0 \le r \le r_1/2$, $0 \le \theta \le \pi/2$, and to be zero outside the set $0 \le r \le r_1$, $0 \le \theta \le \frac{\pi}{2}$. In the region $r_1/2 \le r \le r_1$, $0 \le \theta \le \frac{\pi}{2}$, we define it in such a way that $w_{k_i} \in H^{d+1}$ in $\delta \le r \le r_1$, $0 \le \theta \le \pi/2$, for any $\delta > 0$, eg. by Hermite interpolation in the variable r.

Following [26], we remark that $w_{k_i} \in H^d$ outside some neighborhood of (ξ_i, η_i), $1 \le i \le 4$ and there exist constants $c_0, \ldots, c_{n_i}$ such that $u(r,\theta) - \sum_{j=1}^{n_i} c_j w_{k_j}(r,\theta) = O(r^{2m})$ near (ξ_i, η_i), $1 \le i \le 4$. Combining these remarks with Theorems 4.1 and 4.2, we have as in [26]

Theorem 10.2. Let $d \equiv 2m-1$, where m is a positive integer, $C \subset \mathcal{P}(R)$ be a quasi-uniform set of partitions of R, $p \in C^{2m+1}(R)$ and $q, f \in C^{2m}(R)$. If

$$\tilde{S}_1(2m-1, \rho, z) \equiv \{s(x,y) \mid s(x,y) = \theta(x,y) + \sum_{i=1}^{4} \sum_{j=1}^{n_i} c_{ij} w_{k_j}(x,y),$$

for some c_{ij} and some $\theta(x,y) \in \bigotimes_{i=1}^{2} S_1(2m-1, \Delta_i, z_i)\}$, where $0 \le z_i \le 2m-2$, and u is the solution of (10.1)-(10.2), there exists a positive constant, K, such that

$$(10.3) \qquad \inf_{\theta \in \tilde{S}_1(2m-1, \rho, z)} \|u - \theta\|_{H^1(R)} \le K(\bar{\rho})^{2m-1}$$

for all $\rho \in C$.

Theorem 10.3. Let $d \equiv 2m$, where m is a positive integer, $C \subset \mathcal{P}(R)$ be a quasi-uniform set of partitions of R, $p \in C^{2m+2}(R)$ and $q, f \in C^{2m+1}(R)$.

If $\widetilde{S}_1(2m, \rho, z) \equiv \{s(x,y) \mid s(x,y) = \theta(x,y) + \sum_{i=1}^{4} \sum_{j=1}^{n_i} c_{ij} w_{k_j}(x,y)$, for some c_{ij} and some $\theta(x,y) \in \bigotimes_{i=1}^{2} S_1(2m, \Delta_i, z_i)\}$, where $0 \leq z_i \leq 2m-1$, and u is the solution of (10.1)-(10.2), there exists a positive constant, K, such that

$$(10.4) \qquad \inf_{\theta \in \widetilde{S}_1(2m,\rho,z)} \|u-\theta\|_{H^1(R)} \leq K(\bar{\rho})^{2m},$$

for all $\rho \in C$.

We now turn to the real, eigenvalue problem

$$(10.5) \qquad -\nabla(p(x,y)\nabla u) + q(x,y)u = \lambda t(x,y)u, \quad \text{for}$$

all $(x,y) \in \operatorname{int} R$, subject to the boundary condition

$$(10.6) \qquad u(x,y) = 0, \quad \text{for all } (x,y) \in \partial R,$$

where $p \in C^1(R)$, $q, f \in C(R)$, $p(x,y) > 0$ and $t(x,y) > 0$ for all $(x,y) \in R$, $q(x,y) \geq 0$ for all $(x,y) \in R$, and p, q, and t are analytic in neighborhoods of the four corners of R. Again we need a fundamental result of [44], which we state.

<u>Theorem 10.4.</u> If $u(x,y)$ is an eigenfunction of (10.5)-(10.6) and (ξ,η) is a corner of R, then u has the following asymptotic expansion near (ξ,η): $u(r,\theta) = c_1\{[1-\cos 4\theta]\ln r + \theta \sin 4\theta\}r^4 + C_2\{[1-\cos 4\theta] + \ln r \sin 4\theta\}r^4 + \{\text{terms analytic in } r \text{ and } \theta\} + O(r^5 \ln r)$.

As for the source problem, we construct the singular basis functions, $\{w_{k_i}(x,y)\}_{k_i=1}^{n_i}$, for each corner and adjoin these to subspaces of the form $\bigotimes_{i=1}^{2} S_1(d, \Delta_i, z_i)$, $0 \leq z_i \leq d-1$. Moreover, if $p \in C^{d+2}(R)$, and $q, t \in C^{d+1}(R)$, it is easy to prove, using the results of sections 4 and 9, approximation theory results for the eigenfunctions and eigenvalues of (10.5)-(10.6), which are exact analogues of Theorems 10.2 and 10.3.

The precise details are left to the reader.

§11. Improved Error Bounds

Let $R \equiv \overset{N}{\underset{i=1}{X}} [a_i, b_i] \subset R^N$, $u \in H^p(R)$ be given, C be a quasi-uniform set of partitions of R,

$\mathcal{S}_d \equiv \{ \overset{N}{\underset{i=1}{\otimes}} S(d,\Delta_i, z_i) \mid \overset{N}{\underset{i=1}{\otimes}} S(d,\Delta_i, z_i) \subset H^p(R)$ and $\rho \in C\}$ for each $1 \leq d$, and $\{\mathfrak{m}_S\}_{S \in \mathcal{S}_d}$ be a set of mappings of u into S for all $S \in \mathcal{S}_d$. If $u \in H^{p+t}(R)$ for some $t \geq 1$ and there exists a positive constant, K, such that

$$\|u - \mathfrak{m}_S(u)\|_{H^p(R)} \leq K(\bar{\rho})^t \ , \tag{11.1}$$

for all $S \in \mathcal{S}_d$, we are interested in whether or not such bounds hold in $H^{p+q}(R)$, $1 \leq q \leq t$, and $L^\infty(R)$. For example, if $\mathfrak{m}_S$ is the mapping which assigns to u its Rayleigh-Ritz-Galerkin approximation in S, then such bounds would imply uniform convergence of the approximations. We remark that results analogous to those of this section hold for subspaces of the form $\overset{N}{\underset{i=1}{\otimes}} S_p(d, \Delta_i, z_i)$ and $\theta^p \overset{N}{\underset{i=1}{\otimes}} S(d, \Delta_i, z_i)$, where θ is an appropriate weight function. The details are left to the reader.

<u>Theorem 11.1.</u> If $0 \leq q \leq t$, $u \in H^{p+t}(R)$, and (11.1) holds, then there exists a positive constant, K, such that

$$\|u - \mathfrak{m}_S(u)\|_{H^{p+q}(R)} \leq K(\bar{\rho})^{t-q} \ , \tag{11.2}$$

for all $S \in \mathcal{S}_d$ such that $S \subset H^{p+q}(R)$.

<u>Proof:</u> (11.3) $\|u - \mathfrak{m}_S(u)\|_{H^{p+q}(R)} \leq \|u - v\|_{H^{p+q}(R)} + \|v - \mathfrak{m}_S(u)\|_{H^{p+q}(R)}$

where v is the $L^2(R)$-orthogonal projection of u onto S.

It follows from (11.3), the results of section 4, and the Schmidt inequality, cf. [36], that $\|u - \mathfrak{m}_S(u)\|_{H^{p+q}(R)} \le K(\bar{\rho})^{t-q} + K_1/(\underline{\rho})^q \|v - \mathfrak{m}_S(u)\|_{H^p(R)} \le K(\bar{\rho})^{t-q} + K_1/(\underline{\rho})^q (\|v - u\|_{H^p(R)} + \|u - \mathfrak{m}_S(u)\|_{H^p(R)}) \le K_2(\bar{\rho})^{t-q}$, because of the quasi-uniformity of C. Q.E.D.

We now discuss some uniform convergence properties for the special case of $N = 2$. In general, we may use Theorem 11.1 in conjunction with the Sobolev Embedding Theorem, cf. [50], to obtain somewhat weaker results.

Theorem 11.2. If $N = 2$, $u \in H^{1+t}(R)$, and (11.1) holds with $p = 1$, then there exists a positive constant, K, such that

$$\|u - \mathfrak{m}_S(u)\|_{L^\infty(R)} \le K(\bar{\rho})^t |\ln \bar{\rho}|^{\frac{1}{2}}, \tag{11.3}$$

for all $S \in \mathcal{S}_d$.

Proof: Let $R \equiv [a,b] \times [c,d]$, $w_S(x,y) = u(x,y) - \mathfrak{m}_S u(x,y)$, $\|w_S(x,y)\|_{L^\infty(R)} = M_S$, and $\|u(x,y)\|_{L^\infty(R)} = M$. Then $\|\mathfrak{m}_S u\|_{L^\infty(R)} \le M + M_S$ and by an inequality of A. A. Markov, cf. [36], $\left\|\frac{\partial(\mathfrak{m}_S u)}{\partial y}\right\|_{L^\infty(R)} \le 2(M+M_S)\frac{d^2}{\underline{\rho}}$. Thus,

$$\int_c^d \left(\frac{\partial w_S}{\partial y}\right)^2 dy \le 2\int_c^d \left(\frac{\partial(\mathfrak{m}_S u)}{\partial y}\right)^2 dy + 2\int_c^d \left(\frac{\partial u}{\partial y}\right)^2 dy$$

$$\le \frac{8(b-c)(M+M_S)^2}{(\underline{\rho})^2} d^4 + 2K_0, \text{ where } K_0 = \int_c^d \left(\frac{\partial u}{\partial y}\right)^2 dy.$$

Applying Lemma 1 of [41, pg. 338], we have that there exist two positive constants c_1 and c_2 such that

$$(11.4)\quad M_S \equiv \|w_S\|_{L^\infty(R)} \le c_1(\bar{\rho})^t \left| \ln\left(\frac{8h(d-c)(M+M_S)^2 d^4 + 2(\bar{\rho})^2 K_{(}}{(\underline{\rho})^2 (\bar{\rho})^t} \right.\right.$$

$$+ c_2(\bar{\rho})^t, \quad \text{where } h \equiv [(b-a)^2 + (d-c)^2]^{\frac{1}{2}} .$$

From (11.4) we have that the set of quantities $\{M_S\}_{S \in \mathcal{S}_d}$ is bounded. Taking this into account we can simplify and we obtai (11.3). Q.E.D

We illustrate how Theorems 11.1 and 11.2 may be used by analyzing a method for finding lower bounds for the fundamental (lowest) eigenvalue for second order elliptic differential equations in two variables. Consider the problem

$$(11.5)\qquad \mathcal{L}[u] \equiv -\nabla(p(x,y)\,\nabla u) + q(x,y)u = \lambda t(x,y)u \quad ,$$

for all $(x,y) \in \text{int}\, R \equiv (a,b) \times (c,d)$, subject to the boundary condition

$$(11.6)\qquad u(x,y) = 0, \quad \text{for all } (x,y) \in \partial R \quad ,$$

where $p, q,$ and t are analytic and $p(x,y) > 0$, $t(x,y) > 0$, and $q(x,y) \ge 0$ for all $(x,y) \in R$. If C is a quasi-uniform set of partitions of R, combining Theorems 9.3 and 10.4, we have that there exists a positive constant, K, such that $\|\varphi_1 - \varphi_1(\otimes_{i=1}^{2} S_1(3,\Delta_i,2))\|_{H^1(R)} \le K(\bar{\rho})^3$, for all $p \in C$. From Theorem 11.1, we have that $\|\varphi_1 - \varphi_1(\otimes_{i=1}^{2} S_1(3,\Delta_i,2))\|_{H^t(R)} \le K(\bar{\rho})^{4-t}$, for all $1 \le t \le 3$. Hence, from Theorem 11.2, we have

$$(11.7)\quad \|D^\alpha(\varphi_1 - \varphi_1(\otimes_{i=1}^{2} S_1(3,\Delta_i,2)))\|_{L^\infty(R)} \le K(\bar{\rho})|\ln \bar{\rho}|^{\frac{1}{2}} , \qquad 0 \le |\alpha| \le 2 ,$$

for all $p \in C$.

However, by an extension of an inequality of Barta, cf. [5] and [59], we have that if $\Omega \equiv \{f \in C^2(R) | f(x,y) > 0$ for all $(x,y) \in \text{int } R\}$

$$\lambda_1 = \sup_{f \in \Omega} \{ \inf_{(x,y) \in R} \frac{\mathcal{L}(f)(x,y)}{t(x,y)\varphi(x,y)} \} , \tag{11.8}$$

where an optimizing function is φ_1.

From (11.7) we see that $\theta(x,y) \equiv \varphi_1(\bigotimes_{i=1}^{2} S_1(3,\Delta_i,2)) + K(\bar{\rho})|\ln \bar{\rho}|^{\frac{1}{2}} > 0$ for all $(x,y) \in \text{int } R$, and there exist two positive constants, K_1 and K_2, such that

$$\|\mathcal{L}(\varphi_1) - \mathcal{L}(\theta)\|_{L^\infty(R)} \leq K_1(\bar{\rho})|\ln \bar{\rho}|^{\frac{1}{2}}, \quad \text{for all } p \in C , \tag{11.9}$$

and

$$\|\varphi_1 - \theta\|_{L^\infty(R)} \leq K_2(\bar{\rho})|\ln \bar{\rho}|^{\frac{1}{2}}, \quad \text{for all } \rho \in C . \tag{11.10}$$

Thus, we can obtain a lower bound, $\tilde{\lambda}_1(\rho)$, for λ_1 by taking $\tilde{\lambda}_1(\rho) \equiv \inf_{(x,y) \in R} \frac{\mathcal{L}(\theta_1)(x,y)}{t(x,y)\theta_1(x,y)}$ and it follows from (11.9) and (11.10) that

$$0 \leq \lambda_1 - \tilde{\lambda}_1(\rho) \leq K(\bar{\rho})|\ln \bar{\rho}|^{\frac{1}{2}}, \quad \text{for all } p \in C . \tag{11.11}$$

REFERENCES

[1] Ahlberg, J. H., and E. N. Nilson and J. L. Walsh, The theory of splines and their applications. New York: Academic Press 1967.

[2] Aubin, J. -P., Approximation des espaces de distributions et des opérateurs differentiels. Bull. Soc. Math. France, Mémoire 12 (1967).

[3] Aubin, J. P., Behavior of the error of the approximate solutions of boundary value problems for linear elliptic operators by Galergin's and finite difference methods. Annali della Scoula Normale Superiare di Pisa Classe di Scienze Vol. XXI, Fasc IV, 599-637 (1967).

[4] Babuška, I., and B. Hubbard, Private communication.

[5] Barta, J., Sur la vibration fondamentale d'une membrane Comptes Rendus de l'Acad. des Sci., Paris 204, 472-47 (1937).

[6] Birkhoff, G., and C. de Boor, Error bounds for spline interpolation. J. Math. Mech. 13, 827-836 (1964).

[7] Birkhoff, G., and C. de Boor, Piecewise polynomial interpolation and approximation. Approximation of Functions. (H. L. Garabedian, ed.). pp. 164-190. Amsterdam: Elsevier Publishing Company 1965.

[8] Birkhoff, G., M. H. Schultz, and R. S. Varga, Piecewise Hermite interpolation in one and two variables with applications to partial differential equations. Numer. Math. 11, 232-256 (1968).

[9] Birkhoff, G., C. de Boor, B. Swartz, and B. Wendroff, Rayleigh-Ritz approximation by piecewise cubic polynomials. SIAM J. Numer. Anal. 3, 188-203 (1966).

[10] Boor, C. de and R. E. Lynch, On splines and their minimum properties. J. Math. Mech. 15, 953-969 (196

[11] Bramble, J. H. and B. E. Hubbard, Approximation of sol tions of mixed boundary value problems for Poisson's equation by finite differences. J. Assoc. for Comp. Ma 12, 114-123 (1965).

[12] Bramble, J. H. and B. E. Hubbard, A finite difference analogue of the Neumann problem for Poisson's equatior J. SIAM Numer. Anal. 2, 1-14 (1965).

[13] Bramble, J. and S. Hilbert, Private Communication.

[14] Browder, F. E., Existence and uniqueness theorems for solutions of nonlinear boundary value problems. Proc. Sym. Appl. Math. Amer. Math. Soc. 17, 24-49 (1965).

[15] Céa, J., Approximation variationnelle des problèmes aux limites. Ann. Inst. Fourier (Grenoble) 14, 345-444 (1964).

[16] Ciarlet, P. G., An $O(h^2)$ method for a non-smooth boundary value problem. Aequat. Math. 2, 39-49 (1968).

[17] Ciarlet, P. G., M. H. Schultz, and R. S. Varga, Numerical methods of high-order accuracy for nonlinear two-point boundary value problems. Proceedings of the International Colloquium C.N.R.S., Besançon, France, Sept. 7-lé, 1966 (to appear).

[18] Ciarlet, P. G., M. H. Schultz, and R. S. Varga, Numerical methods of high-order accuracy for nonlinear boundary value problems. I. One dimensional problem. Numer. Math. 9, 394-430 (1967).

[19] Ciarlet, P. G., M. H. Schultz, and R. S. Varga, Numerical methods of high-order accuracy for nonlinear boundary value problems II. Nonlinear boundary conditions. Numer. Math. 11, 331-345 (1968

[20] Ciarlet, P. G., M. H. Schultz, and R. S. Varga, Numerical methods of high-order accuracy for nonlinear boundary value problems. III. Eigenvalue problems. Numer. Math. 12, 120-133 (1968).

[21] Ciarlet, P. G., M. H. Schultz, and R. S. Varga, Numerical method of high-order accuracy for nonlinear boundary value problems. IV. Periodic boundary conditions. Numer. Math. 12, 266-279 (1968).

[22] Ciarlet, P. G., M. H. Schultz, and R. S. Varga, Numerical methods of high-order accuracy for nonlinear boundary value problems. V. Monotone operators. Numer. Math. 13, 51-77 (1969).

[23] Collatz, L., The Numerical Treatment of Differential Equations, 3rd ed. (568 pp.). Berlin-Göttingen-Heidelberg: Springer 1960.

[24] Courant, R., Variational methods for the solution of problems of equilibrium and vibrations. Bull. Amer. Math. Soc. 49, 1-23 (1943).

[25] Douglas, J., Jr. and T. Dupont, The numerical solution of waterflooding problems in petroleum engineering by variational methods. To appear.

[26] Fix, G., Higher-order Rayleigh-Ritz approximations. J. Math. and Mech. 18, 645-658 (1969).

[27] Fix, G., and G. Strang, Fourier analysis of the finite element method in Ritz-Galerkin theory. To appear.

[28] Friedrichs, K. O., and H. B. Keller, A finite difference scheme for generalized Neumann problems. Numerical Solution of Partial Differential Equations. (J. H. Bramb ed.) 1-19. New York: Academic Press Inc., 1966.

[29] Goël, J. J., Construction of basic functions for numerical utilisation of Ritz's method. Numer. Math. 12, 435-447 (1969).

[30] Gould, S. H., Variational Methods for Eigenvalue Problems (275 pp.) Toronto: University of Toronto Pres 1966.

[31] Greenspan D. and S. V. Parter, Mildly nonlinear elliptic partial differential equations and their numerical solution. II. Numer Math. 7, 129-146 (1965).

[32] Herbold, R. J., Consistent quadrature schemes for the numerical solution of boundary value problems by variational techniques. Doctoral Thesis. Case Western Reserve University, 1967.

[33] Herbold, R. J., M. H. Schultz, and R. S. Varga, Quadrature schemes for the numerical solution of boundary value problems by variational techniques. To appear in Aequat. Math.

[34] Harrick, I. I., On the approximation of functions vanishing on the boundary of a region by functions of a special form. Mat. Sb. N.S. 37(79), 353-384 (1955).

[35] Harrick, I. I., Approximation of functions which vanish on the boundary of a region, together with their partial derivatives, by functions of special type. Akad. Nauk SSSR Izv. Sibirk. Otd. 4, 408-425 (1963).

[36] Hille, E., G. Szegö and J. D. Tamarkin, On some generalizations of a theorem of A. Markhoff. Duke Math. Jour. 3, 729-739 (1937).

[37] Hulme, B. L., Interpolation by Ritz approximation. J. Math. Mech. 18, 337-342 (1968).

[38] Jerome, J. W., On the L_2 n-width of certain classes of functions of several variables. J. Math. Anal. Appl. 20, 110-123 (1967).

[39] Jerome, J. W., Asymptotic estimates of the L_2 n-width. J. Math. Anal. Appl. 22, 449-464 (1968).

[40] Jerome, J. W. and R. S. Varga, Generalizations of spline functions and applications to nonlinear boundary value and eigenvalue problems. Theory and Applications of Spline Functions, 103-155. New York: Academic Press Inc., 1969.

[41] Kantorovich, L. V. and V. I. Krylov, Approximate Method of Higher Analysis (681 pp.). New York: Interscience Publishers 1958.

[42] Kellogg, R. B., Difference equations on a mesh arising from a general triangulation. Math. Comp. 18, 203-210 (1964).

[43] Kellogg, R. B., An error estimate for elliptic difference equations on a convex polygon. SIAM J. Numer. Anal. 3, 79-90 (1966).

[44] Lehman, R. S., Developments at an analytic corner of solutions of elliptic partial differential equations. J. Math. Mech. 8, 727-760 (1959).

[45] Levinson, N., Dirichlet problem for $\Delta u = f(P,u)$. J. Math. Mech. 12, 567-575 (1963).

[46] Lorentz, G. G., Approximation of Functions. New York, Holt, Rinehart, and Winston 1966.

[47] Mikhlin, S. G., Variational Methods in Mathematical Physics (584 pp.) New York: The Macmillan Co., 1964.

[48] Mikhlin, S. G. and K. L. Smolitskiy, Approximate Methods for Solution of Differential and Integral Equations (308 pp.). New York: American Elsevier Publishing Co., Inc. 1967.

[49] Minty, George, "Monotone (non-linear) Operators in Hilbert Space. Duke Math. J. 29 (1962).

[50] Morrey, C. B., Multiple Integrals in the Calculus of Variations (506 pp.). New York: Springer-Verlag 1966.

[51] Nečas, J., Les Méthodes Directes in Théorie des Équations Elliptiques. (351 pp.). Paris: Masson et Cie 1967.

[52] Nitsche, Joachim and Johannes C. C. Nitsche, Error estimates for the numerical solution of elliptic differential equations. Arch. Rational Mech. Anal. 5, 293-306 (1960).

[53] Ortega, J. M. and M. L. Rockoff, Nonlinear difference equations and Gauss-Seidel type iterative methods. SIAM J. Numer. Anal. 3, 497-513 (1966).

[54] Parter, S. V., Mildly nonlinear elliptic partial differential equations and their numerical solution. I. Numer. Math. 7, 113-128 (1965).

[55] Perrin, F. M., H. S. Price and R. S. Varga, On higher-order numerical methods for nonlinear two-point boundary value problems. To appear.

[56] Petryshyn, W. V., Direct and iterative methods for the solution of linear operator equations in Hilbert space. Trans. Amer. Math. Soc. 105, 136-175 (1962).

[57] Petryshyn, W. V., On nonlinear P-compact operators in Banach space with applications to constructive fixed-point theorems. J. Math. Anal. Appl. 15, 228-242 (1966).

[58] Polsky, N. I., Projection methods in applied mathematics. Dokl. Akad. Nauk SSSR, 143, 787-790 (1962).

[59] Protter, M. H. and H. F. Weinberger, Maximum Principles in Differential Equations (261 pp.). Englewood Cliffs: Prentice-Hall, Inc., 1967.

[60] Rose, M. E., Finite difference schemes for differential equations. Math. Comp. 18, 179-195 (1964).

[61] Schecter, S., Iteration methods for nonlinear problems. Trans. Amer. Math. Soc. 104, 179-189 (1962).

[62] Schecter, S., Relaxation methods for convex problems. SIAM J. Numer. Anal. 5, 601-612 (1968).

[63] Schoenberg, I. J., Contributions to the problem of approximation of equidistant data by analytic functions. Parts A and B. Quart. Appl. Math. 4, 45-99, 112-141 (1946).

[64] Schultz, M. H. and R. S. Varga, L-splines. Numer. Math. 10, 345-369 (1967).

[65] Schultz, M. H., L^{∞}-multivariate approximation theory. To appear in SIAM J. Numer. Anal.

[66] Schultz, M. H., L^2-multivariate approximation theory. To appear in SIAM J. Numer. Anal.

[67] Schultz, M. H., Rayleigh-Ritz-Galerkin methods for multidimensional problems. To appear in SIAM J. Numer. Anal.

[68] Schultz, M. H., Approximation theory of multivariate spline functions in Sobolev spaces. To appear in SIAM J. Numer. Anal.

[69] Schultz, M. H., L^2-approximation theory of even order multivariate splines. To appear SIAM J. Numer. Anal.

[70] Schultz, M. H., The Galerkin method for nonselfadjoint differential equations. To appear in J. Math. Anal. and Appl.

[71] Schultz, M. H., Error bounds for the Rayleigh-Ritz-Galerkin method. To appear in J. Math. Anal. and Appl.

[72] Schultz, M. H., Multivariate L-spline interpolation. To appear in J. Approx. Theory 2, 127-135 (1969).

[73] Schultz, M. H., Error bounds for polynomial spline interpolation. To appear.

[74] Schultz, M. H., Elliptic spline functions and the Rayleigh-Ritz-Galerkin method. To appear.

[75] Schultz, M. H., Error bounds for the Galerkin method for linear parabolic equations. To appear.

[76] Schultz, M. H., Error bounds for the Galerkin method for nonlinear parabolic equations. To appear.

[77] Sharma, A. and A. Meir, Degree of approximation of spline interpolation. J. Math. Mech. 15, 759-767 (1966).

[78] Swartz, B. and B. Wendroff, Generalized finite difference schemes. Math. of Comp. 23, 37-49 (1969).

[79] Varga, R. S., Matrix Iterative Analysis (322 pp.). Englewood Cliffs: Prentice-Hall 1962.

[80] Varga R. S., Hermite interpolate type Ritz methods for two-point boundary value problems. Numerical Solution of Partial Differential Equations (J. H. Bramble, ed.) 365-373. New York: Academic Press 1966.

[81] Wendroff, B., Bounds for eigenvalues of some differential operators by the Rayleigh-Ritz method. Math. Comp. 19, 218-224 (1965).

[82] Zienkiewicz, O. C., The Finite Element Method in Structural and Continuum Mechanics. London: McGraw Hill 1967.

[83] Zlámal, Miloš, On the finite element method. Numer. Math. 12, 394-409 (1968).

This research supported in part by the
National Science Foundation, GP 11326
and the Chevron Oil Field Research Co.,
La Habra, California

On the Degree of Convergence of Nonlinear Spline Approximation

JOHN R. RICE

1. Introduction. We consider approximation to functions $f(x)$ defined on $[0,1]$ in L_p norms where

$$\|f\|_p = \int_0^1 |f(x)|^p dx]^{\frac{1}{p}} \qquad 1 \leq p < \infty ,$$

$$\|f\|_\infty = \max |f(x)| \qquad x \in [0,1] .$$

We use as approximants spline functions $S^k(\pi;x)$ where

$$\pi = \{t_i | i = 0,1,2,\dots n, \ t_i \leq t_{i+1}\}$$

and $S^k(\pi;x)$ is a polynomial of degree k or less in $[t_i, t_{i+1}]$ (provided $t_{i+1} > t_i$) and $S^k(\pi;x)$ has $k-q$ $(q \leq k+1)$ continuous derivatives at t_i if $t_{i-1} < t_i = t_{i+1} \dots = t_{i+q-1} < t_{i+q}$. That is to say that the knots t_i are allowed to coalesce and we, in fact, consider extended splines in the terminology of [3]. A point with q knots coalesced is called a q-tuple knot. Such a function is a spline of degree k with n knots at the partition π. The collection of such splines is denoted by

$$S_n^k = \{S^k(\pi;x) | \pi = \{t_i | i=0,1,2,\dots,n\}\} .$$

We consider the L_p distance

$$\text{dist}_p(f, A) = \inf_{a \in A} \|f - a\|_p$$

of the function f from the set A . We study in particular the behavior of

$$\text{dist}_p(f, S_n^k)$$

for various functions f as n increases. We say that the degree of convergence of L approximations from S^k to f(x) is r(n) if

$$\text{dist}_p(f, S_n^k) = O(r(n)) .$$

It is known already in the case of linear spline approximation (i.e. the knots are determined in some a priori manner) that spline convergence properties are different - and often better - than that of classical approximation schemes. For example, in [1] a linear projector P_π: $f \to S^k(\pi, x)$ is constructed whose norm is bounded independent of π . Further it is shown that for $f \in C^{(k)}[0,1]$

$$\text{dist}_\infty(f, P_\pi f) = O(|\pi|^k \omega(f^{(k)}; |\pi|))$$

where $\omega(g, \delta)$ is the modulus of continuity of g for the length δ and $|\pi| = \max(t_{i+1} - t_i)$ for $i = 0, 1, 2, \ldots, n-1$. Such linear projectors do not exist for polynomial and similar approximation problems.

Some previous computational experiments [2] indicate that the accuracy of approximation by splines with variable knots is much superior to classical methods for certain types of functions.

To illustrate the type and nature of the results possible we first establish a simple, yet striking, result.

Theorem 1. *Suppose* f(x) *is continuous and monotone non-decreasing with* f(0) = 0, f(1) = 1. *Then*

$$\text{dist}_\infty(f, S_n^k) \le \frac{1}{n} \qquad k = 0,1 .$$

Proof: Choose t_i so that $f(t_i) = 1/n$ for $i = 0,1,2,\dots,n$. These knots exist and are uniquely determined. Determine $S^k(\pi;x)$ so that $S^k(\pi;t_i) = f(t_i)$. Then $S^k(\pi;x)$ is also monotone non-decreasing and the result follows immediately.

Corollary. Suppose $f(x)$ is continuous and of bounded variation V on $[0,1]$. Then

$$\text{dist}_\infty(f, S_n^k) \le \frac{V}{n} \qquad k = 0,1 .$$

The proofs of these results are so simple that one could call them mere observations, yet they connect hypothesis and conclusions which are completely incompatible in classical approximation problems.

In the next section we determine the degree of convergence for L_∞ approximations to x^α by S_n^k as a function of k. In the third section we determine the degree of convergence for L_p approximation to x^α, $\alpha > -\frac{1}{p}$ by S_n^k as a function of k. These two sections contain the "basic constructions" of this paper.

In the fourth section we discuss in some detail the approximation of $f(x) = [\log(1/x)]^{-1}$. Nothing is proved, but some experiments are reported upon and the source of the difficulty is examined. A conjecture about the degree of convergence for this (and similar) functions is made. The conjecture is formulated in several equivalent ways, some of which have no apparent connection with spline approximation. For example, we conjecture that if $f(x)$ is Riemann integrable then there is a sequence of Riemann sums with n terms whose associated step functions converge to $f(x)$ like $1/n$.

The "basic constructions" of Sections 2 and 3 are extended in Section 5 to obtain degree of convergence results for a number of general classes of functions.

2. Uniform Approximation of x^α, $\alpha > 0$. A consistent scheme of spline approximation is used throughout this paper. Each

knot of an approximation $S^k(\pi, x)$ is a k-tuple knot. Thus $S^k(\pi, x)$ is only continuous and we determine $S^k(\pi, x)$ by the projector P_π given by de Boor [1]. This apparently crude scheme allows us to establish degrees of convergence which are probably sharp except for the constants involved.

To simplify the notation, we identify the knots t_{ik+j} in the partition

$$\pi = \{t_{ik+j} \mid i = 1, 2, \ldots, n-1;\ j = 0, 1, 2, \ldots, k-1;\ t_{ik+j} = t_{ik+j+1}\}$$

as t_i. Thus $\pi' = \{t_i \mid i = 0, 1, 2, \ldots, n\}$ has, in fact, $k(n-1)+2$ knots, $n-1$ of which are k-tuple. We use the following basic lemma from [1].

Lemma 1. Let $f(x) \in C^{(k+1)}[a, b]$ and $\pi \subset [a, b]$ be given as above. Then if $S^k(\pi;x) = P_\pi f$ we have

$$\max_{x \in [t_i, t_{i+1}]} |f(x) - S^k(\pi;x)| \leq D_k (t_{i+1} - t_i)^{k+1} f^{(k+1)}(\xi)$$

where $\xi \in [t_i, t_{i+1}]$ and D_k is an absolute constant.

Set $q = (k+1)/\alpha$ and consider the following partition π_α:

$$(2.1)\qquad \begin{aligned} t_0 &= 0 \\ t_1 &= n^{-q} \\ t_j &= j^q t_1 \qquad j = 2, 3, \ldots n \,. \end{aligned}$$

For this partition we define $S^k(\pi_\alpha;x)$ as in Lemma 1 for $x \in [t_1, 1]$ and $S^k(\pi_\alpha; x)$ is linear in $[0, t_1]$ with $S^k(\pi_\alpha;0) = 0$ $S^k(\pi_\alpha;t_1) = t_1^\alpha$.

Theorem 2. For $\alpha > 0$ we have

$$\operatorname{dist}_\infty(x^\alpha, S_n^k) = O\left(\frac{1}{n^{k+1}}\right).$$

Proof. We observe that with $m = kn$

$$\operatorname{dist}_\infty(x^\alpha, S_m^k) \leq \operatorname{dist}_\infty(x^\alpha, S^k(\pi_\alpha;x))$$

and we establish that for some constant K we have

$$\operatorname{dist}_\infty(x^\alpha, S^k(\pi_\alpha;x)) \leq \frac{K}{n^{k+1}}$$

and the result follows since $O(n^{-(k+1)}) = O(m^{-(k+1)})$.

We have for $x \in [0, t_1]$

$$|x^\alpha - S^k(\pi_\alpha;x)| \leq |x^\alpha| \leq t_i^\alpha = \frac{1}{n^{k+1}} .$$

Set $E_j = \max|x^\alpha - S^k(\pi_\alpha;x)|$ for $x \in [t_j, t_{j+1}]$. Then by Lemma 1

$$\begin{aligned} E_j &\leq D_k(t_{j+1} - t_j)^{k+1} t_j^{\alpha-k-1}(\alpha)(\alpha-1)\ldots(\alpha-k) \\ &\leq K[(j+1)^q - j^q]^{k+1} t_1^{k+1} j^{(\alpha-k-1)q} t_1^{\alpha-k-1} \\ &\leq K[(j+1)^{(q-1)(k+1)} j^{q(\alpha-k-1)}] t_1^\alpha \end{aligned}$$

where K denotes a generic constant whose value changes from time to time and which does not depend on j . Note that $q(\alpha-k-1) = -(q-1)(k+1)$ and thus

$$E_j \leq Kt_1^\alpha = \frac{K}{n^{k+1}} .$$

This concludes the proof.

We note the following

Corollary. For the partition π_α in (2.1) we have for all $\beta > \alpha > 0$

$$\mathrm{dist}_\infty(x^\beta, S^k(\pi_\alpha;x)) = O\left(\frac{1}{n^{k+1}}\right) .$$

3. L_p Approximation of x^α, $\alpha > -\frac{1}{p}$. It follows directly from Theorem 2 that one can obtain L_p approximations to x^α for $\alpha > 0$ from S_n^k with degree of convergence $n^{-(k+1)}$. We now show that this is possible for any value of α such that x^α is in $L_p(0,1)$. As before, we consider a specific set of knots (each of which is, in fact, a k-tuple knot). Define π_α as

$$t_0 = 0$$

$$t_1 = n^{-q}$$

$$t_j = j^q t_1 \qquad j = 2, 3, \ldots, n$$

where $$q = \frac{1+p(k+1)}{1+\alpha p} .$$

Theorem 3. For $\alpha > -\frac{1}{p}$ we have

$$\mathrm{dist}_p(x^\alpha, S_n^k) = O\left(\frac{1}{n^{k+1}}\right) .$$

Proof. We need only show, as in the proof of Theorem 2, that

$$\mathrm{dist}_p(x^\alpha, S^k(\pi_\alpha;x)) \le \frac{K}{n^{k+1}}$$

where $S^k(\pi_\alpha;x)$ is the spline constructed as in Lemma 1 for the partition π_α and function x^α. We have

$$\int_0^{t_1} |x^\alpha - S^k(\pi_\alpha, x)|^p \, dx \le \int_0^{t_1} x^{\alpha p} \, dx = \frac{t_1^{\alpha p+1}}{1+\alpha p} = \frac{1}{(1+\alpha p) n^{1+p(k+1)}} .$$

We also have

$$\int_{t_j}^{t_{j+1}} |x^\alpha - S^k(\pi_\alpha;x)|^p\,dx \le (t_{j+1} - t_j)\,E_j^p$$

where E_j is as defined in the proof of Theorem 2. As in that proof we see that

$$E_j \le K(t_{j+1} - t_j)^{k+1}\, t_j^{\alpha-k-1}$$

and thus

$$(t_{j+1} - t_j)\,E_j^p \le K\, t_1^{1+p(k+1)}((j+1)^q - j^q)^{p(k+1)}\, t_1^{p(\alpha-k-1)}\, j^{p(\alpha-k-1)}$$

$$\le K((j+1)/j)^{p(q-1)(k+1)}\, j^{p(q-1)(k+1)+p(\alpha-k-1)}\, t_1^{1+\alpha p}$$

$$\le K\, t_1^{1+\alpha p}$$

where K is a generic constant.

We have established

$$\mathrm{dist}_p(x^\alpha, S^k(\pi^\alpha;x)) = [\int_0^1 |x^\alpha - S^k(\pi_\alpha;x)|^p\,dx]^{\frac{1}{p}}$$

$$= [\sum_{j=0}^{n-1} \int_{t_j}^{t_{j+1}} |x^\alpha - S^k(\pi_\alpha;x)|^p\,dx]^{\frac{1}{p}}$$

$$\le [\sum_{j=0}^{n-1} K\, t_1^{1+\alpha p}]^{\frac{1}{p}}$$

$$\le K[n t_1^{1+\alpha p}]^{\frac{1}{p}} .$$

Note that $n t_1^{1+\alpha p} = n^{-p(k+1)}$ and hence

$$\mathrm{dist}_p(x^\alpha, S^k(\pi_\alpha;x)) \leq \frac{K}{n^{k+1}} .$$

This concludes the proof.

We note the following

Corollary. For the partition π_α in (3.1) we have for all $\beta > \alpha > -\frac{1}{p}$

$$\mathrm{dist}_p(x^\beta, S^k(\pi_\alpha;x)) = O(\frac{1}{n^{k+1}}) .$$

4. On the Uniform Approximation of $\frac{-1}{\mathrm{Log}\, x}$ and Related Functions

We note the degree of convergence obtained above is independent of α and this leads one naturally to consider the possibility that this degree of convergence holds for even "stronger" singularities than x^α. In this section we present the conclusions of our study of the approximations of $\frac{-1}{\mathrm{Log}\, x}$ and the related functions $(\frac{-1}{\mathrm{Log}(x)})^{-p}$ for $p \geq 2$.

No theorems have been established due to two difficulties. First, for $\frac{-1}{\mathrm{Log}\, x}$, the mean value theorem used in connection with Lemma 1 is not sharp enough. That is to say that this function varies so rapidly that even in extremely small intervals it is essential to know with some accuracy the location of ξ within the interval. This difficulty does not appear to occur for $(\frac{-1}{\mathrm{Log}\, x})^{-p}$ as long as $p \geq k+1$. However, the second difficulty does occur and that is that the correct distribution of knots is unknown. It is perhaps more accurate to say that we are unable to find a mathematical expression for the location of the knots which we can analyze. The nature of the knot distribution is thought to be known (see below), but common mathematical expressions fail to provide it.

With the failure to prove any theorems for these functions we undertook an experimental study to give some evidence about the degree of convergence. These experiments are summarized below and they support the conjecture that the degree of convergence for $\frac{-1}{\operatorname{Log} x}$ from S_n^k is, indeed, $n^{-(k+1)}$. The study of these experiments led to some other conjectures (some are equivalent to one another) which are discussed at the end of this section. Some forms of these conjectures appear to be unrelated to spline approximation and raise interesting questions concerning the spaces of Riemann and Lebesgue integrable functions.

The simple experiment consisted of obtaining knots t_i, $i = 1, 2, \ldots, n-1$; $(t_0 = 0,\ t_n = .25)$ so that the broken line interpolant of $\frac{-1}{\operatorname{Log} x}$ made equal maximum errors between each pair of knots. Attempts to analyze this problem indicated that, in order to obtain n^{-2} convergence,

$$t_1 = e^{-n^2}$$

$$t_j \approx e^{-n^2/\alpha_j}$$

where α_j behaves roughly (but not exactly) like j^2 for small j. For large j, the partition must be more uniform than given by such formulas.

The exact experiment performed is to choose $t_1 = e^{-n^2}$ so that the error in (t_0, t_1) is n^{-2} and then determine the rest of the knots in $(0, .25)$ so as to maintain this error level. We then counted the number of knots to see if it increased linearly with n. The results indicate that

$$\operatorname{dist}_\infty\left(\frac{-1}{\operatorname{Log} x}, S_n^1\right) \approx \frac{2.7}{n^2} .$$

The range of values of n is 4 to 32 and the actual number of knots is plotted in Figure 1. The interval $(0, .25)$ was chosen so as to avoid the difficulty with $\frac{-1}{\operatorname{Log} x}$ for $x = 1$.

These computations are complicated by the size of the numbers involved. Thus for $n = 32$ we have $t_1 = e^{-1024} \approx 2{*}10^{-445}$. These calculations require scaling even on a

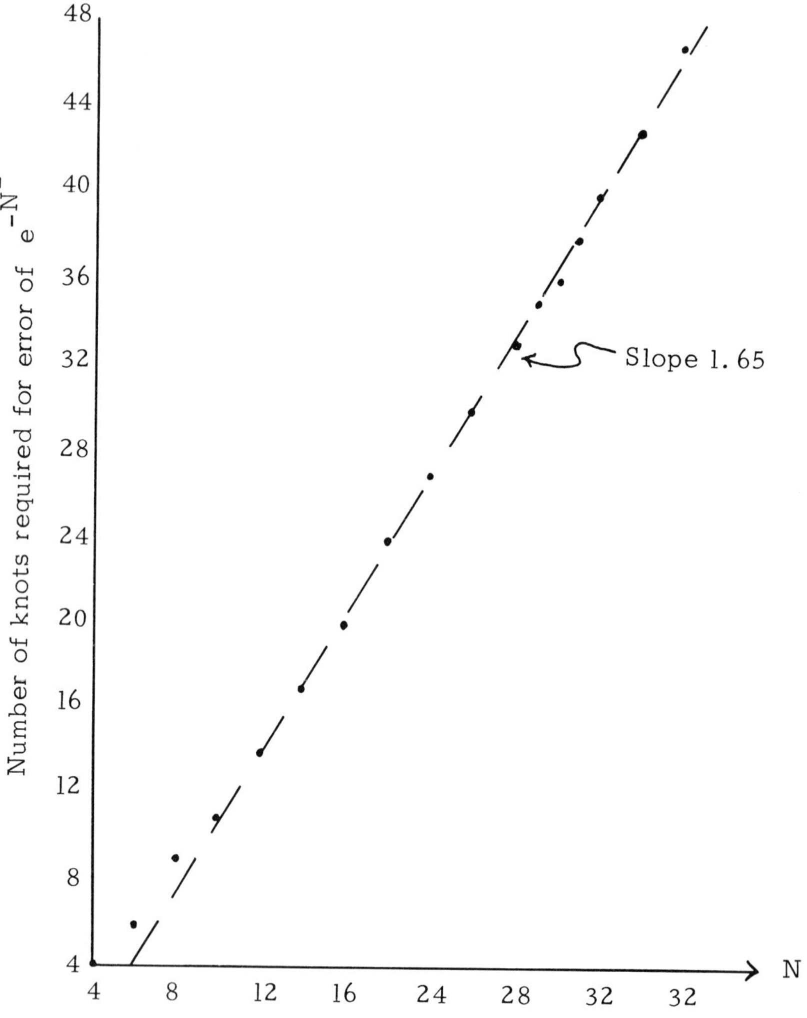

Figure 1: Results of experiments on the degree of convergence for $\frac{-1}{\log(x)}$ on (0, .25)

TABLE 1: The values of r_j and s_n which describe the knot locations

j	r_j	j	r_j	n	s_n	n	s_n
2	3.0	9	42	4	1.67	20	1.94
3	5.95	10	51	8	1.78	22	1.95
4	9.8	11	61	10	1.84	24	1.95
5	14.7	12	72	12	1.88	26	1.96
6	20	13	83	14	1.91	28	1.97
7	26	14	95	16	1.92	30	1.97
8	34	15	108	18	1.93	32	1.98

CDC 6500. Incidentally, these calculations show that this question is almost entirely academic since it is extremely unlikely that one will want to (or be able to) take knots separated by this amount in order to achieve an error of approximation of only .001 .

The knots show an extremely regular behavior which is summarized in Table 1. We give the values of

$$r_j = \mathrm{Log}\,\frac{t_{j+1}}{t_j} \Big/ \mathrm{Log}\,\frac{t_2}{t_1} \qquad j = 2, 3, \ldots, 15$$

and

$$s_n = \frac{1}{n^2}\,\mathrm{Log}\,\frac{t_2}{t_1} \qquad n = 4, 8, 10, 12, \ldots, 32.$$

Thus r_j determines the relation of later knots to the first two and s_n determines the relation of t_2 and t_1. The r_j are valid up to about $j = n-2$ for $6 \le n \le 15$.

On the basis of these experiments and the analysis made we make the following

Conjecture 1: Suppose $f(x)$ is continuous monotone and convex on $[0,1]$, then

$$\mathrm{dist}_{\infty}(f(x), S_n^1) = O(\frac{1}{n^2}) .$$

Note that the convexity here is an essential assumption because it implies $f(x)$ is absolutely continuous. It seems that it is not possible to approximate the standard example of a monotone non-absolutely continuous function by broken lines with this degree of convergence.

It is easily shown that this conjecture is equivalent to the pair of conjectures.

Conjecture 2: Suppose $f(x) \in L_1[0,1]$ and is concave monotone increasing (not necessarily bounded). Then

$$\mathrm{dist}_1(f(x), S_n^0) = O(\frac{1}{n}) .$$

Conjecture 3: Suppose $f(x) \in L_1[0,1]$ and $f(x)$ is monotone increasing (not necessarily bounded). Then

$$\mathrm{dist}_1(f(x), S_n^0) = O(\frac{1}{n}) .$$

These conjectures may be interpreted as indicative of how well integrals of such functions can be approximated by particular Riemann sums. In this same direction we have

Conjecture 4: Suppose $f(x)$ is Riemann integrable on $[0,1]$, then

$$\mathrm{dist}_1(f(x), S_n^0) = O(\frac{1}{n}) .$$

This conjecture states that every Riemann integrable function can be approximated in the L_1 norm by step functions (which define a Riemann sum of n terms) with order $\frac{1}{n}$ (where n is the number of step functions).

5. Extensions to General Classes of Functions. We extend the basic degree of convergence results obtained above in two directions. First, we replace the x^α singularities by a more general variety of singularity. Second, we allow an arbitrary finite number of such singularities.

We make the second extension first as it is the simplest.

Lemma 2. Suppose $[0,1]$ is partitioned by points z_i, $i = 0,1,2,\dots,m$ and suppose for each interval $[z_i, z_{i+1}]$, $i = 0,1,2,\dots,m-1$ there is a spline $S^k(\pi_i;x)$ so that

$$\|f(x) - S^k(\pi_i;x)\|_p = O(\frac{1}{n^{k+1}}) .$$

where the norm is restricted to $[z_i, z_{i+}]$ and n is the number of knots in π_i. Then

$$\text{dist}_p(f(x), S^k_n) = O(\frac{1}{n^{k+1}}) .$$

Proof: We take the partition $\pi = \bigcup_{i=0}^{m-1} \pi_i$ and define

$$S^k(\pi;x) = S^k(\pi_i;x) \qquad x \in (z_i, z_{i+1}) .$$

Then π has at most $n' = m(n+2)$ knots and

$$\text{dist}_p(f, S^k(\pi;x)) \le m\, O(\frac{1}{n^{k+1}}) = O(\frac{1}{(n')^{k+1}}) .$$

This concludes the proof.

We now introduce a classification of functions which includes piecewise smooth functions with singularities of certain types. With a set S in $[0,1]$ we associate the function

$$\omega_S(x) = \text{dist}(x, S)$$

Definition. f(x) is Type (α, k, S) if $f(x) \in \text{Lip}(\alpha)$ and

$$|f^{(k)}(x)| \leq K\,[\omega_S(x)]^{\alpha-k} \qquad x \notin S .$$

f(x) is of Type (α, k) if f(x) is of Type (α, k, S) for some finite set S. The interesting case is when the set S is the set of singularities of the kth derivative of f(x).

Theorem 4. If f(x) is of Type $(\alpha, k+1)$, $\alpha > -\frac{1}{p}$ and $f(x) \in L_p[0,1]$ ($f(x) \in C[0,1]$ for $p = \infty$) then

$$\text{dist}_p(f(x), S_n^k) = O\left(\frac{1}{n^{k+1}}\right)$$

Proof. Since f(x) is of Type $(\alpha, k+1)$, there is a finite set S so that f(x) is of Type $(\alpha, k+1, S)$. Suppose

$$S = \{s_i \mid i=1, 2, \ldots, m\}$$

and set

$$\begin{aligned} z_{2i} &= s_i \\ & & i = 1, 2, \ldots, m \\ z_{2i+1} &= \frac{s_i + s_{i+1}}{2} \end{aligned}$$

where $s_0 = 0$, $s_{m+1} = 1$ (if $s_1 = 0$ or $s_m = 1$ an obvious modification is made). In each interval $[z_i, z_{i+1}]$ $f^{(k+1)}(x)$ has at most one singularity which is z_i or z_{i+1}. For concreteness assume it is z_i.

We now observe $f(y)$, $y = (x-z_i)/(z_{i+1}-z_i)$ is defined on $[0,1]$ and we construct approximations as in Sections 2 and 3 based on the partition (2.1) for $p = \infty$ and (3.1) for $1 \leq p < \infty$. For the case $p = \infty$ we observe, using the notation of the proof of Theorem 2, that for $x \in [0, t_1]$, since $f(x) \in \text{Lip}(\alpha)$,

$$|f(x) - S^k(\pi;x)| \leq |f(x) - f(0) + S^k(\pi;0) - S^k(\pi;x)|$$
$$\leq 2|f(x) - f(0)| \leq K\, t_1^{\alpha} \ .$$

Note that $t_1^{\alpha} = \frac{1}{n^{k+1}}$. Further we have

$$E_j \leq D_k (t_{j+1} - t_j)^{k+1} f^{(k+1)}(\xi)$$

$$\leq D_k (t_{j+1} - t_j)^{k+1} K|\xi|^{\alpha-k-1}$$

$$\leq K(t_{j+1} - t_j)^{k+1} t_j^{\alpha-k-1} \leq K\, t_1^{\alpha} = \frac{K}{n^{k+1}} \ .$$

This establishes that there is an approximation to $f(x)$ on $[z_i, z_{i+1}]$ with the desired accuracy since splines are invariant under the transformation involved.

For $1 \leq p < \infty$ we have a partition π_i so that

$$\int_0^{t_1} |f(x) - S^k(\pi_i;x)|^p\, dx = \int_0^{t_1} |f(x) - f(t_1)|^p\, dx \ .$$

The process of determining $S^k(\pi_i, x)$ and the degree of convergence is independent of a constant addition to $f(x)$ and hence we may assume that $f(t_1) = 0$. Then we have

$$\int_0^{t_1} |f(x) - S^k(\pi_i;x)|^p\, dx \leq K^p \int_0^{t_1} x^{\alpha p}\, dx = \frac{K^p t_1^{\alpha p+1}}{1+\alpha p}$$

$$= \frac{K^p}{(1+p)\, n^{1+p(k+1)}} \ .$$

We also have

$$\int_{t_j}^{t_{j+1}} |f(x) - S^k(\pi_i;x)|^p \, dx \le (t_{j+1} - t_j) E_j^p$$

$$\le K(t_{j+1} - t_j)^{k+1} [f^{(k+1)}(\xi)]^p$$

where $\xi \in [t_j, t_{j+1}]$. Since $f(x)$ is of Type $(\alpha, k+1, 0)$ we have

$$[f^{(k+1)}(\xi)]^p \le K^p |\xi|^{(\alpha-k-1)p} \le K^p t_j^{(\alpha-k-1)p} \quad .$$

The argument is now concluded as in the proof of Theorem 3. Thus we conclude that for each interval $[z_i, z_{i+1}]$ and $\alpha > -\frac{1}{p}$ we have

$$\|f(x) - S^k(\pi_i;x)\|_p = O(\frac{1}{n^{k+1}})$$

where the norm is restricted to the interval $[z_i, z_{i+1}]$. The proof is then complete by the application of Lemma 2.

As an indication of the kinds of functions included in these generalizations we note that

a) $\operatorname{Log} x$ is of Type $(\alpha, k, 0)$ for all $\alpha < 0$

b) $x \operatorname{Log} x$ is of Type $(\alpha, k, 0)$ for all $0 < \alpha < 1$

c) $\sqrt{x} \sin \frac{1}{x^2}$ is in Lip (1/2) but not of Type $(\alpha, 1, 0)$

d) $x^4 \sin \frac{1}{x^2}$ is of Type $(\alpha, 1, 0)$ for all α but not of Type $(\alpha, 2, 0)$.

REFERENCES

1. de Boor, C. , On uniform approximation by splines, J. Approx. Thy. , 1 (1968), 219-235.

2. Esch, R. E. and Eastman, W. L., Computational methods for best approximations, Sperry Rand Research Technical Reports SEG-TR-67-30(1967) and SRRC-CR-68-18 (1968).

3. Rice, J. R., The Approximation of Functions, Vol. 2, Chapter 10, Addison Wesley Publishing Co., 1969.

This work partially supported by NSF grant GP-07163. The first results of this paper were obtained at an Office of Naval Research workshop held at Cornell University in July, 1968.

Error Bounds for Spline Interpolation

RICHARD S. VARGA

§1. Introduction. The object of this paper is to consider various forms of error bounds for spline interpolation in one space variable, along with some applications and extensions. Briefly, the material in §2 concerns the derivation of error bounds for spline interpolation for collections of bounded linear functionals on the Sobolev space $W_2^n[0,1]$ satisfying Property $\mathfrak{R}$, in a sense to be made precise in §2. Some special cases will be given to illustrate the results.

In §3, it is shown how the use of the theory of interpolation spaces leads to error bounds for the more general Besov spaces. In §4, application of the error bounds is made to the study of convergence of discrete variational Green's function to the continuous Green's function defined from two-point boundary value problems. Finally, in §5, Hermite spline functions are considered for the numerical approximation of ordinary differential equations, and improved error bounds are derived.

§2. Property $\mathfrak{R}$ Collections. Consider any ordinary differential operator L of order n of the form

$$(2.1) \qquad L[u] = \sum_{j=0}^{n} a_j(x)\, D^j u(x), \quad D^j \equiv \frac{d^j}{dx^j}, \quad n \geq 1 ,$$

for $u \in C^n[0,1]$, where we assume that $a_j \in C^j[0,1]$ for all $0 \leq j \leq n$, and that $a_n(x) \geq \omega > 0$ on $[0,1]$. In general, let $W_p^m[0,1]$, m a positive integer and $1 \leq p \leq +\infty$, denote the Sobolev space of all real-valued functions f defined on $[0,1]$

such that $D^{m-1}f$ is absolutely continuous with $D^m f \in L_p[0,1]$. It is well known that $W_p^m[0,1]$ is a Banach space. Fixing $p = 2$ in this section, let $\Lambda = \{\lambda_i\}_{i=1}^k$ be a set of k linearly independent bounded linear functionals on $W_2^n[0,1]$. For any $r = (r_1, r_2, \ldots, r_k)$ in real Euclidean space $\mathbb{R}^k$, the minimization problem,

$$(2.2) \quad \begin{cases} \inf\{\|Lg\|_{L_2[0,1]} : g \in K_r^{\perp}\}, & \text{where} \\ K_r^{\perp} \equiv \{g \in W_2^n[0,1] : \lambda_i(g) = r_k \text{ for all } 1 \leq i \leq k\}, \end{cases}$$

possesses a unique solution $s(x)$ in $K_r^{\perp}$ if $\mathfrak{n}(L) \cap K^{\perp} = \{0\}$ (cf. Anselone and Laurent [2] and Jerome and Schumaker [14]; here $K^{\perp}$ denotes those $g \in W_2^n[0,1]$ for which $\lambda_i(g) = 0$, $1 \leq i \leq k$, and $\mathfrak{n}(L)$ denotes the null-space of L. Moreover the collection of all $s(x)$ which solve the minimization problem for some $r \in \mathbb{R}^k$ is a finite-dimensional subspace of $W_2^n[0,1]$ and is denoted by $Sp(L,\Lambda)$. Given any $f \in W_2^n[0,1]$, the unique element s in $Sp(L,\Lambda)$ which solves the minimization problem of (2.2) with $\lambda_i(s) = \lambda_i(f)$ for all $1 \leq i \leq n$, will be called the <u>$Sp(L,\Lambda)$-interpolate</u> of f.

In most applications, the elements λ_i of Λ are usually chosen to be point evaluations of the function or its derivatives through order $n-1$, i.e., $\lambda_i(f) = D^{j_i} f(x_i)$ where $0 \leq j_i \leq n-1$ and $x_i \in [0,1]$. Satisfactory error bounds for $f-s$, where s is the $Sp(L,\Lambda)$-interpolate of f, have been obtained for such Λ (cf. Ahlberg, Nilson, and Walsh [1], and Jerome and Varga [15]). But, as the derivations of the error estimates are based either on Rolle's Theorem or Rayleigh-Ritz inequalities, these known error bounds can be extended to more general Λ. This brings us to

<u>Definition 1</u>. Consider the collection $\{\Lambda_i\}_{i=1}^{\infty}$ where each $\Lambda_i = \{\lambda_{j,i}\}_{j=1}^{k_i}$ is a set of k_i linearly independent bounded linear functionals on $W_2^n[0,1]$, $1 \leq i \leq \infty$. If $K^{\perp}(i)$ denotes those $g \in W_2^n[0,1]$ for which $\lambda_{j,i}(g) = 0$ for all $1 \leq j \leq k_i$, assume that $\mathfrak{n}(L) \cap K^{\perp}(i) = \{0\}$ for each $i \geq 1$. For any $f \in W_2^n[0,1]$, let $s_i(x)$ denote the unique $Sp(L,\Lambda_i)$-interpolate of f. Then, $\{\Lambda_i\}_{i=1}^{\infty}$ satisfies <u>Property $\mathfrak{R}$</u> with respect

to $W_2^n[0,1]$ if, for each $f \in W_2^n[0,1]$, there exist distinct points $\xi_j(i)$ with $0 \le \xi_1(i) < \xi_2(i) < \ldots < \xi_{m_i}(i) \le 1$ with $m_i \ge i$, such that

$$(2.3) \qquad f(\xi_j(i)) = s_i(\xi_j(i)) \quad \text{for all} \quad 1 \le j \le m_i \quad \text{for all} \quad i \ge 1,$$

and, defining $\xi_0(i) = 0$ and $\xi_{m_i+1}(i) = 1$, there exists for each $i \ge 1$ a quantity $\bar{\Delta}_i$, independent of f, such that $\sup_{0 \le j \le m_i} |\xi_{j+1}(i) - \xi_j(i)| \le \bar{\Delta}_i$ for all $i \ge 1$, and

$$(2.4) \qquad \lim_{i \to \infty} \bar{\Delta}_i = 0 .$$

With this definition, we then prove

<u>Theorem 2.1.</u> Let $\{\Lambda_i\}_{i=1}^{\infty}$ be a collection satisfying Property $\mathcal{R}$ with respect to $W_2^n[0,1]$. Then, for any fixed $f \in W_2^n[0,1]$, there exist constants K and K', independent of i, and a positive integer i_0, such that

$$(2.5) \qquad \|D^j(f - s_i)\|_{L_\infty[0,1]} \le K(\bar{\Delta}_i)^{n-j-1/2} \|Lf\|_{L_2[0,1]} \quad \text{for all} \quad 0 \le j \le n-1, \quad i \ge i_0,$$

and

$$(2.6) \qquad \|D^j(f - s_i)\|_{L_2[0,1]} \le K'(\bar{\Delta}_i)^{n-j} \|Lf\|_{L_2[0,1]} \quad \text{for all} \quad 0 \le j \le n, \quad i \ge i_0 .$$

Similarly, if $f \in W_2^{2n}[0,1]$ and for each i, the second integral relation is valid, i.e.,

$$(2.7)\qquad \int_0^1 [L(f-s_i)]^2\,dx = \int_0^2 (f-s_i)L^* Lf\,dx, \quad i \geq 1,$$

then

$$(2.8)\quad \|D^j(f-s_i)\|_{L_\infty[0,1]} \leq K(\bar{\Delta}_i)^{2n-j-1/2}\|L^*Lf\|_{L_2[0,1]},$$

$$0 \leq j \leq n-1, \quad i \geq i_0,$$

and

$$(2.9)\quad \|D^j(f-s_i)\|_{L_2[0,1]} \leq K'(\bar{\Delta}_i)^{2n-j}\|L^*Lf\|_{L_2[0,1]},$$

$$0 \leq j \leq n, \quad i \geq i_0.$$

Proof. Since $\lim_{i\to\infty} \bar{\Delta}_i = 0$, it follows that there exists an integer i_1 such that for $i \geq i_1$, s_i interpolates f in the sense of (2.3) in at least $n+1$ distinct points of $[0,1]$. Using Rolle's theorem, the proof of Theorem 6 of [22] can be directly applied, giving

$$(2.10)\quad \|D^j(f-s_i)\|_{L_\infty[0,1]} \leq \frac{n!\,(\bar{\Delta}_i)^{n-j-1/2}}{\sqrt{n}\; j!}\|D^n(f-s_i)\|_{L_2[0,1]}$$

$$0 \leq j \leq n-1, \quad i \geq i_1.$$

Again because $\lim_{i\to\infty} \bar{\Delta}_i = 0$, there also exists an integer i_2 such that

$$(2.11)\quad \|D^n(f-s_i)\|_{L_2[0,1]} \leq H\|L(f-s_i)\|_{L_2[0,1]} \quad \text{for all } i \geq i_2$$

where H is a positive constant (cf. [22]). Next, for each $i \geq 1$, the $Sp(L, \Lambda_i)$-interpolate s_i of f satisfies $\int_0^1 Ls_i \cdot L(f-s_i)\,dx$ (cf. [2]), which yields the <u>first integral relation</u>:

$$(2.12) \quad \|Lf\|^2_{L_2[0,1]} = \|L(f-s_i)\|^2_{L_2[0,1]} + \|Ls_i\|^2_{L_2[0,1]}, \quad i \geq 1 .$$

Hence, we have from (2.12) that $\|L(f-s_i)\|_{L_2[0,1]} \leq \|Lf\|_{L_2[0,1]}$ for all $i \geq 1$. Thus, combining (2.10) and (2.11) gives the desired result of (2.5) for $i \geq \max(i_1, i_2)$. Similarly, the proof of (2.6) follows that of [22, Theorem 7], and uses Rayleigh-Ritz inequalities instead of Rolle's theorem. The inequalities of (2.7) and (2.8) are established as in the manner of Theorems 8 and 9 of [22].

As an application of the above result, suppose that $L = D^n$, and for each $i \geq n$, define $\Lambda_i = \{\lambda_{j,i}\}_{i=1}^{i}$ by means of the functionals

$$(2.13) \qquad \lambda_{j,i}(f) = \int_{(j-1)/i}^{j/i} f(t)\,dt, \quad i \leq j \leq i .$$

Clearly, the functionals $\{\lambda_{j,i}\}_{j=1}^{i}$ are linearly independent bounded linear functionals on $W_2^n[0,1]$ for all $i \geq n$. Next, $\mathfrak{n}(D^n)$ consists of all polynomials of degree at most $n-1$, and $K^{\perp}(i)$ consists of all function $g \in W_2^n[0,1]$ such that

$$(2.14) \qquad \int_{(j-1)/i}^{j/i} g(t)\,dt = 0, \quad 1 \leq j \leq i .$$

For each $i \geq n$, it is readily verified that $\mathfrak{n}(D^n) \cap K^{\perp}(i) = \{0\}$, and thus, for each $i \geq n$, there exists a unique $Sp(D^n, \Lambda_i)$-interpolate, s_i, of $f \in W_2^n[0,1]$. For $i \geq n$, it is clear that if $f(x) \not\equiv s_i(x)$ in $[\frac{j-1}{i}, \frac{j}{i}]$, then $f(x) - s_i(x)$ must change signs at least once in $[\frac{j-1}{i}, \frac{j}{i}]$; otherwise $\lambda_{j,i}(f-s_i) \neq 0$. Hence, there exist a point $\xi_j(i)$ in $(\frac{j-1}{i}, \frac{j}{i})$ such that $f(\xi_j(i)) = s_i(\xi_j(i))$ for each $1 \leq j \leq i$, $i \geq n$. While these points $\xi_j(i)$ in general depend on f, it does follow that

$$\sup_{0 \leq j \leq i+1} |\xi_{j+1}(i) - \xi_j(i)| \leq \frac{2}{i} \equiv \bar{\Delta}_i$$

for each $f \in W_2^n[0,1]$, and hence $\lim_{i \to \infty} \bar{\Delta}_i = 0$. Consequently $\{\Lambda_i\}_{i=n}^{\infty}$ as defined by (2.13) satisfies property $\mathfrak{R}$, and the error bounds of Theorem 1 are applicable.

Of course, the usual Lg-splines as considered in [14] and [15], as well as the L-splines of [22], are formulated in terms of functionals $\lambda_{i,j}$ which are point evaluations of functions or their derivatives through order $n-1$. Hence $\{\Lambda_i\}_{i=1}^{n}$ for either Lg-splines or L-splines will automatically satisfy property $\mathfrak{R}$, if the partitions π_i of $[0,1]$ defined by these point functionals, are such that $\lim_{i \to \infty} \bar{\pi}_i = 0$. In this sense, Theorem 2.1 generalizes the previously known error bounds for Lg-splines and L-splines.

§3. Besov Spaces. Once one has the error bounds as in Theo[rem] 2.1, one can extend their usefulness via results from the theo[ry] of interpolation spaces. The purpose of this section is to brie[fly] show in a specialized way how this can be done. More detail[ed] results of this nature, as well as considerations of errors of interpolation and best approximation in higher dimensional set-tings, are to be found in Hedstrom and Varga [12].

Let X_0 and X_1 be two Banach spaces with norms $\|\cdot\|_0$ and $\|\cdot\|_1$, respectively, which are contained in a linear Hausdorff space $\mathfrak{X}$, such that the identity mapping of X_i, $i = 0, 1$, in $\mathfrak{X}$ is continuous. If $X_0 + X_1 \equiv \{f \in \mathfrak{X}: f = f_0 + f_1$ where $f_i \in X_i$, $i = 0, 1\}$, then it is known (cf. Butzer and Berens [6, p. 165]) that $X_0 + X_1$ and $X_0 \cap X_1$ are Banach spaces under the norms

$$\|f\|_{X_0 \cap X_1} = \max\{\|f\|_0, \|f\|_1\} ,$$

$$\|f\|_{X_0 + X_1} = \inf\{\|f_0\|_0 + \|f_1\|_1 : f = f_0 + f_1 \text{ with } f_i \in X_i, i=0,1\} .$$

It is understood that the above infimum is taken over all such decompositions, $f = f_0 + f_1$ with $f_i \in X_i$, $i = 0, 1$. Moreover, it follows that

$$X_0 \cap X_1 \subset X_i \subset X_0 + X_1 \subset \mathfrak{X}, \quad i = 0, 1 , \tag{3.1}$$

where the inclusion $A \subset B$ is understood here, and in the rest of this section, to mean that the identity mapping from A into B is continuous. A Banach space $X \subset \mathfrak{X}$ is an <u>intermediate space</u> of X_0 and X_1 if it satisfies the inclusion

$$X_0 \cap X_1 \subset X \subset X_0 + X_1 \subset \mathfrak{X} . \tag{3.2}$$

Peetre (cf. [6] and [21]) has given a real-variable method for constructing intermediate spaces of X_0 and X_1, which we now describe. For each positive t and each $f \in (X_0 + X_1)$, define

$$K(t, f) = \inf\{\|f_0\|_0 + t\|f_1\|_1 : f = f_0 + f_1 \text{ with } f_i \in X_i, i = 0, 1\} .$$

Then, for any θ with $0 < \theta < 1$ and any extended real number q with $1 \leq q \leq +\infty$, let $(X_0, X_1)_{\theta,q}$ be the set of all elements $f \in (X_0 + X_1)$ for which the following norm is finite:

$$\|f\|_{(X_0,X_1)_{\theta,q}} \equiv \begin{cases} \left[\int_0^\infty (t^{-\theta} K(t,f))^q \frac{dt}{t}\right]^{1/q}, & 1 \leq q < +\infty , \\ \sup_{t>0} t^{-\theta} K(t,f) , & q = +\infty . \end{cases}$$

It is known [6, p. 168] that $(X_0, X_1)_{\theta,q}$ is an intermediate space of X_0 and X_1, and thus satisfies the inclusions of (3.2). In particular, $(X, X)_{\theta,q} = X$.

If Y_0 and Y_1 are two Banach spaces continuously contained (with respect to the identity mapping) in the linear Hausdorff space $\mathcal{Y}$, let T denote any linear transformation from $(X_0 + X_1)$ to $(Y_0 + Y_1)$ for which

$$\|Tf\|_i \leq M_i \|f\|_i \quad \text{for all} \quad f \in X_i, \quad i = 0, 1 ,$$

i.e., T is a bounded linear transformation from X_i to Y_i with norm at most M_i, $i = 0, 1$. Then, the following is known (cf. [6, p. 180]).

Theorem 3.1. For $0 < \theta < 1$, $1 \leq q \leq +\infty$, T is a bounded linear transformation from the intermediate space $(X_0, X_1)_{\theta,q}$, whose norm $M \equiv \sup\limits_{\|f\|_{(X_0,X_1)_{\theta,q}}} \|Tf\|_{(Y_0,Y_1)_{\theta,q}}$ satisfies

$$(3.3) \qquad M \leq M_0^{1-\theta} \cdot M_1^{\theta} .$$

With the previous notation, then the Besov space $B_p^{\sigma,q}[0,1]$ is defined as the intermediate space (cf. [3])

$$(3.4) \quad (L_p[0,1], W_p^m[0,1])_{\theta,q} = B_p^{\sigma,q}[0,1] \text{ where } 0 < \sigma = \theta m < m$$

here, $1 \leq p, q \leq +\infty$. It is further known (cf. [6]) for $0 < \theta < 1$ that

$$(3.5) \quad (L_{p_0}[0,1], L_{p_1}[0,1])_{\theta,q} = L_p[0,1] \text{ where } \frac{1}{p} = \frac{1-\theta}{p_0} + \frac{\theta}{p_1}$$

and if $\sigma_0 \neq \sigma_1$, $0 < \theta < 1$, $1 < q_0, q_1 \leq +\infty$, and $1 \leq p \leq \infty$, the

$$(3.6) \quad (B_p^{\sigma_0, q_0}[0,1], B_p^{\sigma_1, q_1}[0,1])_{\theta,q} = B_p^{\sigma,q} \text{ where } \sigma = \theta\sigma_0 + (1$$

and for integer σ_i, either of the spaces $B_p^{\sigma_i, q_i}[0,1]$ in (3.6) can be replaced by $W_p^{\sigma_i}[0,1]$.

We now apply these results to error bounds of Theorem 2.1. Choosing $X_0 = X_1 = \mathfrak{X} = W_2^n[0,1]$, and $Y_0 = L_\infty[0,1] = \mathfrak{Y}$, $Y_1 = L_2[0,1]$, let the linear mapping T on $W_2^n[0,1]$ be defined by

$$Tf = D^j(f - s_i) \,, \tag{3.7}$$

for some fixed j, $0 \le j \le n-1$, where s_i is the $Sp(L, \Lambda_i)$-interpolate of f. Because of previous assumptions on the differential operator L of (2.1), we can write (2.5) and (2.6) of the previous section as

$$\|Tf\|_{L_\infty[0,1]} \le K(\bar{\Delta}_i)^{n-j-1/2} \|f\|_{W_2^n[0,1]} \,,$$

and

$$\|Tf\|_{L_2[0,1]} \le K'(\bar{\Delta}_i)^{n-j} \|f\|_{W_2^n[0,1]} \,.$$

Now, using the result of (3.3) with those of (3.5) for $p_0 = +\infty$, $p_1 = 2$, we obtain

$$\|D^j(f - s_i)\|_{L_q[0,1]} \le K'(\bar{\Delta}_i)^{n-j-\frac{1}{2}+\frac{1}{q}} \|f\|_{W_2^n[0,1]} \,, \tag{3.8}$$

$$0 \le j \le n-1, \quad 2 \le q \le +\infty \,.$$

Similarly, if the error bounds of (2.8) and (2.9), depending on the second integral relation, are valid, then interpolation similarly gives

$$\|D^j(f - s_i)\|_{L_q[0,1]} \le K(\bar{\Delta}_i)^{2n-j-\frac{1}{2}+\frac{1}{q}} \|f\|_{W_2^{2n}[0,1]} \,, \tag{3.9}$$

$$0 \le j \le n-1, \quad 2 \le q \le +\infty \,.$$

Next, from (3.6), we have that $(W_2^n[0,1], W_2^{2n}[0,1])_{\theta,\tau} = B_2^{\sigma,\tau}[0,1]$, where $n < \sigma = 2n - \theta n < 2n$. Thus, interpolating the results of (3.8) and (3.9) gives us

Theorem 3.2. Assuming the error bounds (2.5)-(2.9) of Theorem 2.1, let $f \in B_2^{\sigma,\tau}[0,1]$ where $n < \sigma < 2n$ and $1 \leq \tau \leq +\infty$, and let its $Sp(L,\Lambda_i)$-interpolate be s_i. Then, there exists a constant K, independent of i, such that

$$(3.10)\quad \|D^j(f-s)\|_{L_q[0,1]} \leq K(\bar{\Delta}_i)^{\sigma-j-\frac{1}{2}+\frac{1}{q}} \|f\|_{B_2^{\sigma,\tau}[0,1]},$$

$$0 \leq j \leq n-1, \quad 2 \leq q \leq +\infty .$$

The importance of the error bounds of (3.10) lies in the fact that we now have new error bounds for functions f which are elements of $W_2^n[0,1]$, but not of $W_2^{2n}[0,1]$. In addition, since the exponent of $\bar{\Delta}_i$ in (3.10) doesn't depend on τ, and since

$$(3.11)\qquad B_2^{m,1}[0,1] \subset W_2^m[0,1] \subset B_2^{m,\infty}[0,1],$$

we also have error bounds for spaces intermediate to $W_2^n[0,1]$ which can be larger than intermediate Sobolev spaces. To illustrate this, suppose $n = 2$ in the above discussion. Then, $W_2^4[0,1] \subset W_2^3[0,1] \subset W_2^2[0,1]$, and $W_2^3[0,1] \subset B_2^{3,\infty}[0,1] \subset W_2^2[0,1]$. This means that the error bound of (3.10), with exponent of $\bar{\Delta}_i$ equal to $3-j-\frac{1}{2}+\frac{1}{q}$, is valid not only for $W_2^3[0,1]$, but for $B_2^{3,\infty}[0,1]$ as well. Further results, similar to Theorem 3.2, can be found in [12].

§4. Discrete Variational Green's Functions. As an application of the splines introduced in §2, we consider the boundary value problem

$$(4.1)\qquad -L^*L[u(x)] = f(x), \quad 0 < x < 1,$$

subject to the boundary conditions of

(4.2) $$D^j u(0) = D^j u(1) = 0, \quad 0 \le j \le n-1 ,$$

where L is the n-th order differential operator of (2.1), and L^* is its formal adjoint, i.e.,

(4.3) $$L^*[v(x)] \equiv \sum_{j=0}^{n} (-1)^j D^j \{a_j(x) v(x)\} .$$

With the assumptions on L of §2, it is well known that there exists a Green's function $G(x, \xi)$ associated with the boundary value problem (4.1)-(4.2), defined on $[0,1] \times [0,1]$, such that

(4.4) $$\begin{cases} G(x,\xi) = G(\xi,x) & \text{in the closed unit square,} \\ G(x,\xi) \in C^{2n-2} & \text{in the closed unit square,} \\ G(x,\xi) \in C^{2n} & \text{in the subsets } 0 \le x < \xi \le 1 \text{ and} \\ & 0 \le \xi < x \le 1 \text{ of } [0,1] \times [0,1] . \end{cases}$$

Moreover, for any $0 \le \xi \le 1$, let $G_\xi(x)$ denote the function defined on $[0,1]$ by $G_\xi(x) = G(x, \xi)$. Then, it is also well known that

(4.5) $$\begin{cases} \text{i) } D^{2n-1} G_\xi \in L_\infty[0,1] \text{ and } \sup\{\|D^{2n-1} G_\xi\|_{L_\infty[0,1]} : \\ \qquad\qquad 0 \le \xi \le 1\} = \sigma < +\infty . \\ \text{ii) } D^k G_\xi(0) = D^k G_\xi(1) = 0 \text{ for all } 0 \le k \le n-1 . \\ \text{iii) } \lim_{x \to \xi+} D^{2n-1} G_\xi(x) - \lim_{x \to \xi-} D^{2n-1} G_\xi(x) = \dfrac{(-1)^n}{a_n^2(\xi)}, \text{ and} \\ \text{iv) } L^* L G_\xi(x) = 0 \text{ for all } x \ne \xi \text{ in } [0,1] . \end{cases}$$

Now, given any $f \in C^0[0,1]$, it is classic that the unique solution ϕ of (4.1)-(4.2) can be expressed in terms of the Green's function $G(x,\xi)$ by

$$(4.6) \qquad \phi(x) = \int_0^1 G(x,\xi)\, f(\xi)\, d\xi \; .$$

In addition, if $W^n_{2,0}[0,1]$ denotes the subspace of $W^n_2[0,1]$ which satisfies the boundary conditions (4.2), then ϕ can be characterized as the unique function in $W^n_{2,0}[0,1]$ which minimizes the functional

$$(4.7) \quad F[v] = \int_0^1 \{\sum_{j=0}^n p_j(t)\,(D^j v(t))^2 + 2f(t)v(t)\}dt, \quad v \in W^n_{2,0}[0,$$

Using the Ritz-Galerkin approach, the minimization of F over some finite-dimensional subspace S^M of $W^n_{2,0}[0,1]$ produces a unique ϕ^M in S^M which can, in analogy with (4.6), be described by

$$(4.8) \qquad \phi^M(x) = \int_0^1 G^M(x,\xi) f(\xi)\, d\xi \; .$$

Appropriately, the function $G^M(x,\xi)$ defined on $[0,1]\times[0,1]$ is called the discrete variational Green's function (cf. Ciarlet [7]) for the problem (4.1) - (4.2) and the subspace S^M.

Our purpose in this section is to show how the error boun for spline interpolation can be applied to the problem of estimat the error in $G(x,\xi) - G^M(x,\xi)$, when S^M is a special spline subspace of $W^n_{2,0}[0,1]$.

To make matters precise, we consider the special case o L-splines of Schultz and Varga [22]. If $\pi: 0 = x_0 < x_1 < x_2 < \cdots < x_{N+1} = 1$ is a partition of $[0,1]$, and $\underline{Z} = (z_0, z_1, \ldots, z_{N+1})$ is an incidence vector with positive integer components satisfying $1 \le z_i \le n$ for all $0 \le i \le N+1$, then the spline space $Sp(L, \pi, \underline{Z})$, a subspace of $W^n_2[0,1]$, is simply the particular case of $Sp(L,\Lambda)$ treated in §2 where the elements λ of Λ are all of the form

$$\lambda_{i,j}(f) = D^j f(x_i) \quad \text{for all} \quad 0 \leq j \leq z_i - 1, \quad 0 \leq i \leq N+1 .$$

Choosing $Z_0 = Z_{N+1} = n$ for convenience, the second integral relation needed in (2.7) is valid, as are the other assumptions of Theorem 2.1, and thus, we may use the error bounds of (2.5)-(2.9), with $\bar{\Delta} \equiv \max_{0 \leq i \leq N} |x_{i+1} - x_i|$. We further denote by $Sp_0(L, \pi, \underline{Z})$ the subspace of $Sp(L, \pi, \underline{Z})$ which satisfies the boundary conditions of (4.2). The following is a special case of Ciarlet and Varga [8].

<u>Theorem 4.1</u>. Let $G^M(x, \xi)$ be the discrete variational Green's function associated with the L-spline subspace $Sp_0(L, \pi, \underline{Z})$. Then, there exist positive constants K and K', independent of $\bar{\Delta}$, such that for all $0 \leq \xi \leq 1$,

$$\text{(4.8)} \quad \|D^k(G_\xi^M - G_\xi)\|_{L_\infty[0,1]} \leq K(\bar{\Delta})^{2n-k-3/2} \quad \text{for all } 0 \leq k \leq n-1 ,$$

and

$$\text{(4.9)} \quad \|D^k(G_\xi^M - G_\xi)\|_{L_2[0,1]} \leq K'(\bar{\Delta})^{2n-k-1} \quad \text{for all } 0 \leq k \leq n .$$

In addition, if $G^M(x, \xi)$ is the discrete variational Green's function associated with the Hermite L-spline subspace $Sp_0(L, \pi, \hat{\underline{Z}})$ where $\hat{\underline{Z}} \equiv (n, n, \ldots, n)$, then there exist positive constants K'', independent of $\bar{\Delta}$, such that for all $0 \leq \xi \leq 1$,

$$\text{(4.10)} \quad \|D^k(G_\xi^M - G_\xi)\|_{L_\infty[0,1]} \leq K''(\bar{\Delta})^{2n-k-1} \quad \text{for all } 0 \leq k \leq n-1 .$$

<u>Proof</u>. In [8], it is shown that $G_\xi^M(x)$, the discrete variational Green's function associated with $Sp_0(L, \pi, \underline{Z})$, is in fact the $Sp_0(L, \pi, \underline{Z})$-interpolate of $G_\xi(x)$ for each $0 \leq \xi \leq 1$. Since $D^{2n-1}G_\xi \in L_\infty[0,1]$ by (4.5i), then $G_\xi \in W_\infty^{2n-1}[0,1] \subset W_2^{2n-1}[0,1] \subset B_2^{2n-1,\infty}[0,1]$, using (3.11). Hence, the error

bounds of (4.8) and (4.9) follow as special cases of (3.10) with $q = \infty$ or $q = 2$, $\sigma = 2n-1$, and $\tau = +\infty$. Similarly, the assumption of Hermite L-spline subspaces $Sp_0(L, \pi, \hat{\underline{Z}})$ allows one, as in Birkhoff, Schultz, and Varga [4], to increase the exponent of $\bar{\Delta}$ in (4.8) by 1/2, which gives (4.10).

We further remark that other results, such as the positivity of the discrete variational Green's function $G^M(x, \xi)$, are also considered in [8].

§5. Improved Error Bounds for Ordinary Differential Equations. Applications of spline functions have not been made only to two-point boundary value problems and elliptic partial differential equations. Indeed, spline functions of maximum smoothness were first considered in the numerical solution of ordinary differential equations by Loscalzo and Talbot [18] and [19], and many interesting connections with standard numerical integration techniques have been proved. For example, the trapezoidal rule and the Milne-Simpson predictor-corrector method fall out as special cases of such spline functions applications. Unfortunately, higher-order smooth spline approximations turn out to be unstable, and consequently, the practical use of smooth spline functions to the numerical integration of ordinary differential equations is restricted to cases for whic the resulting method turns out to be classical. For details of the above remarks, we recommend the article by F. R. Loscalzo [16] in Theory and Applications of Spline Functions, edited by T.N.E. Greville.

The main reason why the above-mentioned applications of spline functions to the numerical integration of ordinary differential equations lead to unstable methods is because the result ing numerical approximations are, in a certain sense, too smoo Loscalzo and Schoenberg [16] and [20] have shown that the use of Hermite-splines of lower-order smoothness, to be described below, avoids completely the problem of instability. In [16] and [17], error bounds for the approximate Hermite spline solutions were obtained, but, as we shall show below in Theorem 5.1, the error bounds derived were not in all cases best possib In keeping with the title of this paper, the basic purpose in thi section is to obtain improved error bounds for such Hermite-sp applications.

Although applicable to systems of ordinary differential equations, we consider, for simplicity, the initial value problem

$$(5.1) \qquad Dy(x) = f(x, y(x)), \quad y(0) = y_0 ,$$

where y_0 is specified. We assume that f is continuous on $[0,b] \times \mathbb{R}$, subject to the usual Lipschitz condition

$$(5.2) \qquad |f(x,z) - f(x,\zeta)| \leq L|z-\zeta| \quad \text{for all} \quad x \in [0,b], \quad \text{for all } z, \zeta \in \mathbb{R},$$

where $L > 0$. This assures existence and uniqueness of a solution $y(x)$ of (5.1). For a positive integer n, let $h \equiv b/n$, and let $x_j \equiv jh$, $0 \leq j \leq n$, be the associated knots in $[0,b]$.

To explain the Hermite spline method, consider the $2q+2$ numbers

$$(5.3) \qquad D^j t(0), \quad D^j t(h), \quad 0 \leq j \leq q ,$$

where $t(x) \in C^q[0,h]$. It is clear that, by means of Hermite interpolation, there is a unique polynomial $s(x)$ of degree at most $2q+1$, written $s \in \pi_{2q+1}$, such that

$$(5.4) \qquad D^j t(0) = D^j s(0), \; D^j t(h) = D^j s(h), \quad 0 \leq j \leq q .$$

If, however, we <u>insist</u> that $s \in \pi_{2q}$, i.e., s is of degree at most $2q$, then the following $2q+2$-nd divided difference of $s(x)$ must necessarily vanish (cf. [17, Lemma 3.1]):

$$(5.5) \qquad H_q(s;0;h) \equiv (-1)^q \frac{(q!)^2}{(2q)!} h^{2q+1} s(\overbrace{0,0,\ldots,0}^{q+1}, \overbrace{h,h,\ldots,h}^{q+1}) .$$

This divided difference can also be expressed as the sum

$$(5.6) \qquad H_q(s;0;h) = -\sum_{k=0}^{q} C_{k,q} \{D^k t(0) + (-1)^{k+1} D^k t(h)\} ,$$

where

$$C_{k,q} \equiv \frac{1}{k!} \frac{q!\,(2q-k)!}{(q-k)!\,(2q)!}, \qquad 0 \le k \le q \,. \tag{5.7}$$

Later, we shall show how (5.6) and (5.7) connect the Hermite spline method with Padé approximations.

The algorithm for determining the Hermite spline function $p(x)$ which approximates the solution $y(x)$ of (5.1) can be described as follows. Given the functions

$$g_{\ell-1}(x, u(x)) \equiv D_x^{\ell-1} f(x,u(x)), \quad 1 \le \ell \le q \,, \tag{5.8}$$

where it is assumed for simplicity that $f \in C^{2q+1}([0,b] \times \mathbb{R})$, and $u(x)$ is any element in $C^q[0,b]$, define $p_1(x)$ in $[0,h]$ as the polynomial of degree at most $2q$ such that

$$\begin{cases} D^k p_1(0) = g_k(0, p_1(0)), & 1 \le k \le q \,, \\ D^k p_1(h) = g_k(h, p_1(h)), & 1 \le k \le q \,, \\ p_1(0) = y_0, \quad H_q(p_1;0;h) = 0 \,. \end{cases} \tag{5.9}$$

Using (5.6), it follows from (5.9) that $p(h)$ necessarily satisfies

$$\begin{aligned} p_1(h) &= y_0 + \frac{h}{2}\{g_1(0, y_0) + g_1(h, p_1(h))\} \\ &\quad + \dots + C_{qq} h^q \{g_q(0, y_0) + (-1)^{q+1} g_q(h, p_1(h))\} \,, \end{aligned} \tag{5.10}$$

and the method of successive substitutions suggests itself for the determination of $p_1(h)$. By means of the contraction mapping theorem, it can be shown (cf. [17]) that there is an $h_0 > 0$

such that for all $0 < h \leq h_0$, there is a unique solution $p_1(h)$ of (5.10).

Now, having found $p_1(x)$ in $[0, h]$, determine similarly $p_2(x) \in \pi_{2q}$ in $[h, 2h]$ such that

$$(5.11) \quad \begin{cases} D^k p_2(h) = g_k(h, p_1(h)), & 1 \leq k \leq q, \\ D^k p_2(2h) = g_k(h, p_2(2h)), & 1 \leq k \leq q, \\ p_2(h) = p_1(h); \; H_q(p_2; 0; h) = 0 . \end{cases}$$

In this way, the Hermite spline function $p(x)$, with $p(x) \equiv p_j(h)$ in $[(j-1)h, jh]$, $1 \leq j \leq [b/h]$, is a polynomial of degree at most $2q$ on each subinterval, and, by construction, $p(x) \in C^q[0, b]$. We remark that the solution of (5.10), (5.11), etc., produces only the numbers $p(jh)$, $1 \leq j < [b/h]$, and thus, the Hermite spline method can be viewed simultaneously as a single-step method (cf. [13, p. 209]). Once $p(h)$, say, is determined, the values $D^k p(h)$, $1 \leq k \leq q$, are obtained by evaluating $g_k(h, p(h))$ from (5.9), and $p(x)$ in $[0, h]$ can then be found by Hermite interpolation.

To appraise the errors in the Hermite spline method, the results of Loscalzo [17], utilizing the previous hypotheses, give us that for all $0 < h \leq h_1$, there exists a constant K_1, independent of h, such that

$$(5.12) \quad |D^k y(jh) - D^k p(jh)| \leq K_1 h^{2q}, \quad 0 \leq k \leq q, \; 0 \leq j \leq [b/h],$$

where $y(x)$, of class $C^{2q+2}[0, b]$, is the solution of (5.1), and $p(x)$ is its Hermite spline approximation. Next, let $w(x)$ be the Hermite-interpolation of $y(x)$, i.e.,

$$(5.13) \quad D^k w(jh) = D^k y(jh), \quad 0 \leq k \leq q, \quad 0 \leq j \leq [b/h],$$

and $w(x) \in \pi_{2q+1}$ on each subinterval $[(j-1)h, jh]$, $1 \leq j \leq [b/h]$.

Because $y \in C^{2q+2}[0,b]$, the interpolation error for $w(x)$ satisfies (cf. [4, Theorem 2])

$$(5.14) \qquad \|D^k(y-w)\|_{L_\infty[0,b]} \le K_2 h^{2q+2-k}, \quad 0 \le k \le q,$$

where K_2 is independent of h. Because of (5.13), it follows that

$$(5.15) \qquad |D^k w(jh) - D^k p(jh)| \le K_1 h^{2q}, \ 0 \le k \le q, \ 0 \le j \le [b/h].$$

To complete the picture, we now state a known result, a minor extension of a result of Swartz [23, Lemma 2] (Swartz's inequalities):

Lemma. If there exists a constant K, independent of h, and an integer α such that

$$(5.16) \qquad \max\{|D^k s(0)|, |D^k s(h)|\} \le Kh^{\alpha-k} \text{ for all } 0 \le k \le q,$$

where $s \in \pi_{2q+1}$, then there exists a K', independent of h, such that

$$(5.17) \qquad \|D^k s\|_{L_\infty[0,h]} \le K' h^{\alpha-k}, \qquad 0 \le k \le q.$$

To apply this Lemma, we see from (5.15) that $w(x) - p(x)$, an element of π_{2q+1} on each subinterval $[(j-1)h, jh]$, satisfies the inequalities of (5.16) with $\alpha = 2q$. Hence, applying (5.17) on each subinterval $[(j-1)h, jh]$ of $[0,b]$ gives

$$(5.18) \qquad \|D^k(w-p)\|_{L_\infty[0,b]} \le K' h^{2q-k}, \quad 0 \le k \le q,$$

where K' is independent of h. Then, applying the triangle inequality to (5.14) and (5.18) gives us

Theorem 5.1. Assuming $f(x,y) \in C^{2q+1}([0,b] \times \mathbb{R})$, let $y(x)$ be the unique solution of (5.1) in $[0,b]$. Then, there exist constants $K > 0$ and $0 < h_0 \leq b$ such that for all $0 < h \leq h_0$, the Hermite spline approximation $p(x)$ of $y(x)$, defined in (5.9), satisfies

$$(5.19) \qquad \|D^k(y-p)\|_{L_\infty[0,b]} \leq K h^{2q-k}, \quad 0 \leq k \leq q .$$

We remark that the special case $k = 0$ of (5.19) was obtained earlier by Loscalzo [17], but his error bounds for the higher derivatives of $p(x)$ were weaker than those of (5.19).

To conclude our discussion of the Hermite spline method for ordinary differential equations, we first remark that Loscalzo [17] showed that the Hermite spline method (of order q) is A-stable in the sense of Dahlquist for any q, [9], [10] for any q, i.e., if this method is applied to the particular ordinary differential equation

$$(5.20) \qquad Dy(x) = \lambda y(x), \quad y(0) = 1, \quad \mathrm{Re}\lambda < 0 ,$$

then its approximation $p(x)$ satisfies

$$(5.21) \qquad \lim_{n \to \infty} p(nh) = 0 \quad \text{for } \underline{\text{any}} \ h > 0 .$$

It is easy to verify from (5.6) that the Hermite spline method of order q applied to (5.20) gives $p((n+1)h) = \zeta_h \, p(nh)$, where $p(0) = 1$, and

$$(5.22) \qquad \zeta_h \equiv \left(\sum_{k=0}^{q} C_{q,k} \lambda^k h^k \right) \left(\sum_{k=0}^{q} (-1)^k C_{q,k} \lambda^k h^k \right)^{-1} .$$

However, from the definition of $C_{q,k}$ in (5.7), it turns out that ζ_h is just the q-th diagonal Padé rational approximation of $e^{\lambda h}$ (cf. [24, p. 269]). Consequently, since $\mathrm{Re}\,\lambda < 0$, then $|\zeta_h| < 1$ (cf. [5]). In this regard, it has been more generally shown in the thesis by Ehle [11] that the diagonal and the first two subdiagonals of the Padé table of the exponential function give rise to such A-stable approximations of (5.20), but the connection between Padé approximations and Hermite spline methods for ordinary differential equations appears to be new.

REFERENCES

1. J. H. Ahlberg, E. N. Nilson, and J. L. Walsh, The Theory of Splines and their Applications, Academic Press, New York, 1967.

2. P. M. Anselone and P. J. Laurent, "A general method for the construction of interpolating or smoothing spline-functions", Numer. Math. 12 (1968): 66-82.

3. O. V. Besov, "On some families of functional spaces. Imbedding and extension theorems", (in Russian), Dokl. Akad. Nauk SSSR 126 (1959): 1163-1165.

4. G. Birkhoff, M. H. Schultz, and R. S. Varga, "Piecewise Hermite interpolation in one and two variables with applications to partial differential equations", Numer. Math. 11 (1968): 232-256.

5. G. Birkhoff and R. S. Varga, "Discretization errors for well-set Cauchy problems, I", J. Math. and Physics 44 (1965): 1-23.

6. P. I. Butzer and H. Berens, Semi-groups of Operators and Approximation, Springer-Verlag, New York, 1967.

7. P. G. Ciarlet, "Discrete variational Green's function I", Aequationes Mathematicae (to appear).

8. P. G. Ciarlet and R. S. Varga, "Discrete variational Green's functions. II. One dimensional problem", Numer. Math. (to appear)

9. G. Dahlquist, "A special stability problem for linear multistep methods", BIT 3 (1963) : 27-43.

10. G. Dahlquist, "Stability questions for some numerical methods for ordinary differential equations", Proceedings of Symposia in Applied Mathematics, Amer. Math. Soc., Providence, Rhode Island, vol. 15 (1963), pp. 147-158.

11. B. L. Ehle, "On Padé approximations to the exponential function and A-stable methods for the numerical solution of initial value problems", (120 pp.). Doctoral Thesis, University of Waterloo 1969.

12. G. W. Hedstrom and R. S. Varga, "Applications of Besov spaces to spline approximation", J. Approx. Theory (to appear).

13. P. Henrici, Discrete Variable Methods in Ordinary Differential Equations, John Wiley and Sons, New York, 1962.

14. J. W. Jerome and L. L. Schumaker, "On Lg-splines", J. Approx. Theory 2 (1969): 29-49.

15. J. W. Jerome and R. S. Varga, "Generalizations of spline functions and applications to nonlinear boundary value and eigenvalue problems," Theory and Applications of Spline Functions (edited by T. N. E. Greville), Academic Press, New York, 1969: 103-155.

16. F. R. Loscalzo, "An introduction to the application of spline functions to initial value problems", Theory and Applications of Spline Functions, edited by T. N. E. Greville, Academic Press, New York, 1969, 37-64.

17. F. R. Loscalzo, "On the use of spline functions for the numerical solution of ordinary differential equations", Tech. Summary Report No. 869, Math. Research Center, University of Wisconsin, Madison, May 1968.

18. F. R. Loscalzo and T. D. Talbot, "Spline function approximations for solutions of ordinary differential equations", Bull. Amer. Math. Soc. 73 (1967): 438-442.

19. F. R. Loscalzo and T. D. Talbot, "Spline function approximations for solutions of ordinary differential equations", SIAM J. Numer. Anal. 4(1967): 433-445.

20. F. R. Loscalzo and I. J. Schoenberg, "On the use of spline functions for the approximation of solutions of ordinary differential equations", Tech. Summary Report No. 723, Math. Research Center, University of Wisconsin, Madison, January, 1967.

21. J. Peetre, "Espaces d'interpolation, généralisations, applications", Rend. Sem. Mat. Fis. Milano 34 (1964): 133-164.

22. M. H. Schultz and R. S. Varga, "L-splines", Numer. Math. 10 (1967): 345-369.

23. B. Swartz, "$O(h^{2n-2-\ell})$ bounds on some spline interpolation errors", Bull. Amer. Math. Soc. 74 (1968): 1072-1078.

24. R. S. Varga, Matrix Iterative Analysis, Prentice-Hall, Inc., Englewood Cliffs, New Jersey, 1962.

This research was supported in part by AEC Grant AT(11-1) - 1702.

Multipoint Expansions of Finite Differences

A. MEIR AND A. SHARMA

1. About two decades ago, Kloosterman [2] gave an expansion of the r-th divided difference of a function in terms of the derivatives at one of the end points of the interval. More recently we [4] established an analogous result for the symmetric differences in terms of the derivatives at the mid-point of the interval. These results were used to obtain improvements on certain inequalities of Boas and Polya. An earlier result of Obreshkov [5] on quadrature may be considered as a two point expansion of the first difference of a function in terms of the derivatives of the function at both end points of the interval. More recently in a paper on monosplines, Schoenberg [7] has considered an equivalent of Obreshkov's formula (so-called Hermite's quadrature formula) and has shown that it is best in the sense of Sard [6].

The object of this paper is to obtain explicit formulae for the expansion of higher order differences in terms of derivatives of the function at two specified points. The method of proof is based on an interpolation scheme and could be used mutatis mutandis to obtain other types of multi-point expansions for higher order differences. We shall also show that in certain cases our formulae are best in the sense of Sard, thereby extending the above result of Schoenberg on Hermite's quadrature formula.

2. Notations and Results. For a given positive h, let $\Delta_h = e^{hD} - 1$ and $\delta_h = e^{h(D/2)} - e^{-h(D/2)}$, where D

denotes the differential operator and let $\Delta_h^k = \Delta_h(\Delta_h^{k-1})$ and $\delta_h^k = \delta_h(\delta_h^{k-1})$ for $k = 2, 3, \ldots$. For a fixed positive integer r, let the sequence of numbers $P_k(r)$ be defined by the identi

(2.1) $$\left(\frac{e^x - 1}{x}\right)^r = \sum_{k=0}^{\infty} P_k(r) x^k ,$$

and the constants $A_k(r)$ by

(2.2) $$\left(\frac{\sinh x}{x}\right)^r = \sum_{k=0}^{\infty} A_k(r) x^{2k} .$$

As is easy to see, the following recursion relations hold for these numbers ([2], [4]):

$$P_k(r+1) = \sum_{j=0}^{k} P_j(r) P_{k-j}(1), \quad P_j(1) = \frac{1}{(j+1)!}, \quad P_0(r) = 1 .$$

$$A_k(r+1) = \sum_{j=0}^{k} A_j(r) A_{k-j}(1), \quad A_j(1) = \frac{1}{(2j+1)!}, \quad A_0(r) = 1 .$$

For later convenience we introduce the following notation for $k = 0, 1, \ldots$ and $\lambda > 0$:

(2.2a) $$p_k = k!\, r^{-k} P_k(r), \quad a_k = (2k)!\, \lambda^{-2k} A_k(r) .$$

For non-negative integers m and n and positive integers $r \leq s \leq m + r$, let $B_k \equiv B_k(m, n, r, s)$, $k = 0, 1, \ldots, m-1$ and $C_k \equiv C_k(m, n, r, s)$, $k = 0, 1, \ldots, n-1$ be determined by the identity (as $x \to 0$):

(2.3) $$\left(\frac{e^x - 1}{x}\right)^r = \sum_{k=0}^{m-1} B_k x^k + e^{rx} \sum_{k=0}^{n-1} C_k s^{k+s-r} + O(x^{m+n}) .$$

Further, for a positive λ let $D_k \equiv D_k(m, r, \lambda)$ be defined by the identity (as $x \to 0$):

$$(2.4) \quad \left(\frac{\sinh x}{x}\right)^r = e^{-\lambda x} \sum_{k=0}^{m-1} D_k x^k + e^{\lambda x} \sum_{k=0}^{m-1} (-1)^k D_k x^k + O(x^{2m}) .$$

It will turn out later that the constants B_k, C_k and D_k are uniquely determined by (2.3) and (2.4) and can be expressed as linear combinations of the P_k's and A_k's of (2.1) and (2.2).

We now formulate

Theorem 1. Suppose $f(x) \in C^{m+n+r}[a, a+rh]$. Then

$$(2.5) \quad h^{-r} \Delta_h^r f(a) = \sum_{k=0}^{m-1} B_k h^k f^{(k+r)}(a) + \sum_{k=0}^{n-1} C_k h^{s-r+k} f^{(s+k)}(a+rh)$$

$$+ E_1 h^{m+n} f^{(m+n+r)}(\xi_1) ,$$

where ξ_1 is a suitable point in $(a, a+rh)$ and E_1 is a constant depending only on m, n, r, s but not on f or h.

In case $s = r$, we have

$$(2.5a) \quad C_j \equiv C_j(m, n) = \frac{r^j}{j!} \sum_{k=j}^{n-1} \binom{-m}{k-j} \Delta^k p_m, \quad j = 0, 1, \ldots, n-1 ,$$

$$(2.5b) \quad B_j = (-1)^j C_j(n, m) , \quad j = 0, 1, \ldots, m-1 ,$$

$$(2.5c) \quad E_1 = \frac{r^{m+n}}{(m+n)!} \Delta^n p_m .$$

Furthermore, in case $s = r + m$, we have

$$(2.5d) \quad B_j = P_j(r), \qquad j = 0, 1, \ldots, m-1$$

$$(2.5e)\qquad C_j = \sum_{k=0}^{j} \binom{-m}{j-k} \Delta^k p_m, \quad j = 0,1,\ldots,n-1$$

$$(2.5f)\qquad E_1 = \frac{r^{m+n}}{(m+n)!} \sum_{k=0}^{n} (-1)^{n-k} \binom{m+n}{m+k} p_{m+k}\,.$$

For symmetric differences we state

Theorem 2. For a fixed positive λ let $\Lambda = \max(r,\lambda)$. Suppose $f(x) \in C^{2m+r}[a-\frac{\Lambda h}{2}, a+\frac{\Lambda h}{2}]$ then

$$(2.6)\quad h^{-r}\delta_h^r f(a) = \sum_{k=0}^{m-1} \left(\frac{h}{2}\right)^k D_k\{f^{(k+r)}(a-\frac{\lambda h}{2}) + (-1)^k f^{(k+r)}(a+\frac{\lambda h}{2})\}$$

$$+ E_2 \left(\frac{h}{2}\right)^{2m} f^{(2m+r)}(\xi_2)$$

where $a - \frac{\Lambda h}{2} < \xi_2 < a + \frac{\Lambda h}{2}$ and

$$(2.7)\qquad E_2 = \frac{\lambda^{2m}}{(2m)!} \Delta^m a_0\,.$$

3. Two Lemmas. In order to prove the above theorems we shall need the following lemmas.

Lemma I. [Aitken and Ziauddin]. Let $p_0 < p_1 < \ldots < p_{n-1}$ and $q_0 < q_1 < \ldots < q_{n-1}$ be positive integers satisfying $q_j \leq p_j$ $(j = 0,\ldots,n-1)$. Then

$$\operatorname{Det}\left\{\binom{p_i}{q_j}\right\}_{i,j=0,\ldots,n-1} \geq \frac{1}{q_0!\ldots q_{n-1}!}\,.$$

The proof of this lemma can be found in ([6], p. 47).

Lemma II. Let $\{\beta_k(t)\}_0^{m-1}$ and $\{\gamma_k(t)\}_0^{n-1}$ be polynomials of degree $\leq r+m+n-1$ defined by

$$(3.1)\qquad \begin{cases} \beta_k(a+ih) = 0, & i = 0,1,\dots,r-1 \\ \beta_k^{(r+j)}(a) = \delta_{kj}, & j = 0,1,\dots,m-1 \\ \beta_k^{(s+j)}(a+rh) = 0, & j = 0,1,\dots,n-1 \end{cases}$$

$$(3.2)\qquad \begin{cases} \gamma_k(a+ih) = 0, & i = 0,1,\dots,r-1 \\ \gamma_k^{(r+j)}(a) = 0, & j = 0,1,\dots,m-1 \\ \gamma_k^{(s+j)}(a+rh) = \delta_{jk}, & j = 0,1,\dots,n-1 . \end{cases}$$

Then we have

$$(3.3)\qquad h^{-r-k}\beta_k(a+rh) = B_k,\quad h^{-s-k}\gamma_k(a+rh) = C_k$$

where B_k's and C_k's are given by (2.3).

Proof: From (3.1) and (3.2) we see that the polynomials $\beta_k(t)$ and $\gamma_k(t)$ are the fundamental polynomials of the interpolation problem given by the incidence matrix $E = (e_{ij})$ where

$$\begin{aligned} e_{i0} &= 1, && 0 \le i \le r-1 \\ e_{0j} &= 1, && r \le j \le r+m-1 \\ e_{rj} &= 1, && s \le j \le s+n-1 \end{aligned}$$

and $e_{ij} = 0$ for all other values of i and j, $0 \le i \le r$, $0 \le j \le r+m+n-1$. Denoting the fundamental polynomials of Lagrange interpolation on the nodes $a, a+h, \dots, a+(r-1)h$ by $\{\ell_k(t)\}_0^{r-1}$, it follows from (3.1) and (3.2) that if $f \in C^{m+n+r}$, then the polynomial $Q(t)$ given by

$$(3.4) \qquad Q(t) = \sum_{k=0}^{r-1} f(a+kh)\ell_k(t) + \sum_{k=0}^{m-1} f^{(r+k)}(a)\beta_k(t) + \sum_{k=0}^{n-1} f^{(s+k)}(a+rh)\,\gamma_k(t)$$

is the unique polynomial satisfying the conditions $\underset{\sim}{Q}^{(j)}(a+ih) = f^{(j)}(a+ih)$ whenever $e_{ij} = 1$. Hence from a theorem of Birkhoff [1], we have

$$(3.5) \qquad f(a+rh) - \underset{\sim}{Q}(a+rh) = E\,h^{m+n+r} f^{(m+n+r)}(\xi)$$

where $a < \xi < a+rh$ and E is a constant independent of f and h. It is easy to see that $\ell_k(a+rh) = (-1)^{r-k-1}\binom{r}{k}$ and thus from (3.4) and (3.5) we have

$$(3.5) \qquad f(a+rh) - \underset{\sim}{Q}(a+rh) = E\,h^{m+n+r} f^{(m+n+r)}(\xi)$$

where $a < \xi < a + rh$ and E is a constant independent of f and h. It is easy to see that $\ell_k(a+rh) = (-1)^{r-k-1}\binom{r}{k}$ and thus from (3.4) and (3.5) we have

$$(3.6) \qquad \Delta_h^r f(a) = \sum_{k=0}^{m-1} f^{(r+k)}(a)\,\beta_k(a+rh) + \sum_{k=0}^{n-1} f^{(s+k)}(a+rh)\,\gamma_k(a+rh) + E\,h^{m+n+r} f^{(m+n+r)}(\xi) .$$

In order to determine the quantities $\beta_k(a+rh)$ and $\gamma_k(a+rh)$ we introduce the auxiliary polynomials $\{U_k(x)\}$ and $\{V_k(x)\}$ of degree $m+n-1$ satisfying the conditions:

$$\begin{cases} U_k^{(j)}(0) = \delta_{jk}, & k,j = 0,1,\ldots,m-1 \\ U_k^{(s-r+j)}(1) = 0, & k = 0,1,\ldots,m-1, \quad j = 0,1,\ldots,n-1 \end{cases}$$

$$\begin{cases} V_k^{(j)}(0) = 0, & k = 0,1,\ldots,n-1, \quad j = 0,1,\ldots,m-1 \\ V_k^{(s-r+j)}(1) = \delta_{jk}, & k,j = 0,1,\ldots,n-1 . \end{cases}$$

Then from (3.1) and (3.2) it is readily seen that

$$(3.7) \quad \begin{cases} \beta_k^{(r)}(x) = (rh)^k \, U_k\left(\frac{x-a}{rh}\right), & k = 0,1,\ldots,m-1 \\ \gamma_k^{(r)}(x) = (rh)^{s-r+k} \, V_k\left(\frac{x-a}{rh}\right), & k = 0,1,\ldots,n-1 . \end{cases}$$

Now it is known that for every $g \in C^r[a, a+rh]$

$$(3.8) \qquad \Delta_h^r g(a) = \frac{h^r}{(r-1)!} \int_0^r g^{(r)}(a+ht)\, K(r,t)\, dt$$

where

$$K(r,t) = \sum_{j=0}^{r} (-1)^{r-j} \binom{r}{j} (j-t)_+^{r-1} .$$

Hence from the first r conditions of (3.1), from (3.7) and (3.8) with $g(x) = \beta_k(x)$, we obtain

$$(3.9) \qquad \begin{aligned} \beta_k(a+rh) &= \Delta_h^r \beta_k(a) \\ &= \frac{h^{r+k} r^k}{(r-1)!} \int_0^r U_k\left(\frac{t}{r}\right) K(r,t)\, dt . \end{aligned}$$

Similarly,

$$(3.10) \qquad \gamma_k(a+rh) = \frac{h^{s+k} r^{s-r+k}}{(r-1)!} \int_0^r V_k\left(\frac{t}{r}\right) K(r,t)\, dt .$$

Setting $f(x) = e^x$, in (3.6) we have for every $h > 0$,

$$(3.11)\quad (e^h - 1)^r = \sum_{k=0}^{m-1} \beta_k(a+rh) + e^{rh} \sum_{k=0}^{n-1} \gamma_k(a+rh) + O(h^{m+n+r}) .$$

Comparing (3.11) and (2.3) and keeping in mind that (3.9) and (3.10) imply that $h^{-r-k}\beta_k(a+rh)$ and $h^{-s-k}\gamma_k(a+rh)$ are constants independent of h, the relations (3.3) follow from the uniqueness of the representation (2.3), which will be shown immediately following (3.14).

Proof of Theorem 1. The proof of the first part of Theorem 1 now follows from lemma II and formula (3.6) with $E_1 = E$.

On using (2.1) and (2.3) and comparing coefficients we conclude that the constants B_k, C_k satisfy the linear system of equations:

$$(3.12a)\qquad B_k = P_k(r), \quad k = 0,1,\ldots,s-r-1$$

$$(3.12b)\qquad B_k + \sum_{j=0}^{n-1} \frac{r^{k-s+r-j}}{(k-s+r-j)!} C_j = P_k(r), \quad k = s-r,\ldots,m-1$$

$$(3.12c)\qquad \sum_{j=0}^{n-1} \frac{r^{k-s+r-j}}{(k-s+r-j)!} C_j = P_k(r), \quad k = m, m+1,\ldots,m+n-1$$

Setting again for convenience

$$c_k = r^{r-s-k}(s-r+k)!\, C_k, \quad k = 0,1,\ldots,n-1$$

and using the convention that $\binom{k}{\nu} = 0$ if $\nu < 0$ or $\nu > k$, and the notation of (2.2a), we have from (3.12c):

$$(3.13)\qquad \sum_{j=0}^{n-1} \binom{m+k}{j+s-r} c_j = p_{m+k}, \qquad k = 0,1,\ldots,n-1,$$

which after repeated application of the known identity $\binom{k}{\nu} + \binom{k}{\nu-1} = \binom{k+1}{\nu}$ yields:

$$(3.14) \qquad \sum_{j=0}^{n-1} \binom{m}{j+s-r-k} c_j = \Delta^k p_m, \qquad k = 0,1,\ldots,n-1 .$$

From (3.12a) and (3.12b) it is clear that the B_k's are uniquely determined if the C_k's are known and this in turn depends upon solving the system (3.12). That the determinant of the system (3.13) does not vanish, follows from Lemma I on identifying p_k with $m+k$ and q_j with $s-r+j$. In case (i) $s=r$ or (ii) $s=r+m$, we are able to give an explicit calculable formula for the C_k's. For if $s=r$, on using the identity

$$\sum_{j=0}^{\nu} \binom{m}{j} (-1)^j \binom{m+\nu-j-1}{\nu-j} = \delta_{\nu,0}$$

we obtain (2.5a) from (3.14). When $s=m+r$, we have (2.5e) similarly from (3.14). In order to prove (2.5c) and (2.5f) we set $f(t) = (t-a)^{m+n+r}$ in (2.5). Then it follows from Kloosterman's formula [2] (formula (17)) that the left side of (2.5) is equal to $(m+n+r)!h^{m+n} P_{m+n}(r)$. In view of this formula (2.5) then yields

$$(3.15) \qquad E_1 \cdot (m+n)!\, r^{-m-n} = p_{m+n} - \sum_{k=0}^{n-1} \binom{m+n}{s-r+k} c_k$$

Since

$$\binom{m+n}{j} = \sum_{\nu=0}^{n-1} (-1)^{n-\nu-1} \binom{n}{\nu}\binom{m+\nu}{j}, \qquad j = 0,1,\ldots,n-1$$

$$\binom{m+n}{m+j} = \sum_{\nu=0}^{n-1} (-1)^{n-\nu-1} \binom{m+n}{\nu+m}\binom{m+\nu}{m+j}, \qquad j = 0,1,\ldots,n-1$$

we see easily that (2.5c) and (2.5f) follow from (3.15) and (3.13) for the cases $s = r$ and $s = r + m$ respectively.

This completes the proof of Theorem 1.

An estimate for the constants E_1 in (2.5c) and (2.5f) is of interest. It is known ([2], formula (5)) that

$$(3.16)\qquad P_k(r) = \frac{1}{k!}\int_0^1 \cdots \int_0^1 (t_1 + \cdots + t_r)^k \, dt_1 \cdots dt_r .$$

Thus from (2.5c), if $s = r$

$$E_1 = \frac{r^{m+n}}{(m+n)!}\int_0^1 \cdots \int_0^1 T^m (T-1)^n \, dt_1 \cdots dt_r$$

where $T = r^{-1}\sum_1^r t_j$. Hence

$$|E_1| \le \frac{r^{m+n}}{(m+n)!}\left(\frac{m}{m+n}\right)^m\left(\frac{n}{m+n}\right)^n \le \left(\frac{r}{2}\right)^{m+n}\frac{1}{(m+n)!} .$$

If $s = r + m$, (2.5f) and (3.16) give

$$(3.17)\qquad (-1)^n E_1 = \frac{r^{m+n}}{(m+n)!}\int_0^1 \cdots \int_0^1 \sum_{k=0}^n \binom{m+k-1}{k}(1-T)^{n-k}\, T^m \; dt_1 \cdots dt_r$$

where we have used the identity

$$\sum_{k=0}^n \binom{m+n}{m+k}(-1)^k T^k = \sum_{k=0}^n \binom{m+k-1}{k}(1-T)^{n-k} .$$

To obtain an upper estimate for $|E_1|$ of (3.17) we extend the integration on the right side of (3.17) to the simplex $0 \le t_1 + \cdots + t_r \le r$, $t_i \ge 0$, $i = 1, \ldots, r$. Then using Dirichlet's integral we get

$$|E_1| \le \frac{r^{m+n+r}}{(m+n)!\,(r-1)!} \int_0^1 \sum_{k=0}^{n} \binom{m+k-1}{k} t^{m+r-1} (1-t)^{n-k}\, dt$$

$$\le \frac{r^{m+n+r}}{(m+n)!\,(r-1)!} \binom{m+n-1}{n} \int_0^1 t^{m+r-1} \sum_{k=0}^{n} (1-t)^{n-k}\, dt$$

$$\le \frac{r^{m+n+r}}{(r-1)!} \cdot \frac{1}{m!\,n!\,(m+n)} .$$

For $r=1$, (3.17) can be evaluated directly and yields

$$(-1)^n E_1 = \frac{1}{m!\,n!\,(m+n+1)}$$

which agrees with our earlier formula (11) in [3].

4. Proof of Theorem 2. Using (2.2) and equating coefficients on both sides of (2.4) we obtain that the D_k's satisfy the system of equations

$$\sum_{k=0}^{m-1} D_k \cdot \frac{(-1)^k \lambda^{2j-k}}{(2j-k)!} = \frac{1}{2} A_j(r), \quad j = 0,1,\dots,m-1 .$$

Equivalently, setting $(-1)^k k!\, \lambda^{-k} D_k = d_k$, we have

$$\sum_{k=0}^{m-1} \binom{2j}{k} d_k = \frac{1}{2} a_j, \qquad j = 0,1,\dots,m-1 . \tag{4.1}$$

Multiplying both sides by $(-1)^{\nu-j}\binom{\nu}{j}$ and summing we get

$$\sum_{k=0}^{m-1} d_k \sum_{j=0}^{\nu} (-1)^{\nu-j} \binom{\nu}{j}\binom{2j}{k} = \frac{1}{2} \Delta^{\nu} a_0, \quad \nu = 0,1,\dots,m-1 .$$

However the inner sum on the left side is zero if $\nu > k$, so that the d_k's are the solutions of the upper triangular system of equations

$$(4.2) \qquad \sum_{k=\nu}^{m-1} H_{\nu k}\, d_k = \frac{1}{2}\, \Delta^{\nu} a_0, \qquad \nu = 0,1,\ldots,m-1$$

with $H_{\nu k} = \sum_{j=0}^{\nu} (-1)^{\nu-j} \binom{\nu}{j}\binom{2j}{k}$. Since $H_{\nu\nu} = 2^{\nu}$, $\nu = 0,\ldots,$ $m-1$, the system (4.2) has a unique solution.

The existence of a formula of the form (2.6) would follow on considering an interpolation problem analogously to the method of Lemma II. Hence we omit the details.

To prove (2.7) we set $f(t) = (t-a)^{2m+r}$ in (2.6) and obtain on using ([4], formula (3.1)) after some simplification

$$(4.3) \qquad E_2 = A_m(r) - \frac{2\lambda^{2m}}{(2m)!} \sum_{k=0}^{m-1} \binom{2m}{k} d_k .$$

Since it is easy to see that

$$\sum_{j=0}^{m-1} (-1)^{m-j} \binom{m}{j}\binom{2j}{\nu} = -\binom{2m}{\nu}, \qquad \nu = 0,1,\ldots,m-1$$

we deduce from (4.1) and (4.3)

$$E_2 = \frac{\lambda^{2m}}{(2m)!}\left(a_m + \sum_{j=0}^{m-1} (-1)^{m-j}\binom{m}{j} a_j\right)$$

where a_j's are given by (2.2a). This concludes the proof of Theorem 2.

5. <u>An Optimum Property</u>. If $n \leq m$ and $f \in C^{m+r}[a, a+rh]$, consider the family of linear operators

$$(5.1) \quad L(f) \equiv h^{-r}\Delta_h^r f(a) - \sum_{k=0}^{m-1} B_k^* h^k f^{(k+r)}(a) - \sum_{k=0}^{n-1} C_k^* h^k f^{(k+r)}(a\text{-}$$

depending on the $m+n$ parameters $\{B_k^*\}$ and $\{C_k^*\}$ and such that $L(f) = 0$ if f is a polynomial of degree $\leq m+r-1$. Then by a theorem of Peano [6] for any $f \in C^{m+r}$,

$$L[f] = \frac{1}{(m+r-1)!} \int_a^{a+rh} f^{(m+r)}(t)\, M(t)\, dt$$

where for $a < t < a + rh$,

$$\begin{aligned} M(t) &= L_x[(x-t)_+^{m+r-1}] \\ &= h^{-r} \Delta_h^r (a-t)_+^{m+r-1} \\ &\quad - (m+r-1)! \sum_{k=0}^{n-1} C_k^* h^k \frac{(a+rh-t)^{m-k-1}}{(m-k-1)!} . \end{aligned}$$

The best formula of the form (5.1) in the sense of Sard, in other words the formula for which $\int_a^{a+rh} M^2(t)\,dt$ is a minimum, is obtained if and only if

$$\int_a^{a+rh} M(t) \frac{\partial M(t)}{\partial C_j^*}\, dt = 0, \qquad j = 0, 1, \ldots, n-1 . \tag{5.2}$$

It is not hard to see that the system (5.2) is equivalent to:

$$\sum_{k=0}^{n-1} \frac{C_k^*}{(m-k-1)!} \frac{r^{2m-k-j-1}}{2m-k-j-1} = \sum_{\nu=0}^{m-j-1} (-1)^\nu P_{m+\nu}(r) \frac{r^{m-j-\nu-1}(m-j-1)!}{(m-j-\nu-1)!}, \tag{5.3}$$

$$(j = 0, 1, \ldots, n-1)$$

where the right hand is obtained on evaluating

$$\int_a^{a+rh} (a+rh-t)^{m-j-1} \cdot \Delta_h^r (a-t)_+^{m+r-1}\, dt$$

by using the last formula of §3 in [2] and formula (17) of [2].

Setting

$$c_k^* = k!\,C_k^*\, r^{-k}, \quad p_{m+\nu} = (m+\nu)!\, r^{-m-\nu} P_{m+\nu}(r)$$

and changing variable, (5.3) becomes

$$(-1)^j \frac{(m+j)!}{(m-1)!\, j!} \sum_{k=0}^{n-1} \binom{m-1}{k} \frac{c_k^*}{m+j-k} = \sum_{\nu=0}^{j} (-1)^{j-\nu} \binom{m+j}{m+\nu} p_{m+\nu} ,$$

for $j = 0,1,\ldots,n-1$. Multiplying both sides by $\binom{m+\ell}{m+j}$ and summing on $0 \le j \le \ell$, we have

$$\sum_{k=0}^{n-1} \binom{m+\ell}{k} c_k^* = p_{m+\ell}, \qquad \ell = 0,1,\ldots,n-1 \tag{5.4}$$

since

$$\sum_{j=\nu}^{\ell} (-1)^{j-\nu} \binom{m+j}{m+\nu} \binom{m+\ell}{m+j} = \delta_{\ell\nu} .$$

The system (5.4) is identical with (3.13) if $s = r$, so that $c_k^* = c_k$ whence $C_k^* = C_k$ $(k = 0,1,\ldots,n-1)$.

Also from the assumptions that (5.1) is zero for polynomials of degree $\le m+r-1$, it follows that

$$B_k^* = P_k(r) - \sum_{j=0}^{n-1} \frac{r^{k-j}}{(k-j)!} C_j^*, \quad k = 0,1,\ldots,m-1 ,$$

from which on using (3.12b) we get $B_k^* = B_k$ $(k = 0,1,\ldots,m-1)$. We have thus proved

Theorem 3. Among all linear functionals $L(f)$ of form (5.1) which vanish for polynomials of degree $\leq m+r-1$, the best in the sense of Sard is obtained when

$$C_k^* = \frac{r^k}{k!} \sum_{\nu=k}^{n-1} \binom{-m}{\nu-k} \Delta^\nu p_m, \quad k = 0,1,\ldots, n-1$$

$$B_k^* = (-1)^k \frac{r^k}{k!} \sum_{\nu=k}^{m-1} \binom{-n}{\nu-k} \Delta^\nu p_n, \quad k = 0,1,\ldots,m-1 .$$

6. Formula (2.5) reduces to our earlier result [3] for $r = 1$ and (2.6) yields a result included in [4] (formula (3.1)) if $\lambda = 0$. If $\lambda = r$, then (2.6) is essentially the same as (2.5) with $r = s$ and $m = n$. In this case according to Theorem 3, the formula (2.6) is best in the sense of Sard. We believe that the same assertion is valid even if $\lambda \neq r$.

The method used in proving Lemma 2 and Birkhoff's Theorem [1] could be used to obtain various expansion formulae similar to (2.5) or (2.6). The usefulness of any such formula would however depend on the possibility of explicit evaluation of the coefficients and the remainders involved.

REFERENCES

1. Birkhoff, G. D., General mean value and remainder theorems.... Trans. Am. Math. Soc., 7 (1906), 107-136.

2. Kloosterman, H. D., Derivatives and finite differences. Duke Math. Jour. 17(1950), 169-186.

3. Meir, A. and Sharma, A., An extension of Obreshkov's formula. SIAM Journal Numerical Analysis 5(1968), 488-490.

4. Meir, A. and Sharma, A., Symmetric differences and derivatives. Indagationes Math. 30(1968), 353-360.

5. Obreshkov, N., Neue Quadraturformeln. Abh. Preuss. Akad. Wiss. Math. Nat. K., 5(1940), 1-20.

6. Sard, A., Linear approximation. Math. Surveys, No. 9, Am. Math. Soc., 1963.

7. Schoenberg, I. J., On monosplines of least deviation and quadrature formulae. SIAM Journal, Numerical Analysis 2(1965), 144-170.

8. Whittaker, J. M., Interpolatory function theory. Stechert-Hafner, New York, 1964.

One-Sided L_1-Approximation by Splines of an Arbitrary Degree

ZVI ZIEGLER

Introduction

During the last decade, spline functions have steadily advanced to the front position as an approximation tool. Their qualifications as interpolators, creators of best quadrature formulas and good approximators in the uniform norm have been studied by many authors. For an extensive bibliography of studies published through 1966 see [6].

We propose to investigate the following problem: Let $0 = x_0 < \ldots < x_n = 1$ be a partition of the unit interval with norm $\Delta = \max(x_i - x_{i-1})$. Let f be a function of $C^{(\nu-1)}[0,1]$ such that $f^{(\nu-1)}$ is the integral of a function of bounded variation f_ν. Two ν-th degree spline functions, $a_\Delta(t)$ and $A_\Delta(t)$ whose only knots in $[0,1]$ are the x_i's, are to be constructed in such a way that

$$a_\Delta(t) \le f(t) \le A_\Delta(t) \tag{1}$$

and that the "goodness of fit" to f defined by

$$\int_0^1 (A_\Delta(t) - a_\Delta(t))\,dt \tag{2}$$

have as good an order of vanishing as possible with ν-th degree splines knotted at points of a partition with norm Δ.

A step towards the solution of this problem was taken in a recent paper by A. Meir and A. Sharma [5]. For $\nu = 1$ and for $\nu = 3$ with equidistant knots they constructed splines satisfying (1) for which (2) was of the order $O(\Delta^2)$ and $O(\Delta^4)$ respectively. The proof was computational in character and therefore could not be extended to higher order splines.

We shall show that $O(\Delta^{\nu+1})$ is the best order of vanishing (2) can have, and then construct ν-th degree splines achieving this order. We will indicate how a similar construction can be carried out for arbitrary Tchebycheffian splines without, however, discussing the order of vanishing of the "goodness of fit". This depends heavily on intrinsic properties of the Tchebycheff system, and will be explored elsewhere.

We further note that a full trigonometric analogue is available, as well as extensions of the constructions to achieve the best order of "goodness of fit" for other classes of functions. This will be discussed in a future publication by the author [7].

I with to express my thanks to Professor Schoenberg for very stimulating suggestions which led to a considerable improvement of this paper.

The Approximation Theorems: We first prove a theorem about a bound for the best possible order of approximation by ν-th degree spline functions to the class under consideration on an interval I (I can be finite or infinite). Let $Y = \{y_i\}_{i=-\infty}^{\infty}$ be a sequence of points and assume that the norm of the sequence, defined as $\Delta = \sup(y_{i+1} - y_i)$, is finite.

We introduce the notation:

$A_\nu(I)$ - the class of functions f which are $\nu-1$ times differentiable on I and such that $f^{\nu-1}(t)$ is the integral of a function of bounded variation f_ν.

Theorem 1: Let S_Y^ν denote the class of ν-th degree splines knotted at the sequence Y. Then we have

$$(3)\quad \sup_{f \in A_\nu} \|f - S_Y^\nu\|_p = \sup_{f \in A_\nu} \inf_{s \in S_Y^\nu} \|f - s\|_p \geq C_{\nu p}\, \Delta^{\nu + \frac{1}{p}}$$

where $\| \ \|_p$ is the L_p-norm, $1 \le p \le \infty$, $C_{\nu p}$ is a constant depending on ν and p only, and $1/p$ for $p = \infty$ is taken to mean 0.

Proof: Consider the interval $[-1/2, 1/2]$ and let π_ν be the linear space of polynomials of degree ν. Since x_+^ν (defined, as usual, as $x^{\nu-1} \max(x,0)$) does not belong to π_ν, we have, for $1 \le p < \infty$,

$$(4)\quad 0 < C_{\nu p} = \|x_+^\nu - \pi_\nu\|_p \text{ (on } [-1/2, 1/2]) = \min_{q \in \pi_\nu} \|x_+^\nu - q(x)\|_p \quad \text{(on } [-1/2, 1/2]) .$$

Note next that by a simple change of variables, we obtain, for each $q \in \pi_\nu$,

$$\left[\int_{-d/2}^{d/2} |x_+^\nu - q(x)|^p dx\right]^{1/p} = \left[d^{\nu p+1} \int_{-1/2}^{1/2} |x_+^\nu - \tilde{q}(x)|^p dx\right]^{1/p}$$

where $\tilde{q}(x)$ is also a polynomial of π_ν. Hence

$$(5)\qquad \|x_+^\nu - \pi_\nu\|_p \text{ (on } [-d/2, d/2]) = C_{\nu p} d^{\nu + 1/p} .$$

Let now $s(x)$ be any ν-th degree spline knotted at the y_i's. Choose a subsequence of intervals (y_j, y_{j+1}) such that $(y_{j+1} - y_j) = \Delta_j \to \Delta$ and denote the midpoint of (y_j, y_{j+1}) by z_j. Observing that on (y_j, y_{j+1}) $s(x)$ coincides with a ν-th degree polynomial, we find, using (5), that

$$\|(x - z_j)_+^\nu - s(x)\|_p \text{ (on I)} \ge \|(x - z_j)_+^\nu - s(x)\|_p \text{ (on } (y_j, y_{j+1})) \ge$$
$$\ge C_{\nu p} \Delta_j^{\nu+1/p}, \qquad j = 1, 2, \ldots .$$

Noting that this inequality is true for all $s \in S_Y^\nu$, we deduce

$$\| (x - z_j)_+^{\nu} - S_Y^{\nu} \|_p \geq C_{\nu p} \Delta_j^{\nu + 1/p}, \qquad j = 1, 2, \ldots .$$

Since $(x - z_j)_+^{\nu}$, for all j, belongs to $A_{\nu}(I)$, we obtain the inequality

$$\sup_{f \in A_{\nu}} \| f - S_Y^{\nu} \|_p \geq C_{\nu p} \Delta_j^{\nu + 1/p}, \qquad j = 1, 2, \ldots$$

from which the assertion (3), for $1 \leq p < \infty$, immediately follows. A similar argument works for $p = \infty$.

We have thus shown that for the class A_{ν}, the order of approximation in the L_1-norm by ν-th degree splines cannot be better than $O(\Delta^{\nu+1})$ even when the knots are permitted to coalesce. Moreover, this order is "best" even for the small class of infinitely differentiable functions, in the following precise sense:

Theorem 2: There exists a sequence Y of norm Δ such that

$$\sup_{f \in C^{\infty}(I)} \| f - S_Y^{\nu} \|_p \geq b_{\nu p} \Delta^{\nu+1}, \qquad 1 \leq p \leq \infty$$

where $b_{\nu p}$ is a constant depending on ν and p only.

For the proof of this theorem see [7].

We now show that this best order can be attained with an arbitrarily preassigned set of simple knots (i.e., with the requirement of maximal possible smoothness) and one-sided approximation. Furthermore, the approximating splines (from above and from below) are easy to construct and the operators mapping f onto the approximating splines are linear.

Theorem 3: Let f be a function of $A_{\nu}([0,1])$ and let $0 = x_0 < x_1 < \ldots < x_n = 1$ be a partition of the unit interval with norm Δ. Then there exist ν-th degree splines $a_{\Delta}(t)$ and $A_{\Delta}(t)$ whose only knots in $[0,1]$ are amongst the x_i's satisfying

$$(1) \quad a_\Delta(t) \le f(t) \le A_\Delta(t)$$

and such that

$$(6) \qquad \int_0^1 [A_\Delta(t) - a_\Delta(t)]\,dt \le 2V_\nu \Delta^{\nu+1}$$

where V_ν is the total variation of f_ν .

Proof: We start by proving a lemma from which the theorem will easily follow.

Lemma. For any ξ, $0 \le \xi \le 1$, there exist ν-th degree splines $\gamma_\Delta(t,\xi)$ and $\Gamma_\Delta(t,\xi)$ whose only knots in $[0,1]$ are amongst the x_i's such that

$$(7) \qquad \gamma_\Delta(t,\xi) \le (t-\xi)_+^\nu \le \Gamma_\Delta(t,\xi), \quad -\infty < t < \infty$$

and

$$(8) \qquad \int_0^1 [\Gamma_\Delta(t,\xi) - \gamma_\Delta(t,\xi)]\,dt \le 2 \cdot \nu!\, \Delta^{\nu+1} \quad .$$

Proof of the lemma: If ξ is one of the x_i's then

$$\Gamma_\Delta(t,\xi) = \gamma_\Delta(t,\xi) = (t-\xi)_+^\nu$$

and there is nothing to prove. Assume thus that r is the integer such that

$$(9) \qquad x_r < \xi < x_{r+1} \, .$$

We now extend the system of knots periodically outside $[0,1]$ by defining

$$(10) \qquad x_j = i + x_k, \quad \text{when} \quad j = in + k, \quad 0 \le k \le n \, .$$

For the construction we need Schoenberg's B-splines, and we recall briefly some of their properties (see [1] for further details). The B-spline $M(t; s_1, \ldots, s_{\nu+2})$ is defined as the $\nu+1$-st divided difference of $(\nu+1)(s-t)_+^\nu$ on the points $s_1 < s_2 < \ldots < s_{\nu+2}$. It is known that $M(t; s_1, \ldots, s_{\nu+2})$ is a ν-th degree spline which has simple knots at $s_1, \ldots, s_{\nu+2}$ and those are its only knots. It is positive in $(s_1, s_{\nu+2})$ and vanishes outside this interval. Furthermore, we have

$$(11) \quad M(t; s_1, \ldots, s_{\nu+2}) = (-1)^{\nu+1}(\nu+1)\sum_{j=1}^{\nu+2} \frac{(t-s_j)_+^\nu}{\omega'(s_j)}, \quad \text{where}$$

$$\omega(t) = \prod_{i=1}^{\nu+2}(t-s_i)$$

and

$$(12) \qquad \int_{-\infty}^{\infty} M(t; s_1, \ldots, s_{\nu+2}) = \int_{s_1}^{s_{\nu+2}} M(t; s_1, \ldots, s_{\nu+2})\,dt = 1 .$$

With these properties at hand, we define

$$\Gamma_\Delta(t;\xi) = (t-\xi)_+^\nu + \alpha M(t; x_r, \xi_\nu x_{r+1}, \ldots, x_{r+\nu})$$

where α is so chosen that ξ is not a knot of $\Gamma_\Delta(t;\xi)$. From (11) we read off the value of α obtaining

$$\alpha = \frac{(\xi-x_r)(x_{r+1}-\xi)\ldots(x_{r+\nu}-\xi)}{\nu+1} > 0 .$$

Similarly, we define

$$\gamma_\Delta(t;\xi) = (t-\xi)_+^\nu - \beta M(t; \xi, x_{r+1}, x_{r+2}, \ldots, x_{r+\nu+1})$$

where β is so chosen that ξ is not a knot of $\gamma_\Delta(t;\xi)$, i.e.

$$\beta = \frac{(x_{r+1}-\xi)\ \ldots\ (x_{r+\nu+1}-\xi)}{\nu+1} > 0\ .$$

In view of the non-negativity of the B-splines and the positivity of α and β we have (7). Furthermore, $\gamma(t;\xi)$ and $\Gamma(t;\xi)$ are ν-th degree splines whose only knots are the x_i's and these are simple knots. We further observe, in view of (12), that

$$\int_0^1 [\Gamma_\Delta(t;\xi) - \gamma_\Delta(t;\xi)]\,dt \le \int_{-\infty}^{\infty} [\Gamma_\Delta(t;\xi) - \gamma_\Delta(t;\xi)]\,dt =$$

$$= \alpha + \beta \le 2 \cdot \nu!\ \Delta^{\nu+1}\ ,$$

completing the proof of the Lemma.

Proof of the Theorem: By MacLaurin's formula, we have

$$(13) \qquad f(t) = \sum_{k=0}^{\nu-1} \frac{f^{(k)}(0)\,t^k}{k!} + \frac{t^\nu}{\nu!} f_\nu(0) + \int_0^1 \frac{(t-\xi)_+^\nu}{\nu!}\, df_\nu(\xi)\ .$$

The measure df_ν can be decomposed into

$$df_\nu = df_\nu^+ - df_\nu^-,\quad |df_\nu| = df_\nu^+ + df_\nu^-$$

where df_ν^+, df_ν^- are nonnegative. Denote the polynomial lying outside the integral sign by $P(t)$ and define

$$A_\Delta(t) = P(t) + \frac{1}{\nu!}\Big[\int_0^1 \Gamma_\Delta(t,\xi)\,df_\nu^+(\xi) - \int_0^1 \gamma_\Delta(t,\xi)\,df_\nu^-(\xi)\Big]$$

$$a_\Delta(t) = P(t) + \frac{1}{\nu!}\Big[\int_0^1 \gamma_\Delta(t,\xi)\,df_\nu^+(\xi) - \int_0^1 \Gamma_\Delta(t,\xi)\,df_\nu^-(\xi)\Big]\ .$$

Then, $A_\Delta(t)$ and $a_\Delta(t)$ are obviously ν-th degree splines with simple knots amongst the x_i's . Furthermore, it can be immediately ascertained that (1) is satisfied. Finally, we have

$$\int_0^1 [A_\Delta(t) - a_\Delta(t)]\,dt = \frac{1}{\nu!}\int_0^1 \{ \int_0^1 [\Gamma_\Delta(t,\xi) - \gamma_\Delta(t,\xi)]\,|df_\nu(\xi)| \}\,dt$$

$$= \frac{1}{\nu!}\int_0^1 \{ \int_0^1 [\Gamma_\Delta(t,\xi) - \gamma_\Delta(t,\xi)]\,dt \}\,|df_\nu(\xi)|$$

$$\leq 2V_\nu \cdot \Delta^{\nu+1} .$$

q.e.d.

Remark 1: The existence of ν-th "degree" Tchebycheffian splines (i.e., a piecewise solution of a $\nu+1$-st order differential equation) with knots amongst the x_i's satisfying (1) is established in an analogous way. We need then the generalized B-splines (see [3] and [4]). However, as we have already mentioned, an analogue of (6) cannot be established without imposing some restrictions on the underlying Tchebycheff system.

Remark 2: It is worth noting that for the class of functions under consideration, G. Freud showed in [2] that there exist two n-th degree polynomials $p_n(t)$ and $P_n(t)$ satisfying

$$p_n(t) \leq f(t) \leq P_n(t) \qquad 0 \leq t \leq 1$$

and

$$\int_0^1 \frac{[P_n(t) - p_n(t)]}{\sqrt{t(1-t)}}\,dt \leq \frac{BV_\nu}{n^{\nu+1}} ,$$

where B is a constant. Thus, for one-sided approximation in the L_1-norm there is a striking similarity in the "goodness of fit" between approximation by polynomials of degree n and by ν-th degree splines knotted at the $n+1$ equidistant points $\frac{i}{n}$, $0 \leq i \leq n$.

Remark 3: Using Marsden's identity [4] and proceeding as in the proof of Theorem 2, we can deduce that

$$\|A_\Delta(t) - a_\Delta(t)\|_\infty \leq \frac{2}{\nu+1} V_\nu \Delta^\nu .$$

Theorem 1 implies that for the class of functions under consideration, this is the best order of approximation in the L_∞(= sup) norm. Investigation of other classes will be discussed in [7].

REFERENCES

1. Curry, H. B. and Schoenberg, I. J., On Pólya frequency functions iv: The fundamental spline functions and their limits, J. d'Analyse Math. 17(1966), 71-107.

2. Freud, G., Über einseitige Approximation durch Polynome, Acta. Sci. Math. (Szeged) 16(1055), 12-28.

3. Karlin, S. and Karon, J. M., A variation-diminishing generalized spline approximation method, J. Approx. Th. 1(1968), 255-268.

4. Marsden, M., An identity for spline functions with applications to variation diminishing spline approximation, MRC Tech. Rep. 897, University of Wisconsin, June 1968.

5. Meir, A. and Sharma, A., One sided spline approximation, Studia Sci. Math. Hung. 3(1968), 211-218.

6. Sard, A., Optimal Approximation, J. Funct. Anal. 1(1967), 222-244.

7. Ziegler, Z., On the best order of L_p-approximation by by splines, to appear.

Sponsored by the Mathematics Research Center, United States Army, Madison, Wisconsin under Contract No. DA-31-124-ARO-D-462

Construction of Spline Functions in a Convex Set

P. J. LAURENT

1. Introduction

Let $H^q = H^q[a,b]$ be the Hilbert space of real functions with square integrable q^{th} derivative $(q \geq 1)$. Let us consider n abscissae x_i $(a < x_1 < \ldots < x_n < b)$ and n real numbers $r_1, \ldots, r_n$.

If $n \geq q$, there exists a unique function $\sigma \in H^q$ (interpolating spline function) satisfying $\sigma(x_i) = r_i$, $i = 1, \ldots, n$ and minimizing:

$$\int_a^b (\sigma^{(q)}(x))^2 \, dx, \quad \text{(see I. J. Schoenberg [21]).}$$

Under the same conditions, there exists a unique function $s \in H^q$ (smoothing spline function) minimizing:

$$\int_a^b (s^{(q)}(x))^2 \, dx + \rho \sum_{i=1}^{n} (s(x_i) - r_i)^2, \text{ with } \rho > 0; \text{ (see I. J. Schoenberg [20]).}$$

The functions σ and s are composed of pieces of polynomials of degree $2q-1$. The derivatives of order $2q-1$ of σ and s have in general a discontinuity at each x_i.

The major disadvantage in the practical use of these spline functions σ and s is to introduce a number of discontinuities of derivatives that is systematically equal to the number of abscissae, and this independently from the disposition and the accuracy of the numbers r_i.

In order to correct this default, one can proceed in the following way:

Let us call C the convex set composed of the functions $f \in H^q$ satisfying:

$$\alpha_i \leq f(x_i) \leq \beta_i, \quad i = 1, \ldots, n,$$

where $\alpha_i < \beta_i$ are given real numbers (if the numbers r_i comprise errors $\pm \varepsilon_i$ (experimental or otherwise), we take $\alpha_i = r_i - \varepsilon_i$ and $\beta_i = r_i + \varepsilon_i$).

If $n > q$ and if no polynomial p of degree $q-1$ exists satisfying $\alpha_i \leq p(x_i) \leq \beta_i$, $i = 1, \ldots, n$, then there exists a unique function $\sigma \in C$ which minimizes $\int_a^b (\sigma^{(q)}(x))^2 dx$.

This function is characterized by the following conditions:

(1-a) σ is a polynomial of degree $2q-1$ in each interval $[a, x_1[$, $]x_i, x_{i+1}[$, $i = 1, \ldots, n-1$, $]x_n, b]$.

(1-b) $\sigma^{(j)}(x_i^-) = \sigma^{(j)}(x_i^+)$, $j = 0, 1, \ldots, 2q-2$; $i = 1, \ldots, n$.

(1-c) $\sigma^{(j)}(x_1^-) = \sigma^{(j)}(x_n^+) = 0$, $j = q, q+1, \ldots, 2q-1$.

(1-d) $\alpha_i \leq \sigma(x_i) \leq \beta_i$, $i = 1, \ldots, n$.

(1-e) Setting $\lambda_i = (-1)^q \cdot [\sigma^{(2q-1)}(x_i^+) - \sigma^{(2q-1)}(x_i^-)]$, we ha[ve]

$$\lambda_i \geq 0 \text{ if } \sigma(x_i) = \alpha_i$$
$$\lambda_i \leq 0 \text{ if } \sigma(x_i) = \beta_i$$
$$\lambda_i = 0 \text{ if } \alpha_i < \sigma(x_i) < \beta_i, \qquad i = 1, \ldots, n.$$

If we have $\alpha_i < \sigma(x_i) < \beta_i$, the constraint on σ at the abscis[sa] x_i is not active and σ is one and the same polynomial of degr[ee] $2q-1$ in the interval $]x_{i-1}, x_{i+1}[$.

Thus the number of discontinuities becomes directly related to the difficulty of the problem (configuration of the

$r_i = (\alpha_i + \beta_i)/2$ and accuracy $\varepsilon_i = (\beta_i - \alpha_i)/2$ of the r_i). On the other hand, the calculation of the spline function, solution of this problem, becomes more difficult.

The introduction of this type of spline function is due to M. Atteia [4]. The object of this study is, essentially, to give an efficient algorithm for the calculation of σ. It will be developed in a more general abstract framework which includes the preceding example.

2. Existence

Let X and Y be two real Hilbert spaces and T a continuous linear operator on X onto Y. The kernel (null space) $n(T)$ of T is supposed to be of dimension q. We define a convex subset C of X by the following inequalities:

(2) $$C = \{x \in X : \forall \ell \in L, \ (\ell | x) \leq t(\ell)\}$$

where L is an arbitrary subset of $X (0 \notin L)$ and t a mapping of L into R (real numbers). We suppose that $C \neq \emptyset$. We are looking for an element $\sigma \in C$, satisfying:

(3) $$\|T\sigma\| = \min_{x \in C} \|Tx\| .$$

Such an element σ will be called <u>spline function relative to T in the convex set</u> C. M. Atteia [3] considers various assumptions that imply the existence of σ.

We denote:

(4) $$\tilde{C} = \{x \in X : \forall \ell \in L, \ (\ell | x) \leq 0\}$$

and we make the following assumption:

(H1) $$\tilde{C} \cap n(T) = \{0\} .$$

Thus, we have the existence theorem:

THEOREM 1

Under the assumption (H1), there exists at least one spline function σ relative to T in the convex set C.

Proof:

We set $\Omega = T(C)$ and $\tau = T\sigma$. We must have:

$$\|\tau\| = \min_{y \in \Omega} \|y\| . \tag{5}$$

The subset Ω is convex. If, moreover, Ω is closed, there will exist a unique element $\tau \in \Omega$ satisfying (5): it is the projection of the origin onto Ω. Thus, there will exist $\sigma \in C$ such that $T\sigma = \tau$.

We now prove that Ω is closed: $\Omega = T(C)$ is closed if and only if $\mathfrak{n}(T) + C = T^{-1}(\Omega)$ is closed in X; indeed, the restriction $\tilde{T}$ of T to the subspace $\mathfrak{n}(T)^{\perp}$ is a homeomorphism (cf [8], p. 57) and $\Omega = \tilde{T}((\mathfrak{n}(T) + C) \cap \mathfrak{n}(T)^{\perp})$. Then, we use a theorem due to J. Dieudonné [6]†:

If A and B are two closed non-empty convex subsets of a Hausdorff topological linear space, such that A is locally compact and $A_\infty \cap B_\infty = \{0\}$ (with $A_\infty = \bigcap_{\lambda > 0} \lambda(A - a)$ where $a \in A$) then $A - B$ is closed.

In the present case, $\mathfrak{n}(T)$ is locally compact (as of finite dimension) and we easily prove that:

$$\mathfrak{n}(T)_\infty = \mathfrak{n}(T) \text{ and } C_\infty = \tilde{C} . \tag{6}$$

Thus the assumption (H1) implies that $C + \mathfrak{n}(T)$ is closed.

3. Characterization

We add the following assumptions:

(H2) The subset $L \subset X$ is weakly compact and t is a continuous mapping of L (with the weak topology) into R.

† The idea of using this theorem here, is due to J. L. Joly (Grenoble).

(We also suppose, without loss of generality, that for each $\ell \in L$, we have $D \cap L \subset \{\ell, -\ell\}$ where D denotes the line $\{\lambda \cdot \ell\}$.

(H3) $C \cap \mathfrak{n}(T) = \emptyset$.

(H4) $I = \{x \in X : \forall \ell \in L, \ (\ell | x) < t(\ell)\} \neq \emptyset$.

Remark: The assumptions (H2) and (H4) imply that the interior of C is I. We have the following characterization theorem:

THEOREM 2

Under the assumptions H1, H2, H3, H4, the element $\sigma \in C$ is a spline function (relative to T in C) if and only if:

$$-T^{*} T\sigma \in \overline{CC}(F_{\sigma}) , \tag{7}$$

where

$$F_{\sigma} = \{\ell \in L : (\ell | \sigma) = t(\ell)\} . \tag{8}$$

($\overline{CC}(F_{\sigma})$ denotes the smallest closed convex cone with vertex 0 containing F_{σ} and T^{*}, the adjoint operator of T).

Proof:

Let us consider the function $f(x) = f_1(x) + f_2(x)$ with:

$$f_1(x) = \|Tx\| \quad \text{and} \quad f_2(x) = \begin{cases} 0 & \text{if} \quad x \in C \\ \infty & \text{if} \quad x \notin C . \end{cases}$$

The problem comes to characterize σ such that:

$$f(\sigma) = \min_{x \in X} f(x) . \tag{9}$$

The element σ satisfies (9) if and only if

$$0 \in \partial f(\sigma) , \tag{10}$$

where $\partial f(\sigma) \subset X$ denotes the set of subgradients of f at σ.

As f_1 is convex and continuous and f_2 convex and lower semi-continuous, we have (cf J. J. Moreau [18]):

$$(11) \qquad \partial f(\sigma) = \partial f_1(\sigma) + \partial f_2(\sigma) .$$

By the assumption (H3), we have $\|T\sigma\| \neq 0$, and the set of subgradients of f_1 at σ is reduced to one element:

$$(12) \qquad \partial f_1(\sigma) = \{\frac{T^* T\sigma}{\|T\sigma\|}\} .$$

For the determination of $\partial f_2(\sigma)$, we can use the following theorem due to M. Valadier [22]:
Let L be a compact and (g_ℓ), $\ell \in L$, a family of convex functionals on a Hausdorff topological linear space. Let $g = \sup_{\ell \in L} g_\ell$. If there exists a neighborhood $\mathcal{U}$ of x_0 such that the mapping $[\ell, x] \to g_\ell(x)$ is continuous on $L \times \mathcal{U}$, then

$$\partial g(x_0) = \overline{\text{CO}}\,(\bigcup_{\ell \in F_{x_0}} \partial g_\ell(x_0)) ,$$

where $F_{x_0} = \{\ell \in L : g_\ell(x_0) = g(x_0)\}$.
($\overline{\text{CO}}(A)$ denotes the smallest closed convex subset containing A).
If we set $g_\ell(x) = (\ell | x) - t(\ell)$, the mapping $[\ell, x] \to g_\ell(x)$ is continuous on $L \times X$ ($L \subset X$ with the weak topology and X with the strong topology). If $g(x) = \sup_{\ell \in L} g_\ell(x)$, g is continuous and we have:

$$(13) \qquad C = \{x \in X : g(x) \leq 0\} .$$

According to the theorem of Valadier, as $\partial g_\ell(\sigma) = \{\ell\}$ and $g(\sigma) = 0$, (by the assumption (H3)), we have:

$$(14) \qquad \partial g(\sigma) = \overline{\text{CO}}\,(F_\sigma) .$$

The assumption (H4) implies that g is not minimum in σ ($0 \notin \partial g(\sigma)$); thus, it can be shown that

(15) $$\partial f_2(\sigma) = CC(\partial g(\sigma)) = \overline{CC}(F_\sigma) .$$

The condition (10) is satisfied if and only if there exists $\ell_1 \in \partial f_1(\sigma)$ and $\ell_2 \in \partial f_2(\sigma)$ such that $\ell_1 + \ell_2 = 0$, that is finally if $-T^* T\sigma \in \overline{CC}(F_\sigma)$.

4. Unicity

We add the following assumption:

(H5) Whatever $\ell_1, \ldots, \ell_q \in L(\ell_i \neq \pm \ell_j$, if $i \neq j)$ we have: $(\ell_i | p) = 0$, $i = 1, \ldots, q$ for $p \in \mathfrak{n}(T)$ implies $p = 0$; or equivalently: $\det[(\ell_i | w_j)] \neq 0$ (where $(w_1, \ldots, w_q)$ denotes an arbitrary basis of $\mathfrak{n}(T)$) .

THEOREM 3

Under the assumptions H1, H2, H3, H4, H5, there exists a unique spline function relative to T in C .

Proof:

According to theorem 1, if we have two solutions σ_1 and $\sigma_2 \in C$, we shall have $T\sigma_1 = T\sigma_2 = \tau$, unique projection of 0 on $\Omega = T(C)$. Let us consider $\sigma = \frac{\sigma_1 + \sigma_2}{2}$. This is another solution, hence $T\sigma = \tau$. According to theorem 2, we thus shall have $-T^*T\sigma \in \overline{CC}(F_\sigma)$. But $F_\sigma \subset F_{\sigma_1}$ and $F_\sigma \subset F_{\sigma_2}$: Indeed, if $\ell \in F_\sigma$, we have $(\ell | \sigma) = t(\ell)$; as $(\ell | \sigma_1) \leq t(\ell)$ and $(\ell | \sigma_2) \leq t(\ell)$, $(\ell | \frac{\sigma_1+\sigma_2}{2}) = t(\ell)$ implies that $(\ell | \sigma_1) = (\ell | \sigma_2) = t(\ell)$. Thus, for all $\ell \in F_\sigma$, we have for $p = \sigma_1 - \sigma_2 \in \mathfrak{n}(T)$: $(\ell | p) = 0$. By the assumption (H4), F_σ cannot contain ℓ and $-\ell$ simultaneously. If F_σ contains an infinity of elements, by the assumption (H5), this implies $p = 0$, hence $\sigma_1 = \sigma_2$. If F_σ contains a finite number of elements $\ell_1, \ldots, \ell_m$, the characterization theorem yields $-T^*T\sigma = \sum_{i=1}^{m} \lambda_i \ell_i$ with $\lambda_i \geq 0$. The functional $\sum_{i=1}^{m} \lambda_i \ell_i$ is non-zero and belongs to Range $(T^*) = \mathfrak{n}(T)^\perp$.

The assumption (H5) implies that $m \geq q+1$. As $(\ell_i | p) = 0$, $i = 1, \ldots, m$, we have $p = 0$, hence $\sigma_1 = \sigma_2$.

Remarks

1°/ The assumption (H5) is in fact rather strong. Note that it does not serve for obtaining the unicity of $T\sigma = \tau$. Consequently, without this assumption, two solutions σ_1 and σ_2 are such that $\sigma_1 - \sigma_2 \in \mathfrak{n}(T)$.

2°/ The assumption (H4) is not essential. If we have, in the definition of C, interpolating conditions (that is, conditions of type equality: for instance $k \in L$, $-k \in L$ with $t(-k) = -t(k)$, which comes to write $(k|x) = t(k)$), the assumption (H4) is not any longer satisfied. These conditions can be considered separately without any difficulty. It is even easier to obtain the unicity of the spline function in this case:

Interpolating conditions:

Let us assume L to contain m elements $k_1, \ldots, k_m$ as well as $-k_1, \ldots, -k_m$ and that we have $t(k_i) = -t(-k_i) = r_i$, $i = 1, \ldots, m$. Let us call $L^* = L \sim \{k_1, \ldots, k_m, -k_1, \ldots, -k_m\}$ and let us assume L^* to be, moreover, a weakly compact subset of X. We denote by K the subspace spanned by $k_1, \ldots, k_m$. By writing:

$$K_r^{\perp} = \{x \in X : (k_i | x) = r_i, \quad i = 1, \ldots, m\}$$

which is a translate of the subspace:

$$K^{\perp} = \{x \in X : (k_i | x) = 0, \quad i = 1, \ldots, m\}$$

and

$$C^* = \{x \in X : \forall \ell \in L^*, \quad (\ell | x) \leq t(\ell)\},$$

we have:

$$C = C^* \cap K_r^{\perp}.$$

We denote:

$$I^* = \{x \in X : \forall \ell \in L^*, \ (\ell | x) < t(\ell)\} \ .$$

The assumptions (H4) and (H5) can now be replaced by the following two which are more easily satisfied:

(H4') $$I^* \cap K_r^{\perp} \neq \emptyset$$

(H5') $$\mathfrak{n}(T) \cap K^{\perp} = \{0\} \ .$$

Note that the assumption (H1) is now a consequence of (H5'). The characterization and unicity theorems become:

THEOREM 3':

Under the assumptions H2, H3, H4', H5', there exists a unique spline function $\sigma \in C$ relative to T in C. It is characterized by the condition:

$$-T^* T\sigma + \sum_{i=1}^{m} \alpha_i k_i \in \overline{CC}(F_\sigma) \ ,$$

in which

$$F_\sigma = \{\ell \in L^* : (\ell | \sigma) = t(\ell)\} \text{ and } \alpha_i \in R, \ i = 1, \ldots, m \ .$$

Proof

If σ_1 and σ_2 are two solutions, we have $\sigma_1 - \sigma_2 \in \mathfrak{n}(T)$ and $\sigma_1, \sigma_2 \in K_r^{\perp}$, hence $\sigma_1 - \sigma_2 \in K^{\perp}$. By (H5') we have $\sigma_1 - \sigma_2 = 0$. Let us consider $f(x) = f_1(x) + f_2(x) + f_3(x)$ with

$$f_1(x) = \|Tx\|, \quad f_2(x) = \begin{cases} 0 & \text{if } x \in C^* \\ \infty & \text{if } x \notin C^* \end{cases} \text{ and } f_3(x) = \begin{cases} 0 & \text{if } x \in K_r^{\perp} \\ \infty & \text{if } x \notin K_r^{\perp} \end{cases} .$$

The element σ is the solution iff $0 \in \partial f(\sigma)$.

By (H4'), there exists $x_0 \in X$ such that f_1 and f_2 are finite and continuous at x_0 and f_3 is finite at x_0 (take $x_0 \in I^* \cap K_r^{\perp}$). Thus we have (cf. J. J. Moreau [18]):

$$\partial f(\sigma) = \partial f_1(\sigma) + \partial f_2(\sigma) + \partial f_3(\sigma).$$

As in theorem 3, $\partial f_1(\sigma)$ and $\partial f_2(\sigma)$ are given by (12) and (15). We easily prove that $\partial f_3(\sigma) = K$. We have $0 \in \partial f(\sigma)$ iff there exists $\ell_2 \in K$ and $\ell_3 \in \overline{CC}(F_\sigma)$ such that: $\frac{T^* T\sigma}{\|T\sigma\|} + \ell_2 + \ell_3 = 0$, that is finally iff there exists $\alpha_i \in R$ such that:

$$T^* T\sigma + \sum_{i=1}^{m} \alpha_i k_i \in \overline{CC}(F_\sigma) .$$

5. Examples

Let L be composed of 2n elements:

$$L = \{\ell_1, \ell_2, \ldots, \ell_n, -\ell_1, -\ell_2, \ldots, -\ell_n\} .$$

We set $t(\ell_i) = \beta_i$ and $t(-\ell_i) = -\alpha_i$, $i = 1, \ldots, n$. Thus we have:

(16) $$C = \{x \in X : \alpha_i \leq (\ell_i | x) \leq \beta_i, \quad i = 1, \ldots, n\}.$$

The assumptions H_1, H_2, H_3, H_4, H_5 are supposed to be satisfied (this implies $\alpha_i < \beta_i$, $i = 1, \ldots, n$).
The characterization theorem becomes:

THEOREM 4:

The element $\sigma \in C$ is the spline function (relative to T in C) if and only if:

$$T^* T\sigma = \sum_{i=1}^{n} \lambda_i \ell_i$$

with $\lambda_i \geq 0$ if $(\ell_i|\sigma) = \alpha_i$

$\lambda_i \leq 0$ if $(\ell_i|\sigma) = \beta_i$

$\lambda_i = 0$ if $\alpha_i < (\ell_i|\sigma) < \beta_i$.

We shall consider exclusively examples based on functions of one variable.

$$X = H^q[a,b]\,;\ Y = H^0[a,b]\,;\ Tf = D^q f = f^{(q)} .$$

We take in H^q and H^0 (resp.) the following inner products:

$$(f|g)_q = \sum_{i=0}^{q} \int_a^b f^{(i)}(x) \cdot g^{(i)}(x)\,dx;\ (f|g)_0 = \int_a^b f(x) \cdot g(x)\,dx .$$

a) Let us take the $\ell_i \in H^q$ $(q \geq 1)$ such that:

$$(\ell_i|f)_q = f(x_i), \text{ for all } f \in H^q ,$$

with $a < x_1 < \ldots < x_n < b$.
This is the problem considered in the introduction.
If $n > q$ and if no polynomial p of degree $q-1$ exists satisfying $\alpha_i \leq p(x_i) \leq \beta_i$, $i = 1, \ldots, n$, then the assumptions H_1, H_2, H_3, H_4, H_5 are satisfied.

The element $\ell = \sum_{i=1}^{n} \lambda_i \ell_i$ (in th. 4) belongs to Range $(T^*) = n(T)^{\perp}$. Thus, we have:

$$\sum_{i=1}^{n} \lambda_i (x_i)^j = 0, \qquad j = 0, 1, \ldots, q-1 . \tag{17}$$

By using the Taylor expansion of f :

$$f(x) = f(a) + (x-a)f'(a) + \ldots + \frac{(x-a)^{q-1}}{(q-1)!} f^{(q-1)}(a) + \int_a^b \frac{(x-t)_+^{q-1}}{(q-1)!} f^{(q)}(t)dt , \tag{18}$$

$$(E)_+ = \begin{cases} E & \text{if } E \geq 0 \\ 0 & \text{if } E < 0 \end{cases} ,$$

we obtain

$$(\ell \mid f)_q = \Big(\sum_{i=1}^{n} \lambda_i \ell_i \Big| f\Big)_q = \int_a^b \psi(t) \cdot f^{(q)}(t)\,dt$$

with

$$\psi(t) = \sum_{i=1}^{n} \frac{\lambda_i (x_i - t)_+^{q-1}}{(q-1)!} \quad . \tag{19}$$

Thus, we have for each $f \in H^q$:

$(\ell \mid f)_q = (\psi \mid f^q)_0 = (\psi \mid Tf)_0 = (T^* \psi \mid f)_q$, hence:

$$\ell = T^* \psi \quad . \tag{20}$$

It follows that:

$$T\sigma = \sigma^{(q)} = \psi \ ,$$
$$\sigma^{(q)}(t) = \sum_{i=1}^{n} \frac{\lambda_i (x_i - t)_+^{q-1}}{(q-1)!} \quad . \tag{21}$$

We obtain the values of the discontinuities of σ^{2q-1} at x_i :

$$\sigma^{(2q-1)}(x_i^+) - \sigma^{(2q-1)}(x_i^-) = (-1)^q \lambda_i \ . \tag{22}$$

At the right of x_n, we have:

$$\sigma^{(j)}(x_n^+) = 0, \qquad j = q, q+1, \ldots, 2q-1 \ . \tag{23}$$

The function $\sigma^{(q)}$ being of the form (21), the conditions (17) are equivalent to:

$$\sigma^{(j)}(x_1^-) = 0, \qquad j = q, q+1, \ldots, 2q-1 \ . \tag{24}$$

Finally, according to theorem 4, the function σ is the spline function if and only if the five conditions 1 - a - b - c - d - e are satisfied.

b) Now, let us take the $\ell_i \in H^q$ such that

$$(25) \qquad (\ell_i | f)_q = \int_{a_i}^{b_i} f(x)\,dx$$

with $[a_i, b_i] \subset [a,b]$, $i = 1, \ldots, n$.
In order to simplify the results, we suppose the following disposition:

$$a < a_1 < b_1 < a_2 < b_2 < \ldots < a_n < b_n < b .$$

If $n > q$ and if there is no polynomial p of degree $q-1$ satisfying $\alpha_i \leq \int_{a_i}^{b_i} p(x)\,dx \leq \beta_i$, $i = 1, \ldots, n$, then the assumptions H1, H2, H3, H4, H5 are fulfilled.
As the element $\ell = \sum_{i=1}^{n} \lambda_i \ell_i$ belongs to $n(T)^{\perp}$, we have

$$(26) \qquad \sum_{i=1}^{n} \lambda_i \int_{a_i}^{b_i} x^j\,dx = 0, \quad j = 0, 1, \ldots, q-1 .$$

By using again (18), we have:

$$(27) \qquad (\ell | f)_q = \int_a^b w(t) \cdot f^{(q)}(t)\,dt$$

with

$$(28) \qquad w(t) = \sum_{i=1}^{n} \lambda_i \int_{a_i}^{b_i} \frac{(x-t)_+^{q-1}}{(q-1)!}\,dx .$$

Thus, we have as precedingly:

$$\ell = T^* w$$

and in consequence:

$$T\sigma = \sigma^{(q)} = w$$

(29)
$$\sigma^{(q)}(t) = \sum_{i=1}^{n} \lambda_i \, w_i(t) \quad ,$$

with

(30)
$$w_i(t) = \begin{cases} \dfrac{(-1)^q}{q!}\left[(t-b_i)^q - (t-a_i)^q \right] & \text{if } t < a_i, \ \text{(polyn. of degree q-} \\ \dfrac{(-1)^q}{q!}(t-b_i)^q & \text{if } a_i < t < b_i, \quad \text{(polyn. of degree q} \\ 0 & \text{if } t > b_i \ . \end{cases}$$

Note that we have:

(31)
$$\sigma^{(2q)}(t) = (-1)^q \lambda_i, \quad \text{for all } t \in \left]a_i, b_i\right[\ .$$

As $\sigma^{(2q)}(a_i^-) = \sigma^{(2q)}(b_i^+) = 0$, we have also for the discontinuities of $\sigma^{(2q)}$:

$$\sigma^{(2q)}(a_i^+) - \sigma^{(2q)}(a_i^-) = (-1)^q \lambda_i \ ,$$

$$\sigma^{(2q)}(b_i^+) - \sigma^{(2q)}(b_i^-) = -(-1)^q \lambda_i \ .$$

At the right of b_n we have:

(32)
$$\sigma^{(j)}(b_n^+) = 0, \qquad j = q, q+1, \ldots, 2q \ .$$

The function σ^q being of the form (29), the relations (26) are equivalent to:

$$(33) \qquad \sigma^{(j)}(a_1^-) = 0, \quad j = q, q+1, \ldots, 2q .$$

We finally obtain the following theorem:

The function $\sigma \in H^q$ is the spline function if and only if the following five conditions are satisfied:

(34-a) σ is a polynomial of degree $2q-1$ in $[a, a_1[$, $]b_i, a_{i+1}[$,

$i = 1, \ldots, n-1$, and $]b_n, b]$

σ is a polynomial of degree $2q$ in $]a_i, b_i[$, $i = 1, \ldots, n$.

(34-b) $\sigma^{(j)}(a_i^-) = \sigma^{(j)}(a_i^+)$, $\sigma^{(j)}(b_i^-) = \sigma^{(j)}(b_i^+)$,

$j = 0, 1, \ldots, 2q-1$; $i = 1, \ldots, n$.

(34-c) $\sigma^{(j)}(a_1^-) = \sigma^{(j)}(b_n^+) = 0, \quad j = q, q+1, \ldots, 2q-1$.

$$(34\text{-d}) \qquad \alpha_i \leq \int_{a_i}^{b_i} \sigma(t)\,dt \leq \beta_i, \qquad i = 1, \ldots, n .$$

(34-e) Setting $\lambda_i = (-1)^q \cdot \sigma^{(2q)}(t)$ for $t \in]a_i, b_i[$, we have:

$$\lambda_i \geq 0 \quad \text{if} \quad \int_{a_i}^{b_i} \sigma(t)\,dt = \alpha_i$$

$$\lambda_i \leq 0 \quad \text{if} \quad \int_{a_i}^{b_i} \sigma(t)\,dt = \beta_i$$

$$\lambda_i = 0 \quad \text{if} \quad \alpha_i < \int_{a_i}^{b_i} \sigma(t)\,dt < \beta_i .$$

6. Characterization of $\Omega = T(C)$. Dual problem

LEMMA 1 Assume (H1).

We have the following dual formulae

$$(35) \qquad d = \min_{x\in C} \|Tx\| = \max_{D_Y \supset \Omega} \left(\min_{y \in D_Y} \|y\| \right)$$

$$= \max_{\substack{D_X \supset C \\ D_X /\!/ \mathfrak{n}(T)}} \left(\min_{x\in D_X} \|Tx\| \right)$$

where D_Y and D_X denote closed half-spaces of Y and X (resp.), and $D_X /\!/ \mathfrak{n}(T)$ means that D_X has the equation $\{(\ell | x) \le c\}$ with $\ell \in \mathfrak{n}(T)^{\perp}$.

Proof :

We have in an evident way:

$$\min_{y\in\Omega} \|y\| \ge \max_{D_Y \supset \Omega} \left(\min_{y\in D_Y} \|y\| \right) .$$

Equality is obtained for the half-space $\bar{D}_Y = \{y \in Y: (\tau | y - \tau) \ge 0$ with $\tau = T\sigma$, projection of 0 on Ω.
To each half-space $D_Y = \{y \in Y: (m|y) \le c\}$ corresponds one-to-one the half-space $D_X = \{x \in X: (\ell|x) \le c\}$ with $\ell = T^*m \in \text{Range}(T^*) = \mathfrak{n}(T)^{\perp}$. If $D_Y \supset \Omega$, then $D_X \supset C + \mathfrak{n}(T)$, hence $D_X \supset C$ and conversely. We have $T(D_X) = D_Y$. It follows that we have (35).

We now add the following assumption:

(H6) The subset L contains a finite number of elements

$L = \{\ell_1, \ldots, \ell_n\}$. We again suppose that for $i \neq j$, ℓ_i and ℓ_j are independent unless $\ell_i = -\ell_j$.

Note that the assumption (H6) implies (H2).

Setting $t(\ell_i) = c_i$, C is defined by:

$$C = \{x \in X: (\ell_i|x) \leq c_i, \quad i = 1, \ldots, n\} . \tag{36}$$

Definition:

Let $\mathcal{J}$ be the set of the $J = \{i_1, i_2, \ldots, i_{q+1}\} \subset \{1, 2, \ldots, n\}$ such that $\ell_i \neq -\ell_j$ (for $i, j \in J$, $i \neq j$) and that there exist nonnegative real numbers $\lambda_j^J \geq 0$, $j \in J$, $\sum_{j \in J} \lambda_j^J = 1$ for which:

$$\ell_J = \sum_{j \in J} \lambda_j^J \ell_j \in \mathcal{N}(T)^\perp . \tag{37}$$

Remark: Let res be the orthogonal projector of X onto $\mathcal{N}(T)$. Then

$$J = \{i_1, \ldots, i_{q+1}\} \in \mathcal{J} \text{ iff } 0 \in \mathrm{CO}(\mathrm{res}(\ell_j) \,|\, j \in J) . \tag{38}$$

LEMMA 2: Assume (H1), (H3), (H4), (H5), (H6) .

Then $\mathcal{J}$ is non-empty and for each $J \in \mathcal{J}$, ℓ_J is unique and for all $j \in J$, we have $\lambda_j^J > 0$.

Proof:

By theorems 1 and 2, there exists a spline function σ such that $T\sigma \neq 0$. By theorem 2, we have:

$$T^* T\sigma = \sum_{i \in I_\sigma} \lambda_i \ell_i \tag{39}$$

with $\lambda_i \geq 0$ and

(40) $$I_\sigma = \{i \in \{1, 2, \ldots, n\} : (\ell_i | \sigma) = c_i\} .$$

As $T\sigma \neq 0$, the λ_i are not all zero.
By assumption (H4), I_σ cannot contain i and j such that $\ell_i = -\ell_j$.
Thus, we have:

(41) $$\ell = \sum_{i \in I_\sigma} \lambda_i \ell_i \in \mathcal{R}(T^*) = \mathfrak{n}(T)^\perp$$

hence

(42) $$0 \in \mathrm{CO}(\mathrm{res}(\ell_i) | i \in I_\sigma)$$

in $\mathfrak{n}(T)$ which is of dimension q.
According to the theorem of Caratheodory (cf Eggleston [9]) there exists $\{i_1, \ldots, i_r\} \subset I_\sigma$ with $r \leq q + 1$ such that:

(43) $$0 \in \mathrm{CO}(\mathrm{res}(\ell_{i_k}) | k = 1, \ldots, r), \quad \text{that is:}$$

$$\ell = \sum_{k=1}^{r} \mu_{i_k} \ell_{i_k} \in \mathfrak{n}(T)^\perp \text{ with } \mu_{i_k} > 0 \text{ and } \sum_{k=1}^{r} \mu_{i_k} = 1 .$$

By the assumption (H5) we cannot have $r \leq q$: in fact, if we had $r \leq q$, (by completing ev. up to q with $\mu_i = 0$ for $i = i_{r+1}, \ldots, i_q$) the homogeneous linear system:

$$\sum_{k=1}^{q} \mu_{i_k} (\ell_{i_k} | w_h) = 0, \qquad h = 1, \ldots, q ,$$

would have a non-zero solution, which contradicts (H5); hence $r = q+1$. Setting $J = \{i_1, i_2, \ldots, i_{q+1}\}$ and $\lambda_j^J = \mu_j$ for $j \in J$ we have finally:

(44) $$\ell_J = \sum_{j \in J} \lambda_j^J \ell_j \in \mathfrak{n}(T)^\perp .$$

The assumption (H5) implies the unicity of the λ_j^J.

Notations: For $J = \{i_1, \ldots, i_{q+1}\} \in \mathcal{J}$, we have $\ell_J = \sum_{j \in J} \lambda_j^J \ell_j \in$ Range (T^*). Let

$$\varphi_J = T^{*-1}(\ell_J) \quad \text{and} \quad \delta_J = \sum_{j \in J} \lambda_j^J c_j \tag{45}$$

$$\Omega_J = \{y \in Y: (\varphi_J | y) \leq \delta_J\} . \tag{46}$$

LEMMA 3. Assume (H5), (H6).

If $\ell = \sum_{i=1}^{n} \lambda_i \ell_i \in n(T)^{\perp}$ with $\lambda_i \geq 0$ (non all zero) then there exist $\mu_J \geq 0$ such that $\ell = \sum_{J \in \mathcal{J}} \mu_J \ell_J$.

Proof: Without loss of generality we can suppose that for $\ell_i = -\ell_j$ we have $\lambda_i = 0$ or $\lambda_j = 0$.
Let $I \subset \{1, \ldots, n\}$ be the set of the indices i such that $\lambda_i \neq 0$.
There exists $J \in \mathcal{J}$ such that $J \subset I$: in fact, $0 \in \mathrm{CO}(\mathrm{res}(\ell_i)/i \in I)$, hence, according to the Caratheodory theorem and by using the assumption (H5) as in lemma 2, there exists $J = \{i_1, \ldots, i_{q+1}\} \subset I$ such that $0 \in \mathrm{CO}(\mathrm{res}(\ell_j)/j \in J)$, hence $J \in \mathcal{J}$.
Let us consider:

$$\ell - k\ell_J = \sum_{i \in I} \lambda_i \ell_i - k \sum_{j \in J} \lambda_j^J \ell_j = \sum_{i \in I} \theta_i \ell_i \quad (k \in R) \tag{47}$$

with

$$\theta_i = \begin{cases} \lambda_i & \text{if } i \notin J \\ \lambda_i - k\lambda_i^J & \text{if } i \in J \end{cases} , \quad i \in I . \tag{48}$$

If $k \in R$ is chosen such that:

(49) $$k = \min_{i \in J} (\lambda_i / \lambda_i^J) = \lambda_{i_0} / \lambda_{i_0}^J$$

we shall have $\theta_i \geq 0$ for all $i \in I$ and $\theta_{i_0} = 0$.

Setting $\mu_J = k > 0$ and $I' = \{i \in I : \theta_i \neq 0\}$, we have:

(50) $$\ell = \sum_{i \in I} \lambda_i \ell_i = \sum_{i \in I'} \theta_i \ell_i + \mu_J \ell_J .$$

As I' comprises a number of indices inferior by at least one unit to the one of I, ℓ can be put, by repeating the operation (at the utmost $n-q-1$ times), in the form $\ell = \sum_{J \in \mathcal{J}} \mu_J \ell_J$.

<u>THEOREM 5</u>

Under the assumptions H1, H3, H4, H5, H6, we have:

(51) $$\Omega = \bigcap_{J \in \mathcal{J}} \Omega_J .$$

<u>Proof</u>:

1°/ Let $y \in \Omega$.

We have $y = Tx$ with $x \in C$, hence $(\ell_i | x) \leq c_i$, $i = 1, \ldots, n$. For each $J = \{i_1, \ldots, i_{q+1}\} \in \mathcal{J}$, we have $(\ell_J | x) \leq \delta_J$, hence $(T^* \varphi_J | x) = (\varphi_J | Tx) \leq \delta_J$, that is $y \in \Omega_J$. Thus, we have $y \in \bigcap_{J \in \mathcal{J}} \Omega_J$.

2°/ Let $y \in \bigcap_{J \in \mathcal{J}} \Omega_J$.

The closed convex subset Ω is equal to the intersection of all the closed half-spaces which contain it. An arbitrary half-space $D = \{y : (\varphi | y) \leq \delta\}$ being given containing Ω, it suffice to prove that y belongs to it.

Let $\ell = T^* \varphi \in \mathfrak{n}(T)^{\perp}$. The half-space $\{(\ell | x) \leq \delta\}$ contains C (and even $C + \mathfrak{n}(T)$).
Denoting $A_i = \{x \in X: (\ell_i | x) < c_i\}$, we have:

(52) $$A_1 \cap A_2 \cap \ldots \cap A_n \cap \{(\ell | x) > \delta\} = \emptyset .$$

We use a separation theorem for $n+1$ convex subsets (cf. A. Ja. Dubovickii and A. A. Miljutin [7], C. Lobry [17], P. J. Laurent [12], [13], p. 17): We have (52) iff there exist nonnegative real numbers $\tilde{\lambda}_1, \ldots, \tilde{\lambda}_n, \lambda \geq 0$ (non all zero) such that:

(53) $$\sum_{i=1}^{n} \tilde{\lambda}_i \ell_i - \lambda \ell = 0 ,$$
$$\sum_{i=1}^{n} \tilde{\lambda}_i c_i - \lambda \delta \leq 0 .$$

We have $\lambda \neq 0$: for, by the same separation theorem, $\lambda = 0$ and (53) imply $\bigcap_{i=1,\ldots,n} A_i = \emptyset$, which contradicts (H4).
Dividing by λ and setting $\lambda_i = \dfrac{\tilde{\lambda}_i}{\lambda}$, we obtain:

(54) $$\ell = \sum_{i=1}^{n} \lambda_i \ell_i \in \mathfrak{n}(T)^{\perp} ,$$

$$\sum_{i=1}^{n} \lambda_i c_i \leq \delta .$$

According to lemma 3, we have

(55) $$\ell = \sum_{i=1}^{n} \lambda_i \ell_i = \sum_{J \in \mathcal{J}} \mu_J \ell_J \quad \text{with } \mu_J \geq 0, \quad \text{hence}$$

(56) $$\varphi = \sum_{J \in \mathcal{J}} \mu_J \varphi_J .$$

By (45), (48) and (55) we also have:

(57) $$\sum_{i=1}^{n} \lambda_i c_i = \sum_{J \in \mathcal{J}} \mu_J \delta_J .$$

As $y \in \bigcap_{J \in \mathcal{J}} \Omega_J$, we have $(\varphi_J | y) \leq \delta_J$, hence:

(58) $$\left(\sum_{j \in \mathcal{J}} \mu_J \varphi_J | y\right) \leq \sum_{J \in \mathcal{J}} \mu_J \delta_J = \sum_{i=1}^{n} \lambda_i c_i .$$

Hence, by (54), we have:

(59) $$(\varphi | y) \leq \delta, \quad \text{hence } y \in D .$$

7. Algorithm

Let us suppose $K = \mathcal{L}(\ell_1, \ldots, \ell_n)$, the subspace spanne by the ℓ_i, to be of dimension $m \leq n$.
We can show that the problem then comes to minimize a quadratic degenerate form (null space of dimension q) on a convex polyhedron of R^m. Numerous methods can be applied for obtaining the solution σ : gradient methods, Frank and Wolfe method, relaxation method, etc., (for a comparison of these methods of obtaining σ, see M. Morin [19]). We propose here a dual algorithm deriving from the method of splitting the constraints (cf. J. L. Lions et R. Teman [16], Y. Haugazeau [10], [11], P. J. Laurent and B. Martinet [15]). The modifications lie essentially in the fact that the quadratic form is degenerate.
The algorithm is based on the duality formula (35), (lemma 1) and on the formula (51) (theorem 5).
We construct a sequence of half-spaces D^ν containing Ω, of $y^\nu \in Y$ (the projection of 0 on D^ν: $\|y^\nu\| = \min_{y \in D^\nu} \|y\|$) and of $x^\nu \in X$ such that:

$$\begin{cases} d = \lim\limits_{\nu\to\infty} \|y^\nu\| \\ \tau = \lim\limits_{\nu\to\infty} y^\nu \\ \sigma = \lim\limits_{\nu\to\infty} x^\nu \end{cases}.$$

Let $y^\nu \in Y$ and:

(60) $$D^\nu = \{y \in Y: (y^\nu | y-y^\nu) \geq 0\}.$$

(For $\nu = 0$, we can take $y^0 = 0$).

Let us suppose that we have:

(61-1) $$D^\nu \supset \Omega$$

(61-2) $$(\varphi_{J^\nu} | y^\nu) = \delta_{J^\nu}, \quad \text{with } J^\nu \in \mathcal{J}.$$

By assumption (H5), there exists a unique $x^\nu \in X$ such that

(62) $$Tx^\nu = y^\nu \text{ and } (\ell_j | x^\nu) = c_j, \forall_j \in J^\nu.$$

Note first that $x^\nu \in C$ implies $x^\nu = \sigma$. (Indeed, we would have $y^\nu \in \Omega$ and $(y^\nu | y-y^\nu) \geq 0$, for all $y \in \Omega$, hence $y^\nu = \tau$). We now suppose $x^\nu \notin C$. Then, there exists $\ell_{i^\nu} \in L$ such that:

(63) $$(\ell_{i^\nu} | x^\nu) - c_{i^\nu} = \max_{i=1,\dots,n} [(\ell_i | x^\nu) - c_i] > 0.$$

LEMMA 4

There exists an index $j^\nu \in J^\nu$ such that, if we set

(64) $$J^{\nu+1} = (J^\nu \cup \{i^\nu\}) \sim \{j^\nu\}$$

we have

(65) $$J^{\nu+1} \in \mathcal{J} \ .$$

Proof:

Indeed, we have:

(66) $$0 \in \mathrm{CO}\,(\mathrm{res}\,(\ell_j),\ \ j \in J^{\nu}) \ .$$

There exist numbers $\lambda_j > 0$ such that:

(67) $$\ell = \sum_{j \in J^{\nu}} \lambda_j \ell_j + \lambda_{i^{\nu}} \ell_{i^{\nu}} \in h(T)^{\perp} \ .$$

When following the same way as in lemma 3, we obtain:

(68) $$\ell = \mu_1 \ell_{J_1} + \mu_2 \ell_{J_2}, \quad \text{with } \mu_1, \mu_2 > 0 \text{ and } J_1, J_2 \in \mathcal{J} \ .$$

One of the two subsets J_1 and J_2 contains i^{ν}. We take it as subset $J^{\nu+1}$.

This lemma is analogous to the Stiefel-exchange theorem (Austausch-Satz) for the resolution, by the Remes algorithm, of approximation problems (cf. P. J. Laurent [14]). The index j^{ν} can be obtained analytically by using relation (49).
Let us consider $\Omega_{J^{\nu+1}}$. We have:

(69) $$y^{\nu} \notin \Omega_{J^{\nu+1}} \ .$$

Indeed, we have:

$$(\varphi_{J^{\nu+1}} | y^{\nu}) - \delta_{J^{\nu+1}} = (\ell_{J^{\nu+1}} | x^{\nu}) - \delta_{J^{\nu+1}} = \sum_{j \in J^{\nu+1}} \lambda_j^{J^{\nu+1}} [(\ell_j | x^{\nu}) - c_j] \ .$$

Hence, by (64) and (62):

$$(70) \qquad (\varphi_{J^{\nu+1}} | y^\nu) - \delta_{J^{\nu+1}} = \lambda^{J^{\nu+1}}_{i^\nu} [(\ell_{i^\nu} | x^\nu) - c_{i^\nu}] > 0 \,.$$

Let $y^{\nu+1} \in D^\nu \cap \Omega_{J^{\nu+1}}$ be the projection of 0 on $D^\nu \cap \Omega_{J^{\nu+1}}$. As the projection y^ν of 0 on D^ν does not belong to $\Omega_{J^{\nu+1}}$, we shall have: $y^{\nu+1} \in \text{boundary}(\Omega_{J^{\nu+1}})$, that is:

$$(71) \qquad (\varphi_{J^{\nu+1}} | y^{\nu+1}) = \delta_{J^{\nu+1}} \,.$$

There exists a unique $x^{\nu+1} \in X$ such that

$$(72) \qquad Tx^{\nu+1} = y^{\nu+1} \text{ and } (\ell_j | x^{\nu+1}) = c_j, \forall_j \in J^{\nu+1} \,.$$

We build once again:

$$(73) \qquad D^{\nu+1} = \{y \in Y: (y^{\nu+1} | y - y^{\nu+1}) \geq 0\} \,.$$

As $D^\nu \supset D^\nu \cap \Omega_{J^{\nu+1}} \supset \Omega$, we have:

$$(74) \qquad \|y^\nu\| \leq \|y^{\nu+1}\| \leq \|\tau\| \,.$$

As $y^{\nu+1}$ is the projection of 0 on $D^\nu \cap \Omega_{J^{\nu+1}}$, we have:

$$(75) \qquad (y^{\nu+1} | y - y^{\nu+1}) \geq 0, \forall y \in D^\nu \cap \Omega_{J^{\nu+1}}$$

and as $\Omega \subset D^\nu \cap \Omega_{J^{\nu+1}}$:

$$(76) \qquad (y^{\nu+1} | y - y^{\nu+1}) \geq 0, \forall y \in \Omega, \text{ that is}$$

(77) $$D^{\nu+1} \supset \Omega \ .$$

Thus, we obtained for $y^{\nu+1}$ the same properties ((77) and (71)) as we had for y^{ν} ((61-1) and (61-2)).

8. Convergence of the algorithm

We have the following theorem:

THEOREM 6:

Under the assumptions H1, H3, H4, H5 and H6, the constructed sequence x^{ν} converges strongly to the solution σ .

We first prove two preliminary lemmas (under the same asumptions):

LEMMA 5

If $J \in \mathcal{J}$; $\ell_J = \sum_{j \in J} \lambda_j^J \ell_j \in \mathfrak{N}(T)^{\perp}$; $(\lambda_j^J \geq 0; \sum_{j \in J} \lambda_j^J = 1)$ and $\varphi_J = T^{*-1}(\ell_J)$, then there exists $m > 0$ such that

(78) $$\lambda_j^J / \|\varphi_J\| \geq m$$

where m is independent from $J \in \mathcal{J}$ and $j \in J$.

Proof:

The assumption (H5) implies $\lambda_j^J > 0$, $j \in J$ (see lemma 2) for all $J \in \mathcal{J}$. As $\mathcal{J}$ comprises a finite number of J, there exists $m > 0$ satisfying (78).

LEMMA 6

The sequence $\|x^{\nu}\|$ is bounded.

Proof:

We know that $\|Tx^\nu\| \leq \|\tau\|$ and $(\ell_j|x^\nu) = c_j$, $j \in J^\nu$. Let us set $x^\nu = x_1^\nu + x_2^\nu$ with $x_1^\nu \in \mathfrak{n}(T)$ and $x_2^\nu \in \mathfrak{n}(T)^\perp$. If we call $\tilde{T}$ the restriction of T to $\mathfrak{n}(T)^\perp$, $\tilde{T}$ maps one-to-one $\mathfrak{n}(T)^\perp$ onto Y, hence $\tilde{T}^{-1}$ is continuous. Therefore, the sequence $x_2^\nu = \tilde{T}^{-1}(y^\nu)$ is bounded. Thus, there exists M_2 such that:

(79) $$|(\ell_i|x_2^\nu)| \leq M_2, \quad \forall_i = 1, \ldots, n, \quad \forall\nu \ .$$

As $(\ell_j|x^\nu) = c_j$, $\forall j \in J^\nu$, there exists M_1 such that:

(80) $$|(\ell_j|x_1^\nu)| \leq M_1, \quad \forall j \in J^\nu, \ \forall\nu \ .$$

For each fixed $J \in \mathcal{J}$, by assumption (H5):

(81) $$p_J(x) = \max_{j \in J} |(\ell_j|x)|$$

is a norm on $\mathfrak{n}(T)$.
Thus, for all indices ν such that $J^\nu = J$, we have

(82) $$\|x_1^\nu\| \leq K_J \ .$$

As $\mathcal{J}$ has a finite number of elements, we finally have:

(83) $$\|x_1^\nu\| \leq K, \quad \forall\nu$$

which proves that the sequence $\|x^\nu\|$ is bounded.

Proof of theorem 6:

As the sequence $\|y^\nu\|$ is increasing and bounded by $\|\tau\|$ (see (74)), it converges to $\alpha \leq \|\tau\|$.
As $(y^\nu|y^{\nu+1} - y^\nu) \geq 0$, we have:

(84) $$\|y^{\nu+1} - y^{\nu}\|^2 \leq \|y^{\nu+1}\|^2 - \|y^{\nu}\|^2$$

hence:

(85) $$\lim_{\nu\to\infty} \|y^{\nu+1} - y^{\nu}\| = 0 .$$

As $y^{\nu+1} \in \Omega_{J^{\nu+1}}$, we have by (69):

(86) $$\|y^{\nu} - y^{\nu+1}\| \geq \operatorname*{Inf}_{y\in\Omega_{J^{\nu+1}}}\|y^{\nu} - y\| = [(\varphi_{J^{\nu+1}}|y^{\nu}) - \delta_{J^{\nu+1}}]/\|\varphi_{J^{\nu+1}}\|$$

hence, by (70), (63) and lemma 5:

(87) $$\|y^{\nu} - y^{\nu+1}\| \geq \frac{\lambda_{i^{\nu}}^{J^{\nu+1}}}{\|\varphi_{J^{\nu+1}}\|}[(\ell_{i^{\nu}}|x^{\nu}) - c_{i^{\nu}}] \geq m \cdot \max_{i=1,\ldots,n}[(\ell_i|x^{\nu}) - c_i]$$

and, finally by (85)

(88) $$\overline{\lim_{\nu\to\infty}}\, [(\ell_i|x^{\nu}) - c_i] \leq 0, \quad i = 1, \ldots, n .$$

As the sequence $\|x^{\nu}\|$ is bounded, there is a subsequence x^{ν_k} which converges weakly to $\tilde{x}$.
Since the mapping $x \to \|Tx\|$ is weakly continuous, we have by (74)

(89) $$\|T\tilde{x}\| \leq \|T\sigma\| .$$

By (88), we also have:

(90) $$(\ell_i|\tilde{x}) - c_i \leq 0, \quad i = 1, \ldots, n, \quad \text{that is } \tilde{x} \in C ,$$

which implies:

(91) $$\|T\sigma\| \leq \|T\tilde{x}\| .$$

Thus $\|T\sigma\| = \|T\tilde{x}\|$ and by the unicity of the solution:

(92) $$\tilde{x} = \sigma .$$

The subsequence x^{ν_k} converges weakly to σ. Let us show that the sequence x^{ν} converges strongly to σ:

As the subsequence $\|y^{\nu_k}\|$ converges to $\|\tau\|$, the sequence $\|y^{\nu}\|$ which is increasing, converges to $\|\tau\|$.

Since $(y^{\nu} | \tau - y^{\nu}) \geq 0$, we have:

(93) $$\|\tau - y^{\nu}\|^2 \leq \|\tau\|^2 - \|y^{\nu}\|^2$$

hence:

(94) $$\lim_{\nu \to \infty} y^{\nu} = \tau .$$

If we set $x^{\nu} = x_1^{\nu} + x_2^{\nu}$, with $x_1^{\nu} \in \mathfrak{n}(T)$ and $x_2^{\nu} \in \mathfrak{n}(T)^{\perp}$, the sequence $x_2^{\nu} = \tilde{T}^{-1}(y^{\nu})$ converges strongly to $x_2 = \tilde{T}^{-1}(\tau)$.

Let us consider the set N_J of the ν such that $J^{\nu} = J$ (we can take but the J such that N_J is infinite).

We have:

(95) $$\lim_{\substack{\nu \in N_J \\ \nu \to \infty}} (\ell_j | x_2^{\nu}) \quad (\ell_j | x_2), \qquad j \in J .$$

Since $(\ell_j | x^{\nu}) = c_j$, $j \in J$, $\nu \in N_J$, we also have:

(96) $$\lim_{\substack{\nu \in N_J \\ \nu \to \infty}} (\ell_j | x_1^{\nu}) = \gamma_j = c_j - (\ell_j | x_2), \qquad j \in J .$$

If x_J is the unique element of $\mathfrak{n}(T)$, which satisfies:

$$(\ell_j | x_J) = \gamma_j, \quad j \in J$$

we shall have (with the notation (81)):

(97) $$\lim_{\substack{\nu \in N_J \\ \nu \to \infty}} p_J(x_1^\nu - x_J) = 0$$

Thus we have: $\lim_{\substack{\nu \in N_J \\ \nu \to \infty}} x_1^\nu = x_J$ and consequently:

(98) $$\lim_{\substack{\nu \in N_J \\ \nu \to \infty}} x^\nu = x_2 + x_J \ .$$

By (88), the element $x_2 + x_J$ belongs to C and $\|T(x_2 + x_J)\| = \|Tx_2\| = \|T\sigma\|$, hence $x_2 + x_J = \sigma$. We have finally:

(99) $$\lim_{\nu \to \infty} x^\nu = \sigma$$

REFERENCES

[1] P. M. Anselone and P. J. Laurent: A general method for the construction of interpolating or smoothing spline functions. Num. Math. 12, (1968), 66-82.

[2] M. Attéia: Théorie et applications des fonctions spline en analyse numérique. Thèse, Grenoble (1966).

[3] M. Attéia: Fonctions spline définies sur un ensemble convexe. Num. Math. 12, (1968), 192-210.

[4] M. Attéia: Fonctions spline avec contraintes linéaires de type inégalité. 6e Congrès de l'AFIRO, Nancy, Mai 1967.

[5] C. Carasso: Méthode générale de construction de fonctions spline. Revue Française d'informatique et de recherche opérationnelle. 5, (1967), 119-127.

[6] J. Dieudonné: Sur la séparation des ensembles convexes. Math. Annalen, 163, (1966), 1-3.

[7] A. Ja. Dubovickii and A. A. Miljutin: Problèmes d'optimisation en présence de contraintes. Th. Vychisl. Mat. I. Mat. Fiz., 5 (1965), 395-453.

[8] N. Dunford and J. T. Schwartz: Linear operators, part I, Interscience Pub. (1958).

[9] H. G. Eggleston: Convexity. Cambridge, Univ. Press (1963).

[10] Y. Haugazeau: Sur la minimisation des formes quadratiques avec contraintes. C. R. Acad. Sci. Paris, 264 (1967), 322-324.

[11] Y. Haugazeau: Sur les inéquations variationnelles et la minimisation de fonctionnelles convexes. Thèse, Paris, 7 juin 1968.

[12] P. J. Laurent: Théorèmes de caracterisation en approximation convexe. Colloque "Théorie de l'approximation des fonctions" Cluj (Roumanie) 15-20 Sept. 1967. Mathematica 10 (33), 11, (1968), 95-111.

[13] P. J. Laurent: Cours de théorie de l'approximation. Fascicule 4, Université de Grenoble, 1968-69.

[14] P. J. Laurent: Théorèmes de caractérisation d'une meilleure approximation dans un espace normé et généralisation de l'algorithme de Rémès. Num. Math. 10, (1967), 190-208.

[15] P. J. Laurent et B. Martinet: Méthodes duales pour le calcul du minimum d'une fonction convexe sur une intersection de convexes. Colloque d'optimisation, Nice, 30 Juin - 5 Juillet 1969.

[16] J. L. Lions and R. Teman: Une méthode d'éclatement des opérateurs et des contraintes en calcul des variations. C. R. Acad. Sci. Paris, 263, (1966), 563-565.

[17] C. Lobry: Etude géométrique des problèmes d'optimisation en présence de contraintes. Thèse, Grenoble, 20 Juin 1967.

[18] J. J. Moreau: Fonctionnelles convexes. Séminaire au Collège de France, Paris (1966-67).

[19] M. Morin: Méthodes de calcul des fonctions spline dans un convexe. Thèse, Grenoble (1969), (a paraître).

[20] I. J. Schoenberg: Spline functions and the problem of graduation. Proc. of the Nat. Acad. of Sciences, 52, No. 4., (1964), 947-950.

[21] I. J. Schoenberg: Spline interpolation and the higher derivatives. Proc. of the Nat. Acad. of Sciences, 51, No. 1, (1964), 24-28.

[22] M. Valadier: Quelques propriétés des sous-gradients. Rapport de l'IRIA, (Sept. 1968).

Best Quadrature Formulas and Interpolation by Splines Satisfying Boundary Conditions

SAMUEL KARLIN

In the study of interpolation and approximation by splines, a proliferation of terminology has beset the literature. For example, we encounter natural splines, periodic splines, splines with free knots, g-splines, B-splines, etc. Several of these concepts are distinguished by the nature of the boundary conditions that the splines satisfy.

This paper divides into two principal parts (see also Karlin [5]). Part I reports complete results concerning unique interpolation by splines satisfying rather general boundary conditions. Our study unifies, extends and refines much of the previous work on this topic. Whereas, most related developments focus on interpolating given data exclusively at the knots for special classes of boundary constraints, we have described general criteria on the knots, interpolatory points and prescription of the boundary conditions for unique interpolation by splines.

More specifically, Theorem 1 makes it possible to interpolate at points other than the knots and indeed to determine at exactly which points the interpolation problem is "poised". An interpretation of this result for certain physical systems is pointed out in Section 9, Chapter 10 of Karlin [4].

In Part II best quadrature formulas in the sense of Sard with fixed knots corresponding to splines satisfying boundary conditions are characterized. We generalize the work of Schoenberg [7], [8] but follow his elegant method utilizing the device of integration by parts applied to suitable monosplines.

Detailed proofs and related discussions concerning the assertions of Parts I and II will be published separately. For rather complete references and background on the subject of this paper, we refer the reader to the excellent review articles by Greville [1], Jerome and Varga [2] and Schoenberg [9].

The results summarized in this paper are set forth in the context of ordinary spline polynomials but we hasten to emphasize that these theorems extend to the corresponding case of Techebycheffian splines (for relevant definitions see e. g. Schumaker [10] or Karlin [4, Chap. 10]).†

We close the introduction by fixing some notation. Thus if

$$A = \| A_{ij} \| \quad \text{then} \quad A\begin{pmatrix} i_1, & i_2, & \dots, & i_p \\ j_1, & j_2, & \dots, & j_p \end{pmatrix}$$

denotes the minor of A composed from rows and columns of indices $i_1 < i_2 < \dots < i_p$ and $j_1 < j_2 < \dots < j_p$ respectively. As usual x_+^m represents the function x^m for $x \geq 0$ and 0 for $x < 0$.

---

† The results reported in this work was partly uncovered while the author was visiting M. R. C. between September - December of 1968 and further developed during the authors visit to the Weizmann Institute of Science in Israel between January - March 1969. It is also a great pleasure to acknowledge the numerous beneficial discussions with Professor I. J. Schoenberg which inspired and largely motivated the investigations and results presented below. Part of the material of Section 3 of the subsequent paper of this volume (Karlin [5]) was developed in collaboration with Professor Schoenberg.

BEST QUADRATURE FORMULAS AND INTERPOLATION BY SPLINES

PART I

INTERPOLATION BY SPLINES SATISFYING BOUNDARY CONDITIONS

§1. Total Positivity of Certain Special Kernels

Consider splines of order m (i.e., degree m-1) with r fixed knots $\{\xi_i\}_{i=1}^r$ $(0 < \xi_1 < \xi_2 < \ldots < \xi_r < 1)$ of the explicit form

$$(1) \qquad S(x) = \sum_{i=0}^{m-1} a_i x^i + \sum_{i=1}^{r} c_i (x - \xi_i)_+^{m-1}$$

and satisfying the boundary conditions

$$\mathcal{B}_0 : \sum_{\mu=0}^{m-1} A_{\nu\mu} (D^\mu S)(0) = 0 \qquad \nu = 1, 2, \ldots, p$$

$$\left(D = \frac{d}{dx}\right) .$$

$$\mathcal{B}_1 : \sum_{\mu=0}^{m-1} B_{\lambda\mu} (D^\mu S)(1) = 0 \qquad \lambda = 1, 2, \ldots, q$$

A spline of the kind (1) fulfilling the boundary conditions $\mathcal{B}_0 \cap \mathcal{B}_1$ is said to be of class $\mathcal{S}_{m,r}(\mathcal{B}_0 \cap \mathcal{B}_1)$.

The following requirements are assumed to prevail throughout this part unless stated otherwise.

Postulate I

(i) $p + q \leq m$.

(ii) The $p \times m$ matrix $\tilde{A} = \|A_{\nu\mu}(-1)^\mu\|$ is sign consistent of order p (SC_p) and has rank p (a matrix U is said to be SC_p if all $p \times p$ nonzero subdeterminants of U have a single sign).

(iii) The $q \times m$ matrix $B = \|B_{\nu\mu}\|$ is SC_q and of rank q .

Several concrete illustrations of boundary conditions fulfilling postulate I are indicated at the close of section 2. The objective of this part is to present the solution of the following general interpolation problem. Let $\{x_j\}_1^k$ satisfy $0 < x_1 < x_2 < \ldots < x_k < 1$ and $m+r = k+p+q$. When is it possible to interpolate arbitrarily preassigned values $\{y_j\}_1^k$ at the points $\{x_j\}_1^k$ by a spline of class $\mathcal{S}_{m,r}(\mathcal{B}_0 \cap \mathcal{B}_1)$

The exact criteria when such interpolation is possible is indicated in Theorem 2. A key fact required for the proof of Theorem 2 and of considerable independent importance is the content of Theorem 1. To state the theorem we introduce the following fundamental kernel $K(z,w)$ defined below.

Consider the two ordered sets Z and W consisting of a set of integers and an open interval. Specifically, define

> $Z = \{0, 1, \ldots, m-1, \xi\}$ where ξ ranges over the open unit interval $(0,1)$

and

> $W = \{x, 0, 1, \ldots, m-1\}$ where x traverses $(0,1)$.

(Notice that for the domain Z the integers are arranged to occur prior to ξ values while in W they are placed after the x values.)

Define $K(z,w)$ on $Z \times W$ as follows:

$$
(2) \qquad
\begin{aligned}
K(i,x) &= x^i \\
K(\xi,x) &= (x-\xi)_+^{m-1} \\
K(i,j) &= D^j u_i(x)\big|_{x=1} \qquad (u_i(x) = x^i) \\
K(\xi,j) &= D_x^j \phi(x,\xi)\big|_{x=1} \qquad (\phi(x,\xi) = (x-\xi)_+^{m-1})
\end{aligned}
$$

We can now state the principal result of this paper from which numerous consequences are derived including the solution of the interpolation theorem posed previously.

Theorem 1. (i) The kernel $K(z,w)$ defined in (2) is totally *positive* (TP), *i.e., for any sets* $\{i_\nu, \xi_t\}$ *and* $\{x_t, j_\nu\}$ *satisfying*

$$0 \le i_1 < i_2 < \ldots < i_\alpha \le m-1, \quad 0 < \xi_1 < \xi_2 < \ldots < \xi_\beta < 1$$

and

$$0 < x_1 < x_2 < \ldots < x_\gamma < 1, \quad 0 \le j_1 < j_2 < \ldots < j_\delta \le m-1$$

with $\alpha + \beta = \gamma + \delta$ then

$$K\begin{pmatrix} i_1, i_2, \ldots, i_\alpha, \xi_1, \xi_2, \ldots, \xi_\beta \\ x_1, x_2, \ldots, x_\gamma, j_1, j_2, \ldots, j_\delta \end{pmatrix} \ge 0 \ . \tag{3}$$

(ii) *Strict inequality holds in* (3) *if and only if the indices and variables obey the constraints*.
(Case a; $\alpha \ge \gamma$) ,

$$\begin{aligned} & j_\mu \le i_{\gamma+\mu} \ , && \mu = 1, 2, \ldots, \alpha-\gamma \\ & x_i < \xi_{m-\alpha+i} \ , && i = 1, 2, \ldots, \gamma \end{aligned} \tag{4}$$

(Case b; $\alpha \le \gamma$) ;

$$\begin{aligned} & x_i < \xi_{m-\alpha+i} \ , && i = 1, 2, \ldots, \gamma \\ & \xi_j < x_{\alpha+j} \ , && j = 1, \ldots, \gamma-\alpha \ . \end{aligned} \tag{5}$$

(*Of course these conditions are to apply only when the subscripts are meaningful. Notice that* α *and* δ *are always at most* $m-1$.)

Theorem 1 extends to the case where coincident ξ_t's are permitted (i.e., multiple point knots exist) and coincident x_t values (i.e., the prescription of successive values of the function and some of its higher derivatives). In such a situation the number of elements in each block of coincident knots (or values) cannot exceed m. A slight modification in the conditions of (4) and (5) must be made when m coincidences occur in the combined collection of $\{x_1, \ldots, x_\gamma, \xi_1, \ldots, \xi_\beta\}$ (For its precise statement we refer the reader to Karlin [4; Vol. II].

The proof of Theorem 1 is quite complicated relying on the precise total positivity properties of the fundamental solution $\phi(x,\xi) = (x-\xi)_+^{m-1}$ developed by Karlin and Ziegler [3], see also Karlin [4, Theorems 1.1 and 2.1 of Chap. 10], and several exploitations of the Sylvester determinant identity (see Karlin [4, Chap. 0 and Chap. 10, Section 8]) and a variety of forward and backward induction procedures.

Some special cases of Theorem 1 are worth highlighting.

Example I. Suppose $\alpha = \delta$ and $\gamma = \beta = 0$ then the requirement for strict positivity in (3) reduces to

$$j_\mu \leq i_\mu, \qquad \mu = 1, 2, \ldots, \alpha .$$

Example II. Consider the special case $\alpha = \delta$, $\gamma = \beta$ and $\xi_i = x_i$ for all i. Then (3) is always strictly positive when $\gamma \geq \alpha$ since condition (5) manifestly prevails. When $\gamma < \alpha$ then the condition simply becomes

$$j_\mu \leq i_{\gamma+\mu} \qquad \mu = 1, 2, \ldots, \alpha-\gamma .$$

Remark 1: Theorem 1 in the context of generalized differential operators as studied in Karlin [4, Chap. 10] yields a sharpened version of Theorem 8.2 in [4, Chap. 10]. We will elaborate these applications in Volume II of this cited work.

§2. General Interpolation Criteria and Examples

The next theorem depending on Theorem 1 describes the solution of the interpolation problem posed at the start of Section

Theorem 2. *Let the knots* $\{\xi_t\}_{t=1}^{\beta}$ *and the points* $X = \{x_t\}_{t=1}^{\gamma}$ *be prescribed satisfying* $0 < \xi_1 < \xi_2 < \ldots < \xi_\beta < 1$ *and* $0 < x_1 < x_2 < \ldots < x_\gamma < 1$ where

$$\beta + m = \gamma + p + q \; . \tag{6}$$

Unique interpolation at X *occurs by a spline* $S(x)$ *of class* $\mathcal{S}_{m,\beta}(\mathcal{B}_0 \cap \mathcal{B}_1)$ *provided postulate I holds and then if and only if* $0 \leq i_1 < i_2 < \ldots < i_p \leq m-1$ *and* $0 \leq j_1 < j_2 < \ldots < j_q \leq m-1$ *exist satisfying*

$$A\begin{pmatrix} 1, & 2, & \ldots, & p \\ i_1, & i_2, & \ldots, & i_p \end{pmatrix} \neq 0 \quad \text{and} \quad B\begin{pmatrix} 1, & 2, & \ldots, & q \\ j_1, & j_2, & \ldots, & j_q \end{pmatrix} \neq 0$$

while $\{x_1, \ldots, x_\gamma\}$, $\{\xi_1, \ldots, \xi_\beta\}$, $\{i'_1, \ldots, i'_{m-p}\}$ *and* $\{j_1, \ldots, j_q\}$ *obey the restrictions of* (4) *or* (5). (Here $i'_1, \ldots, i'_{m-p}$ *denotes the complementary set of indices of* $i_1, \ldots, i_p$ *among the collection* $\{0, 1, \ldots, m-1\}$.)

It is worth highlighting some special important examples of boundary conditions $\mathcal{B}_0 \cap \mathcal{B}_1$ fulfilling the conditions of postulate I and the cases of validity of Theorem 2 for them.

Example 1. $q = m-p$ ("a full set of boundary conditions"). If $\gamma = q$ then the stipulation $\beta + m = \gamma + p + q$ implies $\beta = \gamma = q$ and unique interpolation is possible at any sets of prescribed points x's and knots ξ's satisfying (5).

Example 2. $(q = m-p, \; \gamma > q)$. Unique interpolation is possible provided only that $\{\xi_t\}$ and $\{x_t\}$ are specified to satisfy

$$\xi_t < x_{q+t} \qquad t = 1, 2, \ldots, \gamma-q$$

$$x_{q+t-m} < \xi_t \qquad t = 1, \ldots, \beta \; .$$

Example 3. $(q = m-p, \; \gamma < q)$. Unique interpolation holds for all choices of the x's and ξ's provided there exists

$$0 \le i_1 < i_2 < \ldots < i_p \le m-1$$

$$0 \le j_1 < j_2 < \ldots < j_{m-p} \le m-1$$

where

$$A\begin{pmatrix} 1, & \ldots, & p \\ i_1, & \ldots, & i_p \end{pmatrix} \ne 0, \quad B\begin{pmatrix} 1, & \ldots, & m-p \\ j_1, & \ldots, & j_{m-p} \end{pmatrix} \ne 0$$

hold and

$$j_\mu \le i'_{\gamma+\mu}, \quad \mu = 1, 2, \ldots, q-\gamma$$

and where $\{i'_1, \ldots, i'_{m-p}\}$ denotes the set of complementary indices to $\{i_1, \ldots, i_p\}$ in the set $\{0, 1, \ldots, m-1\}$.

The next example embraces a further specialization of wide interest.

<u>Example 4.</u> Let $m = 2n$, $p = q = n$. Let $\mathcal{B}_0 \cap \mathcal{B}_1$ correspond to the simple boundary conditions $S^{(i_1)}(0) = S^{(i_2)}(0) = \ldots = S^{(i_n)}(0) = 0$ and $S^{(j_1)}(1) = S^{(j_2)}(1) = \ldots = S^{(j_n)}(1) = 0$ where $0 \le i_1 < i_2 < \ldots < i_n \le 2n-1$ and $0 \le j_1 < j_2 < \ldots < j_n \le 2n-1$. Suppose $x_t = \xi_t$, $t = 1, \ldots, \beta$. Let $\{i'_1, \ldots, i'_n\}$ denote the complementary indices of $\{i_1, \ldots, i_n\}$ in $\{0, 1, \ldots, 2n-1\}$ According to Theorem 2 unique interpolation holds if and only i

$$j_\mu \le i'_{\beta+\mu}, \quad \mu = 1, 2, \ldots, n-\beta \quad . \tag{8}$$

In particular, if $j_1 = i_1 = 0$, $j_2 = i_2 = 1, \ldots, j_n = i_n = n$ then unique interpolation at the knots holds. On the other if $j_1 = i_1 = n$, $j_2 = i_2 = n+1, \ldots, j_n = i_n = 2n-1$ then inspection of (8) reveals that unique interpolation at the knots is possible if and only if $\beta \ge n$ (i.e., the presence of at least n knots).

Remark 2. For prescribed knots and prescribed derivatives at the ends 0 and 1 the question of well "poised" is settled by the conditions (8). The equivalence of this assertion to the familiar Pólya criteria is easily established.

Remark 3. The corresponding interpolation problem in the case of periodic boundary conditions is easily resolved.[†] In this case when the interpolatory points and knots coincide then unique interpolation constantly prevails. Conditions for unique interpolation in the situation of general knots and points will be treated in a separate publication. The interpolation problem with general mixed boundary conditions is unsettled.

[†] Periodic B.C. means that $S^{(i)}(0) = S^{(i)}(1)$, $i = 0, 1, \ldots, m-1$.

PART II

BEST QUADRATURE FORMULAS WITH FIXED KNOTS

§3. A Special Case

The usual expression of a quadrature formula for a linear functional $L(f)$ defined for continuous functions on $[0,1]$ with preassigned knots $\{\xi_k\}_{k=1}^r$ $(0 < \xi_1 < \xi_2 < \ldots < \xi_r < 1)$ is of the form

$$\sum_{k=1}^{r} a_k f(\xi_k) = Q(f) \quad . \tag{9}$$

Each specification of the coefficients $\{a_k\}$ provides a quadrature formula and the linear functional

$$\Re(f) = L(f) - Q(f) \tag{10}$$

is called the remainder part. A quadrature formula $Q^*(f)$ with corresponding remainder $\Re^*(f)$ is said to be best in the sense of Sard if $\{a_k\}$ is determined in a manner that

(11) $\Re^*(f) = 0$ persists for f a polynomial of degree at most $n-1$

and

$$\min_{\{a_k\}} \sup_{\substack{f \in C^n[0,1] \\ \|f\|_n \leq 1}} |\Re(f)| = |\Re^*(f)| \tag{12}$$

holds where $\|f\|_n$ stands for $\int_0^1 |f^{(n)}(x)|^2 dx$ and the $\min_{\{a_k\}}$ is extended over all admissible quadrature formulas for which the condition (11) is fulfilled. Henceforth we shall concentrate on the special functional $Lf = \int_0^1 f(x)\,dx$. (Adjustments in the statement of the results are appended at the close of Section 4 to take account of more general functionals.)

Schoenberg [7] pointed out a remarkable correspondence between quadrature formulas and monosplines. A monospline $M(x)$ of degree m with knots $\{\xi_i\}_1^r$ is an expression of the form

$$M(x) = \frac{x^m}{m} + \sum_{v=0}^{m-1} \lambda_v x^v + \sum_{i=1}^{r} c_i (x-\xi_i)_+^{m-1} .$$

where λ_v and c_i are real coefficients. Let $m = 2n$ and stipulate f to be of continuity class $C^{(n)}$. We review briefly his analysis. Integration by parts produces the formula

$$(13) \quad \int_0^1 f(x)\,dx = \sum_{j=0}^{n-1} A_j f^{(j)}(1) + \sum_{j=0}^{n-1} B_j f^{(j)}(0) + \sum_{k=1}^{r} a_k f(\xi_k) + (-1)^n \int_0^1 M^{(n)}(x) f^{(n)}(x)\,dx$$

and

$$A_j = (-1)^j M^{(2n-j-1)}(1), \quad B_j = (-1)^{j+1} M^{(2n-j-1)}(0), \quad j = 0,1,\ldots,n-1$$

$$a_v = [M^{(2n-1)}(\xi_v -) - M^{(2n-1)}(\xi_v +)] \qquad v = 1, 2, \ldots, r .$$

If $M(x)$ is determined satisfying the boundary conditions

$$(14) \quad \begin{aligned} M^{(n)}(0) &= M^{(n+1)}(0) = \ldots = M^{(2n-1)}(0) = 0 \\ M^{(n)}(1) &= M^{(n+1)}(1) = \ldots = M^{(2n-1)}(1) = 0 \end{aligned}$$

then (11) reduces to

$$(15) \quad \int_0^1 f(x)\,dx = \sum_{k=1}^{r} a_k f(\xi_k) + (-1)^n \int_0^1 M^{(n)}(x) f^{(n)}(x)\,dx .$$

Comparing with (9) and (10), we set

$$Q(f) = \sum_{k=1}^{r} a_k \, f(\xi_k), \quad \Re(f) = (-1)^n \int_0^1 M^{(n)}(x) \, f^{(n)}(x) \, dx \ .$$

Clearly (11) is satisfied whenever (14) holds and then the converse of the Schwartz inequality implies

$$\sup_{\|f\|_n = 1} |R(f)| = \int_0^1 [M^{(n)}(x)]^2 dx \ .$$

It follows that the best quadrature formula is obtained by minimizing $\int_0^1 [M^{(n)}(x)]^2 dx$ with respect to all monosplines $M(x)$ of degree $2n$ with knots $\{\xi_k\}_1^r$ satisfying (14). To insure (14) we must have $r \geq n$. A standard variational argument shows that the monospline $M_*(x)$ induces the best quadrature formula if and only if the orthogonality relation

$$(16) \qquad \int_0^1 M_*^{(n)}(x) \ S^{(n)}(x) dx = 0$$

holds for splines $S(x)$ of degree $2n$ whose knots are $\{\xi_k\}_1^r$ (i.e., $S(x)$ has the form (1) with $m = 2n$) and which also satisfy the boundary conditions (14). Consulting identity (15) we infer that $M_*(x)$ is characterized as the monospline of degree $2n$ with knots $\{\xi_k\}_1^r$ satisfying the boundary constraints (14) and the further interpolatory conditions

$$M_*(\xi_1) = M_*(\xi_2) = \ldots = M_*(\xi_r) = 0 \ .$$

The interpolation Theorem 2 tells us that $M_*(x)$ is uniquely determined by these stipulations.

§4. Best Quadrature Formulas for Boundary Conditions involving the Vanishing of Certain Derivatives at Both Ends

We develop some extensions of the concept of best quadrature formula allowing quite general boundary conditions.

Indeed, consider quadrature formulas for $\int_0^1 f(x)\,dx$ of the form

$$(17)\quad Q(f) = \sum_{k=1}^{r} a_k f(\xi_k) + \sum_{\mu=1}^{n-p} A_\mu f^{(j'_\mu)}(0) + \sum_{\nu=1}^{n-q} B_\nu f^{(i'_\nu)}(1)$$

where

$$0 \le j'_1 < j'_2 < \dots < j'_{n-p} \le n-1, \quad 0 \le i'_1 < i'_2 < \dots < i'_{n-q} \le n-1 .$$

The problem now is to determine the best quadrature formula in line with the criteria of (11) and (12) where now $Q(f)$ assumes the form (17). Again, suitable integration by parts reveals a one to one correspondence between monosplines $M(x)$ of degree $2n$ with knots $\{\xi_k\}_{k=1}^r$ satisfying the boundary conditions

$$(18)\quad \begin{aligned} M^{(k_1)}(0) &= M^{(k_2)}(0) = \dots = M^{(k_p)}(0) = 0 \\ M^{(\ell_1)}(1) &= M^{(\ell_2)}(1) = \dots = M^{(\ell_p)}(1) = 0 \end{aligned}$$

where

$$n \le k_1 < k_2 < \dots < k_p \le 2n-1$$

$$n \le \ell_1 < \ell_2 < \dots < \ell_q \le 2n-1 ;$$

$$j_\mu = 2n - k_\mu - 1, \qquad \mu = 1, \dots, p$$

$$i_\nu = 2n - \ell_\nu - 1, \qquad \nu = 1, \dots, q$$

and $\{j'_\mu\}_1^{n-p}$ is the complementary set to $\{j_\mu\}_1^p$ in the set $\{0, 1, \dots, n-1\}$ and analogously for $\{i'_\nu\}$ related to $\{i_\nu\}$.

<u>Theorem 3.</u> <u>The quadrature formula of the type</u> (17), <u>best in the sense of Sard, i. e., satisfying</u> (11) <u>and</u> (12), <u>is characterized</u>

through the determination of a monospline $M_*(x)$ of degree $2n$ with knots $\{\xi_k\}_1^r$ satisfying the boundary conditions (18), the adjoint boundary conditions

$$(19)\qquad \begin{aligned} M_*^{(j'_\mu)}(0) &= 0, \qquad \mu = 1, \ldots, n-p \\ M_*^{(i'_\nu)}(1) &= 0, \qquad \nu = 1, \ldots, n-q \end{aligned}$$

and the interpolatory conditions

$$M_*(\xi_k) = 0, \qquad k = 1, 2, \ldots, r .$$

A monospline satisfying these properties is uniquely determined if and only if (see equations (2) and (3)

$$(20)\qquad K\begin{pmatrix} j_1, j_2, \ldots, j_p, k'_1, \ldots, k'_{n-p}, \xi_1, \ldots, \xi_r \\ \xi_1, \ldots, \xi_r, i'_1, \ldots, i'_{n-q}, \ell_1, \ldots, \ell_q \end{pmatrix} > 0$$

where $\{k'_1, \ldots, k'_{n-p}\}$ is the complementary set of indices to $\{k_1, \ldots, k_p\}$ in the collection $\{n, n+1, \ldots, 2n-1\}$. The precise requirements on the indices $\{j_1, \ldots, j_p, k'_1, \ldots, k'_{n-p}\}$ and $\{i'_1, \ldots, i'_{n-q}, \ell_1, \ldots, \ell_p\}$ and r (note the independence of the $\{\xi_k\}$ values) to insure the inequality (20) is given in Theorem 1, especially Example II of Section 1.

As in the previous section, the proof of Theorem 3 is accomplished by verifying the orthogonality relation

$$(21)\qquad \int_0^1 M_*^{(n)}(x)\, S^{(n)}(x)\, dx = 0$$

for all splines $S(x) = \sum_{i=0}^{2n-1} \lambda_i x^i + \sum_{i=1}^{r} c_i (x-\xi_i)_+^{2n-1}$ satisfying the boundary conditions (18).

§5. General Boundary Conditions

Consider monosplines M of degree $2n$ with prescribed knots $\{\xi_k\}_1^r$ satisfying the boundary conditions

$$
(22)\qquad
\begin{aligned}
U_\mu(M) &= \sum_{\nu=0}^{n-1} A_{\mu\nu} M^{(2n-\nu-1)}(0) = 0\,, \quad \mu = 1, \ldots, p \\
V_\nu(M) &= \sum_{\nu=0}^{n-1} B_{\mu\nu} M^{(2n-\nu-1)}(1) = 0, \quad \mu = 1, \ldots, q
\end{aligned}
$$

where $A = \|A_{\mu\nu}\|$ has rank p and $B = \|B_{\mu\nu}\|$ has rank q. By adding further relations to total n for each boundary we can construct in a standard way adjoint boundary linear forms (see Neumark [6]).

$$
(23)\qquad
\begin{aligned}
\tilde{U}_\alpha(M) &= \sum_{\beta=0}^{n-1} \tilde{A}_{\alpha\beta} M^{(\beta)}(0) \qquad \alpha = 1, \ldots, n \\
\tilde{V}_\alpha(M) &= \sum_{\beta=0}^{n-1} \tilde{B}_{\gamma\beta} M^{(\beta)}(1) \qquad \gamma = 1, \ldots, n\,.
\end{aligned}
$$

(The forms $\tilde{U}$ and $\tilde{V}$ are not uniquely determined by the forms (22) but usually there is a natural set of adjoint boundary conditions.) Suppose M is a monospline of degree $2n$ satisfying the boundary conditions (22). Integration by parts with $f \in C^n[0,1]$ produces the identity

$$
(24)\qquad \int_0^1 f(x)\,dx = \sum_{i=1}^{n} U_i(M)\,\tilde{U}_i(f) + \sum_{i=1}^{n} V_i(M)\,\tilde{V}_i(f)
$$

$$
+ \sum_{k=1}^{r} c_k\, f(\xi_k) + (-1)^n \int_0^1 f^{(n)}(x)\, M^{(n)}(x)\,dx\,.
$$

This expression suggests a quadrature formula of the type

$$(25) \quad Q(f) = \sum_{i=p+1}^{n} u_i \tilde{U}_i(f) + \sum_{i=q+1}^{n} v_i \tilde{V}_i(f) + \sum_{k=1}^{r} c_k f(\xi_k)$$

where u_i, v_i and c_k are free real constants. In other words, the quadrature part involves the values of f at ξ_k and certain additional quantities involving special linear combinations of the derivatives of f (up to order n-1) at 0 and 1.

We next formulate the problem of determining best quadrature formulas of the type (25). Suppose $M_*(x)$ is a monospline of degree 2n with knots $\{\xi_k\}_1^r$ satisfying the boundary conditions

$$(26) \quad U_\mu(M_*) = 0, \quad \mu = 1, \ldots, p; \quad V_\nu(M_*) = 0, \quad \mu = 1, \ldots, q ,$$

the adjoint boundary conditions

$$(27) \quad \begin{aligned} &\tilde{U}_\alpha(M_*) = 0, && \alpha = p+1, \ldots, n \\ &\tilde{V}_\gamma(M_*) = 0, && \gamma = q+1, \ldots, n \end{aligned}$$

and vanishing at $\{\xi_k\}_{k=1}^r$, i.e.,

$$(28) \quad M_*(\xi_k) = 0, \qquad k = 1, 2, \ldots, r .$$

If such a monospline exists then by exploiting the integration by parts formula we may establish the orthogonality relation

$$\int M_*^{(n)}(x)\, S^{(n)}(x)\, dx = 0$$

for any spline of degree 2n with knots $\{\xi_k\}_{k=1}^r$ and satisfying the B.C. (26). This discussion leads to the following theorem.

Theorem 4. (a) Suppose a monospline $M_*(x)$ satisfying (26 (27) and (28) exists. The induced quadrature formula is best i the sense of (11) and (12) among all quadrature formulas of the type (25).

(b) If the B.C. (26) and (27) satisfy the criteria of Postulate I and $r \geq n$ then a unique monospline $M_*(x)$ with the desired properties exists.

Example 5. Consider boundary conditions of the explicit form

$$M^{(n+\mu-1)}(0) + (-1)^{\mu+1} c_\mu M^{(2n-\mu)}(0) = 0, \quad c_\mu > 0, \quad \mu = 1, 2, \ldots, p$$

(29)

$$M^{(n+\nu-1)}(1) + (-1)^{\nu} d_\nu M^{(2n-\nu)}(1) = 0, \quad d_\nu > 0, \quad \nu = 1, 2, \ldots, q.$$

It is not difficult to establish that there exists adjoint boundary conditions such that the total set of boundary conditions satisfy the requirements of Postulate I.

The boundary conditions (29) encompass the usual one that occur in physical problems of vibrating segments (see Karlin [4, Chap. 10]).

Remark 4. The results of Theorems 3 and 4 can be interpreted as a characterization of best approximation in the L_n^2 norm to the zero function by monosplines satisfying appropriate boundary conditions with prescribed knots.

Remark 5. The analogs of Theorems 3 and 4 for determining best quadrature formulas associated with linear functionals $L(f) = \int_0^1 f(x)\, w(x)\, dx$ where $w(x)$ is a positive continuous weight function admits the following solution. Merely replace the term $x^{2n}/(2n)!$ in the definition of monospline by the $2n-1$ fold integral

$$W_n(x) = \int_0^x \cdots \int_0^{\xi_2} \int_0^{\xi_1} w(\xi)\, d\xi\, d\xi_1 \cdots d\xi_{2n-2} \ .$$

The correspondence between "monosplines" with quadrature

formulas results as previously. The characterization of best quadrature in terms of monosplines satisfying certain boundary conditions as well as vanishing at the knots holds.

Remark 6. Another criteria for generating good (distinguished from best) quadrature formulas of the form of (24) is based on the concept of computing

$$\min_{\{a_\nu\}} \max_{\substack{f \in \mathcal{P} \\ \|f\|_n = 1}} |\mathfrak{R}(f)| \tag{30}$$

with respect to all quadrature formulas satisfying (11) where $\mathcal{P}$ denotes the set of all $f \in C^n[0,1]$ satisfying prescribed boundary conditions. For example, let $\mathcal{P}$ denote the set of all $f \in C^n$ for which $f^{(i)}(0) = f^{(i)}(1)$, $i = 0, 1, \ldots, n-1$. It can be shown that the solution of (30) is achieved by constructing the unique monospline M_* of degree $2n$ with knots $\{\xi_k\}_1^r$ which fulfills the periodicity condition $M_*^{(i)}(1)$, $i = 0, 1, \ldots, 2n-1$ and vanishes at $\{\xi_k\}$; i.e., $M^*(\xi_k) = 0$, $k = 1, 2, \ldots, r$.

Other choices for $\mathcal{P}$ lead to other interesting classes of good quadrature formulas. Further examples will be included in the paper elaborating the proofs of the theorems stated in this presentation.

REFERENCES

[1] Greville, T. N. E., "Introduction to Spline Functions" in Theory and Applications of Spline Functions, edited by T. N. E. Greville, Academic Press, New York, (1969) 1-37.

[2] Jerome, J. W. and R. S. Varga, "Generalizations of Spline Functions and Applications to Nonlinear Boundary Value and Eigenvalue Problems" in Theory and Applications of Spline Functions, edited by T. N. E. Greville, Academic Press, New York (1969) 103-157.

[3] Karlin, S. and Z. Ziegler, Tchebycheffian Spline Functions, J. SIAM Numer. Anal. 3, (1966) 514-543.

[4] Karlin, Samuel, Total Positivity, Stanford University Press, Stanford California (1968).

[5] Karlin, Samuel, The Fundamental Theorem of Algebra for Monosplines Satisfying Certain Boundary Conditions and Applications to Optimal Quadrature Formulas, this volume (1969).

[6] Neumark, M. A. Linear Differentialoperatorem Akademie Verlag, Berlin (1960).

[7] Schoenberg, I. J., On Monosplines of least deviation and best Quadrature Formulae, J. SIAM Numer. Anal. 2 (1965) 144-170.

[8] Schoenberg, I. J., On Monosplines of Least Square Deviation and Best Quadrature Formulae II, J. SIAM Numer. Anal. 3 (1966) 321-328.

[9] Schoenberg, I. J., "Monosplines and Quadrature Formulae" in Theory and Applications of Spline Functions, edited by T. N. E. Greville Academic Press, New York, (1969) 157-207.

[10] Schumaker, L. I., "Approximation by Splines" in Theory and Applications of Spline Functions, edited by T. N. E. Greville, Academic Press, New York (1969) 65-87.

Research supported in part at the Mathematics Research Center, University of Wisconsin, during a 3 month visit; The Weizmann Institute of Science, Rehovot, Israel during a 3 month visit; and supported in part at Stanford University, Stanford, California under contract N0014-67-A-0112-0015.

It was brought to my attention that Theorem 1 of Schoenberg† cited below is close to Theorem 3 of our manuscript.

† I. J. Schoenberg, "A second look at approximate quadrature formulae and spline interpolation", MRC Technical Report 966, Feb. (1969).

The Fundamental Theorem of Algebra for Monosplines Satisfying Certain Boundary Conditions and Applications to Optimal Quadrature Formulas

SAMUEL KARLIN

Motivated by work of Schoenberg [12] on characterizing optimal quadrature formulas ("optimality" as distinguished from "best in the sense of Sard" allows the knots in addition to the coefficients of the quadrature expression to be regarded as free variables) I was led to the problem of investigating the validity of the fundamental theorem of algebra for monosplines vanishing at prescribed points and also obeying suitable boundary constraints. Section 1 summarizes these refinements on the fundamental theorem of algebra for monosplines with boundary conditions. The proofs appear to be quite intricate as they involve proving that certain nonlinear mappings cover the origin. Aside from the intrinsic utility in determining optimal quadrature formula the fundamental theorem of algebra has independent interest and relevance for the study of best approximation in the sup norm as well as in the solution of certain area problems and elsewhere.

In Section 3 we present the characterization of the optimal quadrature formula. Analysis of the first and second variations of certain nonlinear functionals play a role. The results of Section 3 were partly developed in collaboration with Professor Schoenberg (c. f. Schoenberg [12]). The full details will be published separately. We formulate all the results in the case of ordinary spline polynomials although most of the theorems generalize to the context of Tchebycheffian splines.

§1. Fundamental Theorem of Algebra for Monosplines with Boundary Conditions

A monospline of degree m with knots $\{\xi_k\}_1^r$ ($\xi_1 < \xi_2 < \dots < \xi_r$) is an expression of the form

$$(1) \qquad K(x) = \frac{x^m}{m!} + \sum_{i=0}^{m-1} a_i x^i + \sum_{i=1}^{r} c_i (x-\xi_i)_+^{m-1} .$$

We denote the class of functions of this form by $\mathfrak{m}_{m,r}$ $(m>1)$. (A more general concept of monospline with respect to an extended Tchebycheff system is given in Karlin and Schumaker [5]. All the results reported in this paper extend to Tchebycheffian monosplines with appropriate adjustments in the proofs, e.g., see Karlin [3, §4, Chap. 10].) We regard $\{a_i\}_0^{m-1}$ and $\{c_1, c_2, \dots, c_r, \xi_1, \dots, \xi_r\}$ as $m + 2r = k$ parameters. The fundamental theorem of algebra for monosplines asserts in particular that given any k numbers $t_1 < t_2 < \dots < t_k$, there exists a unique monospline $K \in \mathfrak{m}_{m,r}$ vanishing exactly at the points $\{t_\nu\}_1^k$.

The theorem encompasses also the case of nonsimple zeros provided the following convention for multiplicity of zeros is adopted. Specifically, the multiplicity of a zero t not a knot of K has the standard interpretation. Counting the multiplicity of a zero t equal to a knot of K is the usual one if $K(t) = K'(t) = \dots = K^{(\ell)}(t) = 0$, $K^{(\ell)}(t) \neq 0$ when $\ell \leq m-2$ since K is globally C^{m-2}. On the other hand with $\ell = m-1$ and t a knot then the number of sign changes of $K^{(m-1)}(t)$ at t are to be counted. Explicitly, if $A = K^{(m-1)}(t-)$ and $B = K^{(m-1)}(t+)$ then the knot t is said to be of order

(i) $m-1$ if $AB > 0$,

(ii) m if $AB < 0$,

(iii) m if $AB = 0$ and $B - A > 0$,

(iv) $m+1$ if $AB = 0$ and $B - A < 0$.

We denote by $Z^*(K)$ the number of zeros of K on $(-\infty, \infty)$ following the above convention in evaluating the multiplicity of a zero.

The complete version of the standard fundamental Theorem of algebra involving no boundary conditions is

Theorem A. (Schoenberg [13] and Karlin and Schumaker[5]). Let $K \in \mathfrak{M}_{m,r}$ then $Z^*(K) \leq m + 2r$. Conversely, given $t_1 < t_2 < \ldots < t_\ell$ and positive integers $\{\alpha_i\}_1^\ell$ satisfying $\alpha_i \leq m+1$ for $i = 1, 2, \ldots, \ell$ with $\sum_{i=1}^{\ell} \alpha_i = m + 2r$ then there exists a unique monospline $K \in \mathfrak{M}_{m,r}$ possessing zeros $\{t_i\}$ with corresponding multiplicities $\{\alpha_i\}$. The monospline depends continuously on its set of zeros.

The proof is intricate and relies on some facts concerning variation diminishing properties induced by totally positive transformations, perturbation arguments using the implicit function theorem and the study of certain determinants.

Theorem A serves various applications including characterizing best approximation to certain functions in the sup norm and some area theorems (see Johnson [2], Schumaker [8], and Fitzgerald and Schumaker [1]).

In this paper we describe a significant extension of Theorem A where K obeys certain boundary constraints in addition to vanishing at prescribed zeros. Applications of the fundamental theorem and other generalizations are discussed in Sections 2 and 3.

We first state a special case of the main theorem. For this purpose it is convenient to introduce additional notation.

Let $\mathfrak{M}_{m,r}(\mathfrak{B})$ denote the collection of monosplines $K(x)$ of the form (1) with knots $\{\xi_\nu\}_1^r$ $(0 < \xi_1 < \xi_2 < \ldots < \xi_r < 1)$ satisfying the boundary conditions of the vanishing of certain derivatives at the end points 0 and 1 :

$$(2) \qquad \mathfrak{B} \begin{cases} K^{(\alpha_1)}(0) = K^{(\alpha_2)}(0) = \ldots = K^{(\alpha_p)}(0) = 0 \\ \\ K^{(\beta_1)}(1) = K^{(\beta_2)}(1) = \ldots = K^{(\beta_q)}(1) = 0 \end{cases}$$

where

(3)
$$0 \le \alpha_1 < \alpha_2 < \ldots < \alpha_p \le m-1$$
$$0 \le \beta_1 < \beta_2 < \ldots < \beta_q \le m-1$$

and

(4)
$$p+q \le m \; .$$

Let $\{\alpha'_1, \alpha'_2, \ldots, \alpha'_{m-p}\}$ denote the complementary set of indices of $\{\alpha_1, \ldots, \alpha_p\}$ in $\{0, 1, \ldots, m-1\}$. We postulate for the moment (compare Theorems 1 and 2 below) in line with Example 4 of Karlin [4] the conditions

(5)
$$\beta_1 \le \alpha'_{m-p-q+1}, \; \beta_2 \le \alpha'_{m-p-q+2}, \; \ldots, \; \beta_q \le \alpha'_{m-p} \; .$$

The count of zeros of monosplines will henceforth refer only to the open interval $(0,1)$ and multiplicities of zeros are evaluated by the convention set forth earlier. We denote the number of zeros by this count as $Z^*(K; 0,1)$.

Although Theorem 1 below is a special case of Theorems 2 and 3, because of its occurrence in applications it is worth special emphasis.

Theorem 1. *Let* K *be a monospline in* $\mathfrak{M}_{m,r}(\mathfrak{B})$ *(i.e., satisfying the boundary conditions* (2) *where the indices satisfy* (3), (4) *and* (5) *then*

$$Z^*(K; 0,1) \le m-p-q + 2r \; .$$

Conversely, given distinct $t_1, t_2, \ldots, t_k$ $(t_i \in (0,1))$ *and positive numbers* $\alpha_1, \alpha_2, \ldots, \alpha_k$ *such that* $\alpha_i \le m+1$ *for each* i *with* $\sum_{i=1}^{k} \alpha_i = m-p-q+2r$. *Then there exists a unique monospline* $K \in \mathfrak{M}_{m,r}(\mathfrak{B})$ *possessing zeros* $\{t_i\}$ *with corresponding* multiplicities $\{\alpha_i\}$. The monospline depends continuously on its set of zeros.

Remark: Notice that when $q = 0$ where no boundary conditions at 1 are imposed then (5) is automatic. Furthermore, when $p = 0$ then $\alpha'_{m-i+1} = m-i$ $(i = 0, 1, 2, \ldots, q)$ and (5) certainly holds.

Suppose that (5) is not necessarily satisfied. The next theorem covers this contingency.

Theorem 2. Let the notation in (2), (3) and (4) prevail. Determine γ as the smallest nonnegative integer such that

$$(6) \quad \beta_1 \le \alpha'_{m-p-q+\gamma+1}, \ \beta_2 \le \alpha'_{m-p-q+\gamma+2}, \ \ldots, \ \beta_{q-\gamma-1} \le \alpha'_{m-p-1}, \quad \beta_{q-\gamma} \le \alpha'_{m-p} .$$

(If $\gamma \ge q$ then by convention there restrictions are ipso facto satisfied.) Given $t_1, t_2, \ldots, t_k$ and positive numbers $\alpha_1, \alpha_2, \ldots, \alpha_k$ such that $1 \le \alpha_i \le m+1$ $(i = 1, 2, \ldots, k)$ and $\sum_{i=1}^{k} \alpha_i = m-p-q+2r$ hold with $r \ge \gamma$. Then there exists a unique monospline $K \in \mathfrak{M}_{m,r}(\mathfrak{B})$ possessing zeros $\{t_i\}$ with corresponding multiplicities $\{\alpha_i\}$. The monospline depends continuously on its set of zeros.

We next generalize the boundary conditions (2). Consider

$$(7a) \quad \mathfrak{B}_0: \sum_{\mu=0}^{m-1} A_{\nu\mu} D^{\nu} K(0) = 0 \qquad \nu = 1, 2, \ldots, p$$

$$p+q \le m$$

$$(7b) \quad \mathfrak{B}_1: \sum_{\mu=0}^{m-1} B_{\lambda\mu} D^{\mu} K(1) = 0 \qquad \lambda = 1, 2, \ldots, q$$

where the matrices $\tilde{A} = \|A_{\alpha\mu}(-1)^{\mu}\|$ and $B = \|B_{\beta\mu}\|$ satisfy

(i) $\tilde{A}$ is sign consistent of order p (SC_p) and has rank p (a matrix is said to be SC_p if all $p \times p$ nonzero determinants have a single sign),

(ii) B is SC_q ,

(iii) There exists $0 \le \alpha_1 < \alpha_2 < \dots < \alpha_p \le m-1$ and $0 \le \beta_1 < \beta_2 < \dots < \beta_q \le m-1$ satisfying

$$A\begin{pmatrix} 1, & 2, & \dots, & p \\ \alpha_1, & \alpha_2, & \dots, & \alpha_p \end{pmatrix} \neq 0 \quad \text{and } B\begin{pmatrix} 1, & 2, & \dots, & q \\ \beta_1, & \beta_2, & \dots, & \beta_q \end{pmatrix} \neq 0 \quad \dagger$$

where $\{\alpha'_1, \alpha'_2, \dots, \alpha'_{m-p}\}$ and $\{\beta_1, \beta_2, \dots, \beta_q\}$ obey the inequalities (5).

The class of monosplines (1) satisfying (7) is denoted by $\mathfrak{M}_{m,r}(\mathfrak{B}_0, \mathfrak{B}_1)$.

Theorem 3. *Suppose* (i), (ii) *and* (iii) *hold. Let* K *be a monospline in* $\mathfrak{M}_{m,r}(\mathfrak{B}_0 \cap \mathfrak{B}_1)$. *Then* $Z^*(K;0,1) \le m-p-q-2r$. *Conversely, given distinct* $t_1, t_2, \dots, t_k$ $(t_i \in (0,1))$ *and positive numbers* $\alpha_1, \alpha_2, \dots, \alpha_k$ *such that* $1 \le \alpha_i \le m+1$ $(i = 1, 2, \dots, k)$ *and* $\sum_{i=1}^{k} \alpha_i = m-p-q+2r$. *Then there exists a unique* $K \in \mathfrak{M}_{m,r}(\mathfrak{B}_0 \cap \mathfrak{B}_1)$ *vanishing at* $\{t_i\}$ with associated multi*plicities* $\{\alpha_i\}$. *The monospline depends continuously on its set of zeros.*

The analog of Theorem 2 extends to the situation of boundary conditions of the form (7). We will not formulate the precise result here.

---

† $A\begin{pmatrix} i_1, i_2, \dots, i_p \\ j_1, j_2, \dots, j_p \end{pmatrix}$ stands for the minor of A based on the rows of indices $i_1, i_2, \dots, i_p$ and columns of indices $j_1, j_2, \dots, j_p$.

We say a few words about the proof of Theorem 3 which is delicate and involved. The method of Karlin and Schumaker [5] needs considerable refinement and modification to take account of the imposed boundary conditions. Several applications of the variation diminishing properties of certain totally positive transformations are made. A continuity method decisively relying on the implicit function theorem is carried out. The conclusion of Theorem A serves as the starting point of the continuity method. Induction on the number of boundary conditions is part of the argument. Several interesting ancillary results emerge from the analysis. We cite two.

Proposition 1. If

$$K(x) = \frac{x^m}{m!} + \sum_{i=0}^{m-1} a_i x^i + \sum_{i=1}^{r} c_i (x-\xi_i)_+^{m-1}$$

is a monospline in $\mathfrak{M}_{m,r}(\mathcal{B}_0 \cap \mathcal{B}_1)$ satisfying the conditions of Theorem 3 with a maximal number of zeros, i.e., $Z^*(K;0,1) = m-p-q+2r$, then

$$c_1 < 0,\ c_2 < 0,\ \ldots,\ c_r < 0\ . \tag{8}$$

Proposition 2. Let K be a monospline satisfying the conditions of Theorem 1 with a maximal number of zeros in $(0,1)$. Then

$$\begin{aligned} &M^{(i)}(1) > 0 && \beta_q < i \\ &(-1)^k M^{(i)}(1) > 0 && \beta_{q-k} < i < \beta_{q-k+1} \quad (k=1,2,\ldots,q) \end{aligned} \tag{9}$$

and

$$M^{(i)}(0)M^{(j)}(0) < 0 \text{ for all } i, j \text{ satisfying } \alpha_k < i < \alpha_{k+1} < j < \alpha_{k+2}$$

$$(k = 0, 1, \ldots, p-2;\ \alpha_0 = -1)\ .$$

§2. Extensions, Applications and Discussion

I. As remarked at the start of §1, Theorems 1 - 3 apply mutatis mutandis to the case of Tchebycheffian monosplines satisfying boundary condition of the structure (7) where D^{μ} is replaced by the generalized differential operator

$$L_{\mu} = D_{\mu} D_{\mu-1} \cdots D_1, \quad \mu = 0, 1, \ldots, m-1 \quad (L_0 = I = \text{identity})$$

and

$$(D_i f)(\varphi) = \left(\frac{d}{dx} \frac{1}{w_i(x)}\right) \varphi(x)$$

with $w_i(x)$ positive m-1 times continuously differentiable functions prescribed on $[0,1]$.

II. It is interesting to contrast Theorem 3 involving free knots and the corresponding theorem with fixed knots (Theorem 2 of Karlin [4]). In the case of Theorem 3 of this paper the zeros can be arbitrarily specified while in the fixed knot case there are limitations on the prescriptions of the zeros for a given set of knots.

A general problem of interest would be to have some of the knots fixed and others free. We hope to uncover in future investigations the nature of the fundamental theorem of algebra for this more general setting and specifically delimit criteria guaranteeing the possibility of unique interpolation.

In the characterization of the <u>optimal quadrature formula</u> (see Schoenberg [12] and also §3 of this paper) the following problem arises. Can one construct a monospline $K(x)$ of degree $2n$ with $r \geq n$ $(n \geq 2)$ knots $\{\xi_i\}_1^r$ $(0 < \xi_1 < \xi_2 < \ldots < \xi_r < 1)$, i.e.,

$$K(x) = \frac{x^{2n}}{2n!} + \sum_{i=0}^{2n-1} a_i x^i + \sum_{i=1}^{r} c_i (x-\xi_i)_+^{2n-1}$$

such that K exhibits a double zero at each ξ_i and satisfies the boundary conditions

$$K^{(n)}(0) = K^{(n+1)}(i) = \ldots = K^{(2n-1)}(0) = 0 \tag{10}$$
$$K^{(n)}(1) = K^{(n+1)}(1) = \ldots = K^{(2n-1)}(1) = 0 .$$

The determination of such a monospline can be secured utilizing Theorem 2 and the Brouwer fixed point theorem in the following manner.

Let $\{t_i\}_{i=1}^{r}$ $(0 < t_1 < t_2 < \ldots < t_r < 1)$ be r given points and assign to each point the value $\alpha_i = 2$, $i = 1, 2, \ldots, r$ (i.e., multiplicity 2 as a zero).

We apply Theorem 2 with $m = 2n$ for the boundary conditions (10) to conclude the existence of a unique monospline $M(x)$ with knots $\{\xi_1, \xi_2, \ldots, \xi_r\}$ $(0 < \xi_1 < \xi_2 < \ldots < \xi_r < 1)$ such that

$$M(t_i) = M'(t_i) = 0, \qquad i = 1, 2, \ldots, r$$

and M satisfies (10). Theorem 2 further informs us that

$$\xi_i = \xi_i(t_1, t_2, \ldots, t_r) \qquad i = 1, 2, \ldots, r \tag{11}$$

are continuous functions mapping the open simplex Δ: $\{\bar{t} = (t_1, \ldots, t_r);\ 0 < t_1 < t_2 < \ldots < t_r < 1\}$ into itself. The mapping (11) can be extended continuously to $\bar{\Delta}$ = the closure of Δ such that the boundary has no fixed point. An application of the Brouwer fixed point theorem then affirms the existence of monosplines having the desired properties. The rigorous details of this proof will be elaborated elsewhere. We formally state the version of this result in the setting of the boundary conditions (2).

Theorem 4. Let $r \geq \gamma$ where γ is defined as in Theorem 2 with boundary conditions (2) such that $p + q = m$. There exists a unique monospline $K(x)$ with knots $\{\xi_i\}_i^r$ satisfying $0 < \xi_1 < \xi_2 < \ldots < \xi_r < 1$ such that

$$K(\xi_i) = K'(\xi_i) = 0, \quad i = 1, 2, \ldots, r$$

and K satisfies the boundary conditions (2).

When $m = 2n$ and n is even then K is strictly convex in the neighborhood of each ξ_i and when n is odd K is strictly concave at each ξ_i.

The proof of uniqueness is delicate and indirect. It decisively exploits the result of the extremal characterization described in Theorem 5 below. The proof also invokes several times in different forms the result of Theorem 2.

§3. Optimal Quadrature Formulas

In Section 3 of Karlin [4] the problem of ascertaining "best quadrature formulas in the sense of Sard" with quite general boundary conditions was examined. In that discussion the knots were maintained fixed. Considering the knots as free variables in addition to the coefficients of the quadrature expression we seek now to determine the "optimal quadrature formula" so named by Schoenberg [12].

The precise definition of the optimal quadrature formula is as follows. Among all quadrature expressions with r distinct knots

$$Q(f) = \sum_{k=1}^{r} a_k f(\xi_k) \tag{12}$$

and variables $\{a_1, a_2, \ldots, a_r, \xi_1, \xi_2, \ldots, \xi_r\}$ $(0 < \xi_1 < \xi_2 < \ldots < \xi_r < 1)$ satisfying the condition

$$\int_0^1 f(x) = Q(f) \quad \text{for } f \text{ a polynomial of degree } \le n-1 , \tag{13}$$

determine $\{a_k, \xi_k\}_{k=1}^{r}$ such that

$$\sup_{\substack{\|f\|_n \le 1 \\ f \in C^n[0,1]}} |\mathfrak{R}(f)| = \min \tag{14}$$

where $\mathfrak{R}(f) = \int_0^1 f(x)\,dx - Q(f)$ and $\|f\|_n$ stands for $\int_0^1 [f^{(n)}(x)]^2 dx$

It is instructive and of independent interest to generalize the prescription of the problem. To this end, consider quadrature formulas of the form

$$(15)\qquad Q(f) = \sum_{k=1}^{r} q_k f(\xi_k) + \sum_{\mu=1}^{n-p} A_\mu f^{(j'_\mu)}(0) + \sum_{\nu=1}^{n-q} B_\nu f^{(i'_\nu)}(1)$$

where the indices

$$0 \le j'_1 < j'_2 < \dots < j'_{n-p} \le n-1;\ 0 \le i'_1 < i'_2 < \dots < i'_{n-q} \le n-1$$

are prescribed.

The optimal quadrature formula in this more general setting is attained by solving the variational problem (14) where the min is extended over the variables

$$\{a_1, \dots, a_r, \xi_1, \dots, \xi_r, A_1, \dots, A_{n-p}, B_1, \dots, B_{n-q}\}$$

where $Q(f)$ is defined by (15) subject to (13).

Recall that the best quadrature formula (15) where $\{\xi_k\}_1^r$ are fixed and provided $r \ge n$ is induced by constructing the unique monospline

$$(16)\ M_*(x; \xi_1, \dots, \xi_r) = \frac{x^{2n}}{2n!} + \sum_{i=0}^{2n-1} \lambda_i x^i + \sum_{k=1}^{r} c_k (x-\xi_k)_+^{2n-1}$$

satisfying the boundary conditions $(p+q \le n)$

$$(17)\qquad \begin{aligned} M^{(j'_1)}(0) &= M^{(j'_2)}(0) = \dots = M^{(j'_{n-p})}(0) = 0 \\ M^{(i'_1)}(1) &= M^{(i'_2)}(1) = \dots = M^{(i'_{n-q})}(1) = 0 \end{aligned}$$

and

$$(18)\qquad \begin{aligned} M^{(k_1)}(0) &= M^{(k_2)}(0) = \ldots = M^{(k_p)}(0) = 0 \\ M^{(\ell_1)}(1) &= M^{(\ell_2)}(1) = \ldots = M^{(\ell_q)}(1) = 0 \end{aligned}$$

and which vanishes at the knots, i.e.,

$$(19)\qquad M(\xi_1) = M(\xi_2) = \ldots = M(\xi_r) = 0$$

where $k_\mu = 2n-j_\mu-1$ $(\mu = 1, \ldots, p)$, $\ell_\nu = 2n-i_\nu-1$ $(\nu = 1, 2, \ldots, q)$ and $\{j'_\mu\}_1^{n-p}$ and $\{j_\mu\}_1^p$ are complementary sets of indices in $\{0, 1, \ldots, n-1\}$ and similarly for $\{i'_\nu\}_1^{n-q}$ and $\{i_\nu\}_1^q$ (see Theorem 3 of Karlin [4]). The minimal error (i.e., the minimum value of (14)) in the fixed knot case is

$$(20)\qquad \int_0^1 [M_*^{(n)}(x; \xi_1, \ldots, \xi_r)]^2\, dx = F(\xi_1, \xi_2, \ldots, \xi_r) .$$

The optimal quadrature formula is then achieved by minimizing $F(\xi_1, \xi_2, \ldots, \xi_r)$ with respect to $(\xi_1, \ldots, \xi_r)$ satisfying $0 < \xi_1 < \xi_2 < \ldots < \xi_r < 1$. It is conceivable that in calculating $\inf F(\xi_1, \ldots, \xi_r)$ certain of the knots coalesce or stream to the endpoints 0 and/or 1. This is actually not the case. The following theorem provides a complete characterization of the optimal quadrature formula.

Theorem 5. There exists $\{\xi_k^*\}_1^r$ satisfying $0 < \xi_1^* < \xi_2^* < \ldots < \xi_r^* < 1$ such that

$$F(\xi_1^*, \xi_2^*, \ldots, \xi_r^*) = \inf_{\{\xi_k\}_{k=1}^r} F(\xi_1, \xi_2, \ldots, \xi_r)$$

where $F(\xi_1, \ldots, \xi_r)$ is defined in (20) and the corresponding "optimal monospline" $\widetilde{M}(x; \xi_1^*, \ldots, \xi_r^*)$ satisfies the boundary conditions (17) and the adjoint boundary conditions (18), the interpolatory conditions (19) and

$$(21) \qquad \tilde{M}'(\xi_1^*) = \tilde{M}'(\xi_2^*) = \dots = \tilde{M}'(\xi_r) = 0 \ .$$

The optimal monospline is uniquely determined. (Notice that the total set of conditions (17), (18), (19) and (21) equals $2n + 2r$, the number of free parameters in the expression of the monospline.)

In the special case $p = q = 0$ the minimization of $F(\xi_1, \xi_2, \dots, \xi_r)$ is equivalent to the problem of calculating the minimun least square deviation of splines of degree $n-1$ with r variable knots to the function $x^n/n!$. Partial contributions to the solution of this problem were made recently by Powell [7]. His result (Theorem 10) is subsumed in our Theorem 5. Powell also proposed some computational algorithms for determining $\tilde{M}$ in the special case that he considered.

A comment about the proof of Theorem 5 may be in order. It is an easy matter to establish that the minimum of $F(\xi_1, \xi_2, \dots, \xi_r)$ is attained for a monospline M^* satisfying the boundary constraints (17) and (18) with possible multiple knots $\hat{\xi}_i$ $(0 < \hat{\xi}_1 < \hat{\xi}_2 < \dots < \hat{\xi}_k < 1)$ of respective multiplicity m_i where $\sum_{i=1}^{k} m_i = r$. A simple variational argument (compare with (23) below) shows that $M^*(x; \xi_1, \dots, \xi_k)$ has a zero at $\hat{\xi}_i$ of order $m_i + 1$. If $m_{i_0} > 1$ for some i_0 then by judicious application of Theorem 2 we can construct a monospline M^{**} with the same zeros as $M^*(x)$ satisfying the boundary conditions (17) and (18) such that the value of $F(\xi_1, \dots, \xi_r)$ is strictly diminished. This contradicts the property that M^* induces an optimal quadrature formula. The elaboration of this argument is deleicate. The proof of the uniqueness assertion is more involved.

It is not difficult to show that any monospline with r distinct knots $(\xi_1^*, \xi_2^*, \dots, \xi_r^*)$ obeying (17), (18), (19) and (21) induces a strict minimum of $F(\xi_1, \dots, \xi_r)$. We do this in two stages by examining the first variation and then the second variation of F in the neighborhood of $(\xi_1^*, \xi_2^*, \dots, \xi_r^*)$.

Let $M(x; \xi_1, \ldots, \xi_r)$ be any monospline satisfying (17) and (19). Integration by parts produces the identity (see also equation (13) in Karlin [14])

$$(22)\quad \int_0^1 M^{(n)}(x, \xi_1, \ldots, \xi_r)\, \tilde{M}^{(n)}(x; \xi_1^*, \ldots, \xi_r^*)dx$$
$$= (-1)^n \left[\int_0^1 \tilde{M}^{(n)}(x; \xi_1^*, \ldots, \xi_r^*)dx - \sum_{\nu=1}^{n} c_\nu(\xi_1, \ldots, \xi_r)\, \tilde{M}(\xi_\nu; \xi_1^*, \ldots, \xi_r^*) \right]$$

where we have used the fact that $\tilde{M}$ satisfies the full set of boundary conditions (17) and (18). Differentiating with respect to ξ_k and evaluating at $\xi_1^*, \xi_2^*, \ldots, \xi_r^*$ gives

$$(23)\quad \int_0^1 \left[\frac{\partial}{\partial \xi_k} M^{(n)}(x, \xi_1^*, \ldots, \xi_r^*) \right] \tilde{M}^{(n)}(x, \xi_1^*, \ldots, \xi_r^*)dx$$
$$= (-1)^{n-1} \left[c_k(\xi_1^*, \ldots, \xi_r^*)\, \tilde{M}'(\xi_k^*; \xi_1^*, \ldots, \xi_r^*) + \sum_{\nu=1}^{n} \frac{\partial c_\nu}{\partial \xi_k}(\xi_1^*, \ldots, \xi_r^*)\, \tilde{M}(\xi_\nu^*; \xi_1^*, \ldots, \xi_r^*) \right] = 0$$

all terms vanishing because of (19) and (21).

In particular, this analysis establishes that $(\xi_1^*, \ldots, \xi_r^*)$ is a stationary point of $F(\xi_1, \ldots, \xi_r)$.

We next calculate the second variation of $F(\xi_1, \ldots, \xi_r)$ at the point $(\xi_1^*, \ldots, \xi_r^*)$. Thus, for $i \neq j$

$$(24)\quad \frac{\partial^2 F}{\partial \xi_i \partial \xi_j} = 2 \int_0^1 \frac{\partial^2 M^{(n)}(x; \xi_1, \ldots, \xi_r)}{\partial \xi_i\, \partial \xi_j} M^n(x; \xi_1, \ldots, \xi_r)dx$$
$$+ \int_0^1 T_i(x; \xi_1, \ldots, \xi_r) T_j(x; \xi_1, \ldots, \xi_r)dx = I_1 + I_2$$

where

$$T_i(x;\xi_1, \ldots, \xi_r) = \frac{\partial M^{(n)}(x;\xi_1, \ldots, \xi_r)}{\partial \xi_i}, \qquad i = 1, 2, \ldots, r .$$

We infer that the first integral I_1 evaluated at $(\xi_1^*, \ldots, \xi_r^*)$ is zero since I_1 approaches

$$\int_0^1 \frac{\partial^2 \widetilde{M}^{(n)}(x; \xi_1^*, \ldots, \xi_r^*)}{\partial \xi_i \, \partial \xi_j} \widetilde{M}^{(n)}(x; \xi_1^*, \ldots, \xi_n^*)\, dx$$

$$= (-1)^{n-1} \Bigg[c_1(\xi_1^*, \ldots, \xi_n^*) \widetilde{M}'(\xi_i^*) + c_j(\xi_1^*, \ldots, \xi_r^*) \widetilde{M}'(\xi_j^*)$$

$$+ \sum_{\nu=1}^{r} \frac{\partial^2 c_\nu}{\partial \xi_i \, \partial \xi_j} (\xi_1^*, \ldots, \xi_r^*) \widetilde{M}(\xi_\nu^*) \Bigg] = 0$$

with each term vanishing by virtue of (19) and (21). By an analogous calculation, we deduce

$$\text{(25)} \qquad \frac{\partial^2 F}{\partial \xi_i^2} = \int_0^1 [T_i(x; \xi_1, \ldots, \xi_r)]^2 dx + (-1)^{n-1} c_i \widetilde{M}''(\xi_i^*; \xi_1^*, \ldots, \xi_r^*) .$$

Referring to Proposition 1, Section 1, we know that $c_i < 0$ for each i and according to Theorem 4 for n even $\widetilde{M}''(\xi_i^*) > 0$ while for n odd $\widetilde{M}''(\xi_i^*) < 0$. In view of these facts it follows that the matrix

$$\text{(26)} \qquad \Gamma = \left\| \frac{\partial^2 F}{\partial \xi_i \partial \xi_j} \right\|_{i,j=1}^{r} \quad \text{evaluated at } (\xi_1^*, \ldots, \xi_r^*)$$

is a sum of a Gramian matrix plus a strict diagonal positive definite matrix. Thus, the matrix Γ is positive definite and therefore $F(\xi_1, \xi_2, \ldots, \xi_r)$ exhibits a strict local minimum at $(\xi_1^*, \ldots, \xi_r^*)$.

This argument has proved that every monospline with r knots obeying the full set of boundary conditions (17) and (18) and the interpolatory requirements (19) and (21) induces a strict local minimum of $F(\xi_1, \ldots, \xi_r)$. Theorem 5 asserts that the global minimum is induced by a monospline of the stated kind.

The proof of global uniqueness uses the results indicated above and further perturbation arguments with reliance on the fundamental theorem of algebra for monosplines delineated in Theorem 2.

§4. Extensions and Remarks

(i) Schoenberg has highlighted the fascinating fact that many of the classical interpolation formulas are optimal in the sense of Theorem 5 for suitable boundary conditions. This includes the Euler-Maclaurin quadrature formula and the associated Bernoulli monospline. Some optimal monosplines are calculated for the case of optimal cubic monospline where the knot are equally spaced and one simple example of an optimal quintic monospline with 2 knots is exhibited (see Schoenberg [12]).

(ii) A straightforward extension of the results of Section 3 to the case of Tchebycheffian monosplines and their corresponding optimal quadrature formulas are available.

(iii) It seems likely that an analog of the theory of this section can be developed for the case of monosplines with multi knots and the associated optimal quadrature formula for them. This is somewhat akin to the Turàn type quadrature formula briefly touched on in Schoenberg [12].

(iv) For the special case of boundary conditions (10) the determination of K can be reduced to a moment problem (see Karlin and Studden [6, Chap. 4, Sec. 9]).

REFERENCE

[1] FitzGerald, Carl H. and L. Schumaker, A differential equation approach to interpolation at extremal points, MRC Tech. Summ. Rpt. No. 731, Mathematics Research Center, U.S. Army, University of Wisconsin, Madison, Wisconsin, 1967.

[2] Johnson, R. S., On monosplines of least deviation, Trans. Amer. Math. Soc. 96 (1960), 458-477.

[3] Karlin, S., Total Positivity, Stanford University Press, Stanford, California, 1968.

[4] Karlin, S., Best quadrature formulas and interpolation by splines satisfying boundary conditions, this volume.

[5] Karlin, S. and L. Schumaker, The fundamental theorem of algebra for Tchebycheffian monosplines, J. Analyse Math., 20 (1967), 233-270.

[6] Karlin, S. and W. Studden, Tchebycheff Systems with Applications in Analysis and Statistics, Interscience, New York, 1966.

[7] Powell, M. J. D., On Best L_2 Spline Approximations, Numerische Mathematik, Differentialgleichungen, Approximationstheorie, Sonderdruck aus ISNM Vol. 9 (1968), 317-339.

[8] Schumaker, L., Uniform approximation by Chebyshev spline functions, SIAM J. Numer. Anal. 5 (1968), 647-656.

[9] Schumaker, L., Approximation by Splines, Theory and Applications of Spline Functions (1969), 65-85.

[10] Schoenberg, I. J., On monosplines of least deviation and best quadrature formulae, SIAM J. Numer. Anal. 2 (1965), 144-170.

[11] Schoenberg, I. J., On monosplines of least square deviation and best quadrature formulae II, SIAM J. Numer. Anal. 3 (1966), 321-328.

[12] Schoenberg, I. J., "Monosplines and quadrature formulae" in Theory and Applications of Spline Functions, edited by T. N. E. Greville, Academic Press, New York (1969), 157-207.

[13] Schoenberg, I. J., "Spline Functions, Convex Curves and Mechanical Quadratures", Bull. Amer. Math. Soc. 64 (1958), 352-357.

Research supported in part at the Mathematics Research Center, University of Wisconsin, during a 3 month visit; The Weizmann Institute of Science, Rehovot, Israel during a 3 month visit; and supported in part at Stanford University, Stanford, California under contract N0014-67-A-0112-0015.

Index